Cellular and Molecular Aspects of Plant-Nematode Interactions

Edited by

C. FENOLL
Department of Biologia,
University of Autonoma de Madrid,
Madrid, Spain

F. M. W. GRUNDLER
Institut für Phytopathologie,
Universität Kiel,
Kiel, Germany

and

S. A. OHL
MOGEN International nv,
Leiden, The Netherlands

KLUWER ACADEMIC PUBLISHERS
DORDRECHT / BOSTON / LONDON

A C.I.P. Catalogue record for this book is available from the Library of Congress.

ISBN 0-7923-4637-8

Published by Kluwer Academic Publishers,
P.O. Box 17, 3300 AA Dordrecht, The Netherlands.

Sold and distributed in the U.S.A. and Canada
by Kluwer Academic Publishers,
101 Philip Drive, Norwell, MA 02061, U.S.A.

In all other countries, sold and distributed
by Kluwer Academic Publishers,
P.O. Box 322, 3300 AH Dordrecht, The Netherlands.

Printed on acid-free paper

All Rights Reserved
© 1997 Kluwer Academic Publishers
No part of the material protected by this copyright notice may be reproduced or
utilized in any form or by any means, electronic or mechanical,
including photocopying, recording or by any information storage and
retrieval system, without written permission from the copyright owner.

Printed in the Netherlands

CONTENTS

PROLOGUE
    Peter C. SIJMONS   1

INTRODUCTION

ROOT PARASITIC NEMATODES: AN OVERVIEW
    Urs WYSS   5

A: CELLULAR AND MOLECULAR ASPECTS OF THE INTERACTION

ROOT ANATOMY AND DEVELOPMENT, THE BASIS FOR NEMATODE PARASITISM
    Ben SCHERES, Peter C. SIJMONS, Claudia van den BERG, Heather McKHANN, Geert de VRIEZE, Viola WILLEMSEN and Harald WOLKENFELT   25

PLANT SIGNALS IN NEMATODE HATCHING AND ATTRACTION
    Roland N. PERRY   38

INVASION AND MIGRATION BEHAVIOUR OF SEDENTARY NEMATODES
    Nicola von MENDE   51

THE BIOLOGY OF GIANT CELLS
    Teresa BLEVE-ZACHEO and Maria T. MELILLO   65

THE STRUCTURE OF SYNCYTIA
    Wladyslaw GOLINOWSKI, Miroslaw SOBCZAK, Wojciech KUREK and Grazyna GRYMASZEWSKA   80

NEMATODE SECRETIONS
    John T. JONES and Walter M. ROBERTSON   98

PHYSIOLOGY OF NEMATODE FEEDING AND FEEDING SITES
    Florian M.W. GRUNDLER and Annette BÖCKENHOFF   107

CELL CYCLE REGULATION IN NEMATODE FEEDING SITES
Godelieve GHEYSEN, Janice DE *ALMEIDA ENGLER and Marc* VAN *MONTAGU*   120

REGULATION OF GENE EXPRESSION IN FEEDING SITES
Carmen FENOLL, Fabio A. ARISTIZABAL, Soledad SANZ-ALFEREZ and Francisca F. Del CAMPO   133

B: NATURAL RESISTANCE

NATURAL RESISTANCE: THE ASSESSMENT OF VARIATION IN VIRULENCE IN BIOLOGICAL AND MOLECULAR TERMS
Carolien ZIJLSTRA, Vivian C. BLOK and Mark S. PHILLIPS   153

GENETIC AND MOLECULAR STRATEGIES FOR THE CLONING OF (A)VIRULENCE GENES IN SEDENTARY PLANT-PARASITIC NEMATODES
Phillipe CASTAGNONE-SERENO, Pierre ABAD, Jaap BAKKER, Valerie M. WILLIAMSON, Fred J. GOMMERS and Antoine DALMASSO   167

BREEDING FOR NEMATODE RESISTANCE IN SUGARBEET: A MOLECULAR APPROACH
Michael KLEINE, Daguang CAI, Rene M. KLEIN-LANKHORST, Niels N. SANDAL, Elma M.J. SALENTIJN, Hans HARLOFF, Sirak KIFLE, Kjeld A. MARCKER, Willem J. STIEKEMA and Christian JUNG   176

RESISTANCE TO ROOT-KNOT NEMATODES IN TOMATO
Tsvetana B. LIHARSKA and Valerie M. WILLIAMSON   191

BIOCHEMISTRY OF PLANT DEFENCE RESPONSES TO NEMATODE INFECTION
Giuseppe ZACHEO, Teresa BLEVE-ZACHEO and Maria T. MELILLO   201

C: ENGINEERED RESISTANCE

ENGINEERING RESISTANCE AGAINST PLANT PARASITIC NEMATODES USING ANTI-NEMATODE GENES
Paul R. BURROWS and Dirk DE WAELE   217

ENGINEERING PLANT NEMATODE RESISTANCE BY ANTIFEEDANTS
  Michael J. McPHERSON, Peter E. URWIN, Catherine J. LILLEY and Howard J. ATKINSON   237

ANTI-FEEDING STRUCTURE APPROACHES TO NEMATODE RESISTANCE
  Stephan A. OHL, Frédérique M. van der LEE and Peter C. SIJMONS   250

TOWARDS PLANTIBODY-MEDIATED RESISTANCE AGAINST NEMATODES
  Willem J. STIEKEMA, Dirk BOSCH, Annemiek WILMINK, Jan M. DE BOER, Alexander SCHOUTEN, Jan ROOSIEN, Aska GOVERSE, Gert SMANT, Jack STOKKERMANS, Fred J. GOMMERS, Arjen SCHOTS and Jaap BAKKER   262

EPILOGUE
  Paul R. BURROWS   273

INDEX   275

AUTHORS   279

PROLOGUE

Peter C. SIJMONS
Agrotechnological Research Institute (ATO-DLO), P.O. Box 17, 6700 AA Wageningen, The Netherlands

This book marks the end of a four year European Union Concerted Action Program on "Resistance Mechanisms against Plant-parasitic Nematodes". The objectives of the Concerted Action were, and I quote from the original application, "to increase European collaboration and to coordinate research on plant-parasite nematodes, with a focus on engineering nematode-resistant crop varieties".

In comparison to other fields in phytopathology, plant-nematode research is a minor area of research in every EU country. As a result, national feedback and exchange of scientific information amongst different institutes within a member state is limited. Financial support is usually focussed at particular (local) crop-nematode interactions with limited allowance for travel and international exchange. These factors, as well as the economical impact of nematode damage to crop plants, together with a rising public pressure to decrease the use of nematicides, as formulated by EU directives, made it apparent that European research on plant-parasitic nematodes needed a boost. The availability, just prior to the program, of a non-economic plant as a suitable model host for laboratory studies (*Arabidopsis*), could suddenly provide the common denominator to compare and exchange scientific date on a transnational basis. But, more importantly, the sheer number of scientists working with *Arabidopsis* were giving an unprecedented momentum to plant science. Nematologists could now board this fast moving train, overcoming at least part of the technical restraints when working with such a complex host-parasite system. This was parallel to the development of transformation protocols for all of the major crop species. It was therefore apparent at the time the Concerted Action was initiated that major breakthroughs in the area of nematode resistance could be expected in the coming years. This was recognized by the European Commission too, and a substantial grant was secured at the end of 1992. This was allowed for large scale annual meetings of all 16 participating groups (Leiden, NL 1993; Rothamsted, UK 1994; Kiel, D 1995; Toledo, ES 1996), including invited speakers, students and technicians, thus bringing a wealth of fresh ideas and enthusiasm in these meetings. Very direct contacts between those working at the bench could be established, a prerequisite to make collaboration work. Participants were able to visit each other's lab without almost any constraints, and more than a hundred short scientific missions were sponsored by the program, disseminating new protocols and results quickly.

Although the ability to travel and to have large scale meetings on a fully sponsored basis was sometimes regarded as a very luxurious program, I consider the levy of every

ECU spent through the Concerted Action high. The impact of this grant would not have been higher when actual research was being sponsored, a much more costly setup for the EU. It would have involved fewer groups and less mobility, and eventually EU collaboration would still be limited, the prime objective of the project. But now, after 4 years of extensive exchanges, personal relations have evolved and a closely knit EU network will definitely continue after the end of the program, thus ensuring collaborative progress in the area of plant-nematode interactions for a long time. Already several new EU-projects have been initiated through contacts made in the Concerted Action, focussed on more applied aspects of research and based on the considerable progress that has been made in the past few years. Thus for probably less than 0.5% of personal research budgets, the Concerted Action has had a strong influence on the scientific progress of participating groups, indeed a high levy per ECU spent.

The action coincided with exciting times in plant nematology. This field can now no longer be considered the last wagon on the train of phytopathology or be described as a very black box. Although major issues are far from resolved, enormous progress has been made when compared to preceding decades. More and more results are published in high impact journals, attracting scientists from different fields and thus adding to our understanding of this unique and highly evolved relationship between a parasite and its host. Driven by the demand to produce nematode resistant crops, several strategies emerged around the world, already culminating in field trials in 1995. It has become clear over the past 4 years that nematodes are a pest that can indeed be tackled through the use of transgenic crops. The obligate parasitism makes it a more sensitive target for manipulation than, *e.g.* fungi or bacteria. I am therefor optimistic that fully resistant commercial crops will be in the field before 2000 and that this strategy will provide a feasible alternative for the use of nematicides.

At this point I would like to thank the EU-Commission for their decision to sponsor our Concerted Action, and all scientists that made this program successful. It has been a privilege and a pleasure to act as coordinator for such a group.

Introduction

ROOT PARASITIC NEMATODES: AN OVERVIEW

Urs WYSS
Institut für Phytopathologie, Universität Kiel, Hermann-Rodewald-Str. 9, 24118 Kiel, Germany

Abstract

This overview emphasizes the present state of knowledge on the feeding behaviour of root parasitic nematodes and associated cell responses. Several unresolved questions related to this topic could recently be answered with the aid of high resolution video-enhanced contrast light microscopy, which enables direct observations of live nematodes within roots.Unlike in other plant pathogens, such as fungi, where cell wall surface molecules elicit a cascade in plant gene regulation, nematodes inject copious amounts of salivary secretions from oesophageal gland cells to transform root cells and tissues for their own reproductive success. Hence this chapter covers several examples of root cell responses to salivary secretions. As shown, *Xiphinema index* may be a good model. It has a huge dorsal gland cell and the effects of its secretions can be studied in detail. However, from an economic point of view, nearly all efforts in understanding the molecular bases for feeding structure establishment are currently devoted to the economically most important cyst and root-knot nematodes. In order to compare their feeding structures with those of other nematodes with different parasitic strategies, this introductory chapter gives also a general overview on the wide array of food cell modifications induced and maintained. These modifications are entirely nematode specific, irrespective of the host plant parasitized. Each is successful in its own to support development and reproduction of the nematode species involved. From an evolutionary point of view the uninucleate giant cells of some non-cyst-forming nematodes are for instance more primitive than the syncytia of cyst-forming nematodes, which ensure a higher reproductive capacity. Root-knot nematodes are generally regarded to have evolved the most advanced form of root parasitism. Multinucleate giant cells may therefore be considered to represent the most advanced form of nurse cell systems. Hopefully, with rapid advances in molecular biology, we will finally find an answer to the intriguing question how nematode secretions regulate plant gene expression for their own benefit and how the specificity of the nurse cells or nurse cell systems induced by them can be explained.

1. Introduction

Nematodes have established themselves in nearly all possible ecological niches, despite their relatively simple body structure and uniform organization. Those dwelling in the soil live either from bacteria, fungi, nematodes, insects or from roots of plants. Root parasitic nematodes exploit the root as their only source of nutrients and may spend their whole life cycle outside it, feeding from the surface or deeper tissues, others have evolved the capacity to invade the root and to feed from cortical or stelar cells. All of them are obligate biotrophic parasites as they require nutrients from living cells, which they modify by salivary secretions prior to food ingestion. In many cases the food cells are transformed into highly specialized feeding structures to support nematode development and reproduction. Hence these feeding structures can be called nurse cells when they consist of single or few cells, or nurse cell systems, when many cells are incorporated. Cell and tissue modifications induced by root parasitic nematodes have been recently reviewed [e.g. 14, 18, 50, 51, 56]. The aim of the present overview is to summarize and categorize current knowledge of the feeding behaviour of root parasitic nematodes and associated cell responses. On the basis of some selected examples the reader of this book should become acquainted with some fundamental facts before indulging into details on cellular and molecular aspects in plant-nematode interactions.

The phylum Nemata with currently about 20 000 described species is divided into two classes: Adenophorea and Secernentea with 12 and 6 orders, respectively [38]. Root parasitism has only evolved in the order Dorylaimida (dorylaimid nematodes) within the class Adenophorea and in the order Tylenchida (tylenchid nematodes) within the class Secernentea. Dorylaimid nematodes can cause serious root damage when they occur in large populations, however, they are economically more important as vectors of several soilborne viruses, which can only be transmitted by these nematodes. Tylenchid nematodes are unable to transmit viruses but many of them seriously affect plant growth by inducing profound alterations in the anatomy and function of the roots as well as in the structure of the cells from which they feed.

2. Dorylaimid Nematodes

Most nematodes in the order Dorylaimida live predacious from other nematodes. Only those grouped into the two families Trichodoridae and Longidoridae are obligate root parasites. They are exclusively migratory ectoparasites (Fig. 1; 1A-C) and they either browse along the root surface (trichodorid nematodes with the main genera *Trichodorus*, *Paratrichodorus*) or feed for longer periods at specific sites from deeper root tissues (longidorid nematodes with the main genera *Longidorus*, *Xiphinema*). Some species of the Trichodoridae transmit Tobraviruses, whereas those of the Longidoridae transmit Nepoviruses.

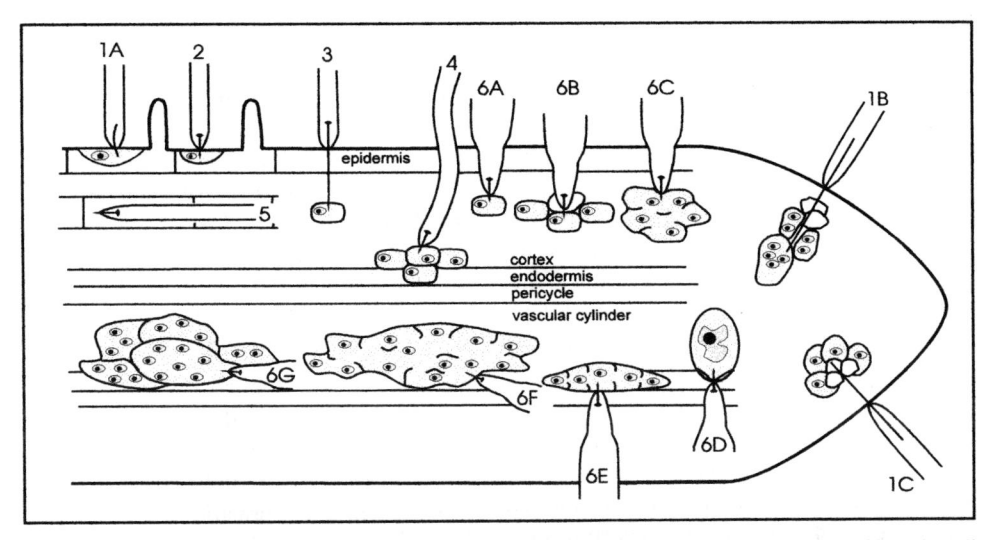

Figure 1. Examples of feeding sites of some selected root parasitic nematodes with their specific root cell responses: 1A-C: **DORYLAIMID NEMATODES:** all migratory ectoparasites. 1A: *Trichodorus* spp.; 1B: *Xiphinema index;* 1C: *Longidorus elongatus.* 2 - 6: TYLENCHID NEMATODES: 2: Migratory ectoparasite: *Tylenchorhynchus dubius* 3: Sedentary ectoparasite: *Criconemella xenoplax* 4: Migratory ecto-endoparasite: *Helicotylenchus* spp. 5: Migratory endoparasite: *Pratylenchus* spp. 6: Sedentary endoparasites: 6A: *Trophotylenchulus obscurus;* 6B: *Tylenchulus semipenetrans* ; 6C: *Verutus volvingentis* ; 6D: *Cryphodera utahensis* ; 6E: *Rotylenchulus reniformis;* 6F: *Heterodera* spp. ; 6G: *Meloidogyne* spp.

2.1. TRICHODORID NEMATODES

All trichodorid nematodes possess a ventrally curved stylet for cell wall perforation. It is composed of two parts, an anterior solid tip and a posterior part, to which stylet protractor muscles are attached. Behind the mouth opening the stoma wall contains strengthening rods with associated short sets of muscles [46]. The oesophagus is slender in the anterior region and has a triangular lumen (food canal). It gradually swells to form a flask-shaped basal bulb that contains five secretory gland cells, one dorsal and a pair of anterior and posterior subventral cells. The ducts of all glands open into the food canal within the basal bulb. The bulb also contains radial muscles that contract during pumping for both food ingestion and salivation [46].

Trichodorid nematodes usually aggregate near the tips of rapidly growing roots, where they often feed gregariously on epidermal cells and root hairs, occasionally also on subepidermal cells. The feeding behaviour is unique among nematodes as the stylet is continuously thrust into perforated cells during salivation and food ingestion. Feeding on individual cells lasts only a few minutes and is composed of distinct phases [46, 62]. Before root cell perforation, the nematodes explore the surface of the wall with their lips. When a suitable spot has been detected, the strengthening rods are drawn into close contact with the wall, which is then perforated by forceful stylet thrustings. Thereafter, during salivation, just the tip of the stylet is repeatedly inserted through the perforation hole. Saliva, produced in one or more of the five secretory glands and injected into the cell, hardens rapidly around the strengthening rods to form a feeding tube [70]. In combination with repeated stylet insertions, the secretions incite a rapid and massive accumulation of cytoplasm at the perforation site. The cell's nucleus is drawn to this aggregation; it swells and quickly loses its refractive index [68]. When most of the cytoplasm has accumulated, the nematodes thrust their stylets deeply into the aggregated mass. These forceful thrusts widen the plug of saliva beneath the perforation hole of the cell wall and most likely pierce the tonoplast. The resulting mixture with cell sap facilitates the removal of a large amount of cytoplasm through the feeding tube concurrent with each stylet retraction. Whenever the nucleus is pierced by the deep stylet thrusts, its contents are also removed. Upon departure, the feeding tube remains firmly anchored in the cell wall. Cell death soon becomes evident by the coagulation of remaining cytoplasm. However, occasionally cells survive, when feeding is suddenly interrupted at the beginning of salivation. Virus transmission may be effective only under these rather rare conditions [46].

Trichodorids are the only root parasitic nematodes capable of removing cytoplasm together with its organelles, such as mitochondria and plastids, because the lumen of the feeding tube is wide enough to permit this. In spite of their remarkable strategies for root cell exploitation, they are little advanced in mode of parasitism and copious amounts of cytoplasm are required for reproduction. Cytoplasm of several hundred cells has to be withdrawn to produce just a few eggs within one week.

2.2. LONGIDORID NEMATODES

Longidorids are relatively large in body size (up to 12 mm long) and possess a hollow axial stylet, about 100 - 300 µm long, that enables them to feed deep within plant roots (Fig. 1; 1B-C). The stylet is composed of an anterior spear (odontostyle) that penetrates the root cells and of a posterior part (odontophore) to which stylet protractor muscles are attached. The odontophore also contains nerve tissue adjacent to the food canal and is hence thought to have a gustatory function [46] to enable food discrimination deep within the root. The oesophagus consists of a long, narrow convoluted anterior tube that connects the stylet with a cylindrical muscular basal bulb. This bulb is about 100 µm long and contains a triradial pump chamber to which radial muscles for food ingestion are attached. It also harbours three secretory gland cells, one dorsal and two subventral.

2.2.1. *Genus Xiphinema*
Some *Xiphinema* spp. may feed for many hours or even several days from cells within the differentiated vascular cylinder. Others, for instance *X.index* and *X.diversicaudatum*, feed exclusively on root tips, which they transform into attractive galls, containing groups of much enlarged multinucleate cells (Fig. 1; 1B). These nurse cells arise by synchronous mitoses without cytokinesis [e.g. 48, 71]. They are metabolically highly active and essential for nematode reproduction.

The best examined species is *Xiphinema index,* the vector of grapevine fanleaf virus. The very large dorsal secretory gland cell fills nearly the whole bulb and has an elaborate duct system. During feeding large amounts of salivary secretions are flushed forward through the ducts to a main collecting duct that opens into the food canal just anterior to the pump chamber [45, 46]. The orifice of this duct is opened by dilator muscles at definite feeding phases [69, 72]. The ducts of the two subventral glands open midway into the bulb chamber, and the secretions of these comparatively small glands are most likely ingested during feeding [69, 72].

On root tips of fig seedlings, grown in monoxenic agar culture, *X.index* feed on a column of cells, feeding on each cell for a few minutes only and then progressing into successively deeper layers. Secretory fluids of the dorsal gland are injected shortly after a cell wall has been perforated and also during brief pauses between spells of food ingestion [45, 69]. With the aid of video-enhanced interference contrast microscopy, cell responses to injected salivary fluids could be observed and analyzed [69, 72]. The thin walls of meristematic cells are easily perforated by the stylet tip, which then stays protruded in the perforated cell a few μm deep throughout feeding. A few seconds after wall perforation, secretory fluids from the dorsal gland are injected, which quickly liquefy the cytoplasm around the inserted stylet tip. Then the nematodes start to ingest the modified nutrients by a pumping action of the basal bulb. The ducts of the dorsal gland become filled again with secretory fluids, which are injected even more forcefully after a brief pumping period. Within about one minute of food ingestion, interspersed by one to two salivary injections, most of the cytoplasm and also the nucleoplasm are liquefied. The nuclear envelope is obviously not affected, whereas the nucleolus is quickly degraded. Food withdrawal from each meristematic cell lasts about two minutes, then the stylet is pushed through the emptied cell towards the next deeper cell, the contents of which are liquefied and ingested in the same way. After maximal protrusion, the stylet is retracted and soon afterwards it is inserted into the root tip at another site. The responses described are typical for all attacked meristematic cells [69, 72]. Occasionally feeding nematodes become inactive for many minutes immediately after cell wall perforation. Secretory fluids are then seen to emanate through the stylet tip orifice, forming a plug-like structure [72]. During the extensive periods of continuous ingestion that follow such an inactive phase, the plug impedes the injection of salivary secretions for some time and thus retards the liquefaction of the cell contents.

The secretory fluids, produced in the dorsal oesophageal gland and injected into root tip cells have obviously a double function: they quickly liquefy the protoplast of directly attacked cells to facilitate food ingestion and at the same time modify neighbouring cells to become multinucleate and metabolically highly active. One day after the first attack, necrotic cells are surrounded by binucleate cells. These cells are degraded by *X. index* in the same way as uninucleate meristematic cells, but the time necessary for their liquefaction now lasts longer, hence food ingestion, interspersed by salivary secretions, now lasts several minutes [69, 72]. It is even more prolonged, when the nematodes feed on the much enlarged multinucleate cells inside root tip galls.

2.2.2. *Genus Longidorus*

Longidorus spp. feed apparently exclusively on root tips, transforming them into terminal galls. They insert their stylets nearly fully protracted before they start salivating and feeding at the same spot for several hours. Cells that have been fed on become necrotic and are surrounded by hypertrophied, metabolically active cells (Fig. 1; 1C), the nuclei of which do not divide. In root tip galls of *Lolium perenne*, induced by *L.*

elongatus, a cluster of empty cells, surrounded by modified uninucleate cells, becomes progressively larger as feeding progresses [22]. The walls of empty cells adjacent to modified cells are holed, possibly as a result of local cell wall dissolution, indicating a route of food withdrawal from the modified cells [44]. Slightly different cell responses have also been described [e.g. 76].

3. Tylenchid Nematodes

The order Tylenchida is divided into three suborders [38]. All root parasitic nematodes belong to the suborder Tylenchina, which includes nine families and numerous genera [38]. The economically most important root parasitic tylenchids, cyst and root-knot nematodes, belong to the family Heteroderidae.

3.1. THE FEEDING APPARATUS

All tylenchid nematodes possess a uniform feeding apparatus (Fig. 2). The stylet has a narrow lumen, through which salivary secretions are injected into punctured cells and liquids of their modified cytoplasm are ingested. Close to the stylet tip the lumen has a ventrally located orifice, which is usually not larger than about 100 nm in internal diameter. Hence intact larger cell organelles, for instance mitochondria and plastids, cannot be withdrawn during food ingestion. At its posterior end the stylet is composed of three knobs to which protractor muscles are attached. The oesophagus is divided into a nonmuscular procorpus, a muscular metacorpus with a metacorpal bulb, containing a pump chamber and a posterior nonmuscular glandular region. The oesophageal lumen (food canal) is at first narrow and circular in cross section, but within the pump chamber of the metacorpal bulb it is triradiate and continues in this form right back to the intestine. During food ingestion the radial muscles of the metacorpal bulb contract several times per second, dilating the pump chamber at each contraction. Constraining muscles in front and behind the bulb stay contracted to stabilize the pulsating bulb.

The nonmuscular glandular region is composed of three secretory cells, one dorsal and two subventral. Each cell produces copious amounts of salivary proteins that are sequestered into secretory granules. The granules are then transported through a cytoplasmic extension (duct) into a terminal ampulla, where they are collected prior to the release of their contents into a valve with a complex membrane structure [16]. The valve of the dorsal duct joins the food canal just behind the stylet knobs, whereas the valves of the two subventral glands join the triradiate food canal just behind the pump chamber (Fig. 2). Due to the structure of the pump with a narrow rigid circular lumen in front and an expandable triradiate lumen behind, secretory fluids of the subventral glands flow most probably into the intestine, where they may assist digestion [28, 65, 74]. However, it still remains to be clarified whether this is indeed their only role. Artificially induced oral release of subventral granule proteins was demonstrated by specific monoclonal antibodies in cyst [21] and root-knot nematodes [15], and hence it was speculated that they may play a role in the initial stages of root infection.

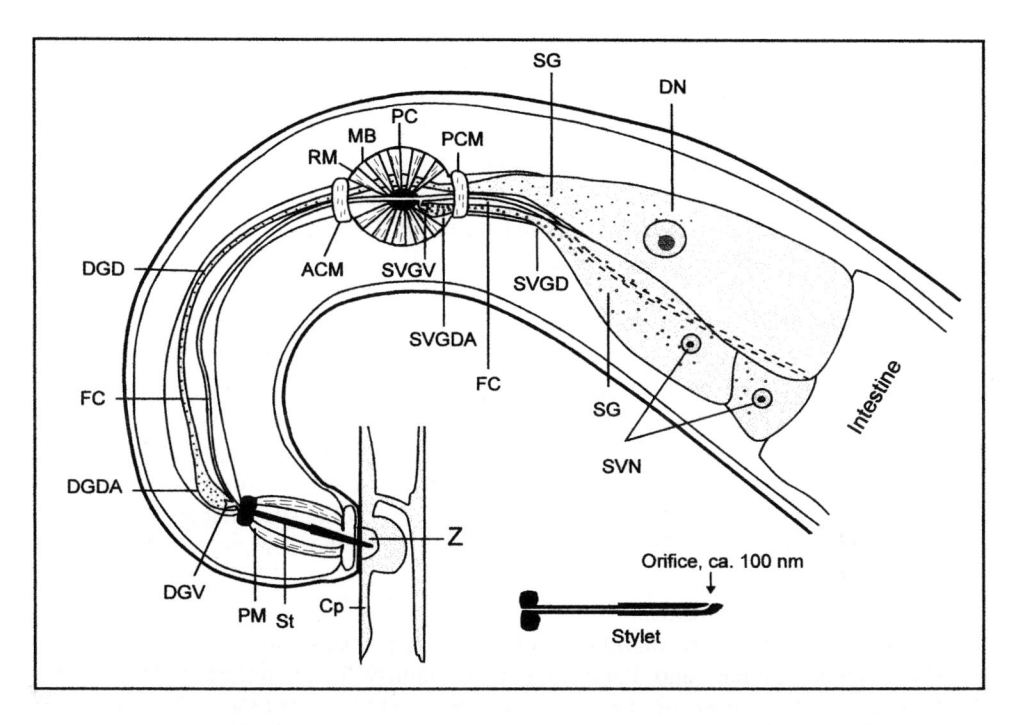

Figure 2. Feeding apparatus of a tylenchid nematode (here *Tylenchorhynchus dubius*, feeding on a root hair). Abbreviations: ACM, anterior constraining muscles of metacorpal bulb; Cp, cytoplasm of root hair; DGD, dorsal gland duct; DGDA, dorsal gland duct ampulla; DGV, dorsal gland valve; DN, dorsal gland nucleus; FC, food canal; MB, metacorpal bulb; PC, pump chamber of metacorpal bulb; PCM, posterior constraining muscles; PM, stylet protractor muscles; RM, radial muscles; SG, secretory granules; SVGD, subventral gland duct; SVGDA, subventral gland duct ampulla; SVGV, subventral gland valve; SVN, nuclei of the two subventral glands; St, stylet; Z, zone of modified cytoplasm around inserted stylet tip.

Secretory fluids from the dorsal gland are definitely injected through the stylet into root cells. They are obviously involved in the regulation of gene activity by yet unknown mechanisms in order to initiate and maintain specific nurse cell systems. They may also modify the cytoplasm around the inserted stylet tip to facilitate food ingestion. With certainty they are involved in the formation of feeding tubes. These tubes are essential tools for food withdrawal and are probably produced by all sedentary endoparasites [47, 64, 73]. The ultrastructure of feeding tubes was described in detail for the reniform nematode, *Rotylenchulus reniformis* [42] and root-knot [29] as well as cyst nematodes [53]. Feeding tubes are also produced by migratory ecto-endoparasites, which feed for many days on a single cortical cell [35].

3.2. PARASITIC STRATEGIES

In spite of the uniform structure of their feeding apparatus, tylenchid nematodes have evolved many different parasitic strategies to exploit the plant root as a food source. The strategies can be classified into the following main categories:

3.2.1. *Migratory Ectoparasitism*

As in dorylaimid nematodes, each member of this category stays vermiform throughout its life cycle and feeds at a selected site for only a limited period of time. Those feeding on epidermal cells, such as *Tylenchorhynchus dubius* (Fig. 2), possess relatively short stylets and cause only a moderate perturbance of the protoplast within the cells they attack. After cell wall perforation, secretions from the dorsal gland are injected, forming a clear zone of cytoplasm around the inserted stylet tip. During salivation and ingestion cytoplasm accumulates at this zone. Video-enhanced contrast light microscopy showed that the 'clear zone' apparently consists only of salivary secretions, which may act as a cytoplasmic filter or may help to predigest the surrounding cytoplasm [63].

A number of nematode genera within this category have long, needlelike stylets, eg. species of the genus *Belonolaimus*, with which they exploit deeper root tissues. Although *Belonolaimus* causes serious root damage, including extensive necrosis, very little is known about cellular responses to feeding. *Hemicycliophora* spp. usually feed on root tips, transforming them into gall-like swellings. Prolonged feeding of *H.typica* on emerging root tips of rice seedlings produces cavities at feeding sites that are surrounded by intact cells with partially dissolved cell walls, similar to those of syncytia [5].

3.2.2. *Sedentary Ectoparasitism*

Members of this group are mainly encountered in the subfamilies Criconematinae (family Criconematidae) and Paratylenchinae (family Tylenchulidae). Within the Paratylenchinae, *Paratylenchus* spp. have been observed to ingest food for many days at the same site on epidermal cells and root hairs without causing any serious disturbance of the cell's protoplast [43]. Within the Criconematinae *Criconemella xenoplax* (Fig. 1; 3) is the best examined species. It feeds continuously for many days from a single cell in the outer root cortex in various hosts examined. Nutrients are withdrawn from this cell via a zone of modified cytoplasm in intimate contact with the stylet tip orifice. This is the only part of the inserted stylet that is not covered by callose depositions [31]. The plasmalemma of the food cell surrounds these depositions and becomes tightly appressed to the wall of the stylet orifice, creating a hole for an unimpeded uptake of nutrients [30]. Feeding tubes have not been detected, but fibrillar material appears to form a barrier in the cytosol around the stylet orifice in order to limit movement of cell organelles towards it. Plasmodesmata between the food cell and surrounding cells are modified in a unique manner, obviously to facilitate solute transport to the food cell [30].

3.2.3. *Migratory Ecto-Endoparasitism*

Nematodes of this category remain vermiform throughout development and can feed ectoparasitically on roots. In most cases, however, they invade the roots partially to feed on cortical or outer stelar cells. Members of this category belong frequently to the subfamily Hoplolaiminae (family Hoplolaimidae). *Helicotylenchus* spp. were observed to feed for many days on a single cell in cereal roots. The food cell (Fig. 1; 4), usually located close to the stele, is surrounded by a few metabolically active cells with dense cytoplasm but without nuclear enlargement. Feeding tubes, surrounded by a membraneous network, were detected in the food cell [35]. An extensive membraneous network, thought to originate from the plasmalemma, was also seen to surround the stylet tip and feeding tubes of *Scutellonema brachyurum* in cortical cells in potato roots [49].

3.2.4. *Migratory Endoparasitism*

Migratory endoparasites are found in members of the subfamily Pratylenchinae (family Pratylenchidae), of which *Pratylenchus* and *Radopholus* are the economically most important genera. *Pratylenchus* spp. generally have a wide host range, invade or leave the root at any developmental stage, and in nearly all cases restrict their parasitism to cortical cells (Fig. 1; 5). Younger stages, however, prefer to feed on root hairs. The feeding behaviour of *P. penetrans* on root hairs and cortical cells has been studied with the aid of video-enhanced contrast light microscopy [77, 78]. Extended feeding on cortical cells leads to a loss in turgor pressure and a gradual increase in the size of the nucleus, followed by cell death, often several hours after the nematode has moved away [78]. Many cortical cells are killed by this nematode during intracellular migration. In alfalfa roots, this destruction is not limited to cortical cells. It extends also to adjacent, uninvaded cells. Tannin deposition occurs, membrane integrity is lost, and cell organelles degenerate. This effect is most pronounced in the endodermis but can also extend into stelar cells [58].

3.2.5. *Sedentary Endoparasitism*

Nematodes within this category are the most advanced specialists and include the economically most important root-knot and cyst nematodes. Various other genera of less economic importance can also be assigned to this category. Associated with sedentariness is female obesity and sexual dimorphism. The female is always sedentary, whereas the male stays or becomes veriform again. Several authors, e.g [37] classify sedentary endoparasites into different categories, depending on the localization of the food cells within the root tissue. However, as females become obese and stay permanently fixed to their feeding structures within the root, this classification is not pursued here. In many cases, especially concerning *Meloidogyne* and *Nacobbus* spp., the roots respond with hyperplastic cell responses around the permanent feeding structures, producing galls, in which the females become embedded.

Sedentary nematodes invade roots and induce specific nurse cells or nurse cell systems as a source of nutrients at a particular developmental stage. The stages involved in feeding site induction are either freshly hatched second-stage juveniles (J_2; e.g. *Globodera*, *Heterodera*, *Meloidogyne* spp.) or adult females (e.g. *Nacobbus*, *Rotylenchulus*, *Tylenchulus* spp.). Locomotory musculature is lost after feeding site induction and the body volume increases soon after feeding begins.

4. Parasitic Strategies of Sedentary Endoparasites

The thin and translucent roots of some cruciferous plants provide excellent facilities to study the parasitic behaviour of some cyst and root-knot nematodes inside the root, especially when light microscopy is enhanced by video contrast [63]. One of these cruciferous plants, *Arabidopsis thaliana*, has now become the most important model plant for *in vivo* observations of the parasitic behaviour [25, 26, 52, 66, 67] as well as for molecular studies to identify genes or gene regulating mechanisms involved in the induction and maintenance of nurse cell systems by cyst and root-knot nematodes.

4.1. CYST NEMATODES

The parasitic behaviour of *Heterodera schachtii* has so far most intensively been studied among cyst nematodes, because cruciferous plants, including *A.thaliana*, are good hosts of this species. Freshly hatched J_2 invade the roots predominantly in the zone of elongation. Continuous vigorous stylet thrusts, stimulated by root diffusates [24], finally rupture the wall of epidermal cells and the J_2 invade the first cortical cells, from where they progress by intracellular migration to the vascular cylinder. Although the two subventral glands, including their ampullae, are then packed with secretory granules, intracellular migration is obviously not assisted by salivary secretions. The walls of cortical and endodermal cells are cut open by highly coordinated stylet thrusts that produce a line of merging holes to form a slit, through which the J_2 pass into the neighbouring cell [74, 75]. In the vascular cylinder the destructive behaviour changes into subtle exploration when the initial cell for the developing nurse cell system is selected. All cyst nematodes induce and maintain syncytia (see later) as their permanent food source, hence the first cell chosen is generally termed initial syncytial cell (ISC). Once the wall of an ISC has been perforated by careful stylet thrusts, the stylet tip stays protruded in the ISC for 6-8 h [65]. During this preparation period no obvious changes in the protoplast of the ISC can be recognized, and the metacorpal bulb does not pump. The only visible events concern a gradual decrease in the density of secretory granules in the subventral glands, accompanied by an increase of granules in the dorsal gland, Also a few defecations are evident before the J_2 start to ingest food for the first time. Therefore it has been suggested that secretions of the subventral glands may be used to mobilize lipid reserves, while the intestine is transformed into an absorptive organ [1, 65]. J_2 that have completed this preparation period are no longer capable of leaving the root.

First visible changes in the ISC, such as increase in cytoplasmic streaming and enlargement of the ISC's nucleus, are apparent a few hours after the J_2 have started feeding. Feeding occurs in repeated cycles, each consisting of three distinct phases (I-III). During phase I, which is by far the longest and which increases with time, nutrients are withdrawn from the ISC by a continuous rapid pumping action of the metacorpal bulb. The constraining muscles in front and behind the bulb stay contracted throughout pumping and thus they impede a forward flow of secretory granules within the duct of the dorsal gland cell. Phase II is characterized by stylet retraction and reinsertion, and phase III by a continuous forward movement of secretory granules, especially from the dorsal gland, with the stylet tip staying inserted in the ISC. The three feeding phases, described in detail for a *H. schachtii* J_2 in an early infective stage [74] are maintained throughout development, also in the adult female stage [65]. These phases and also the preparation period are most likely common to all cyst nematodes, as they have also been described for J_2 and J_3 juveniles of *Globodera rostochiensis* feeding in tomato roots [54].

Initial stages of syncytium development, i.e. partially dissolved cell walls, are clearly visible 24 h after the J_2 have commenced feeding. About 12 h later feeding tubes become distinct in light microscopic studies [65]. At the ultrastructural level, however, these tubes are already detected at an earlier stage [53]. *In vivo o*bservations show that the feeding tube is surrounded by a zone of modified cytoplasm (Fig. 3), which keeps larger cell organelles away [65, 74, 75]. Electron microscopic studies confirm that this zone is free of larger organelles but rich in tubular endoplasmic reticulum [53, 73]. The feeding tube stays permanently attached to the tip of the stylet in phase I (Fig. 3). When the stylet is retracted at the beginning of phase II, the feeding tube becomes detached from the stylet tip. Soon afterwards the stylet is again inserted and a new feeding tube is formed during phase III by secretions of the dorsal gland cell.

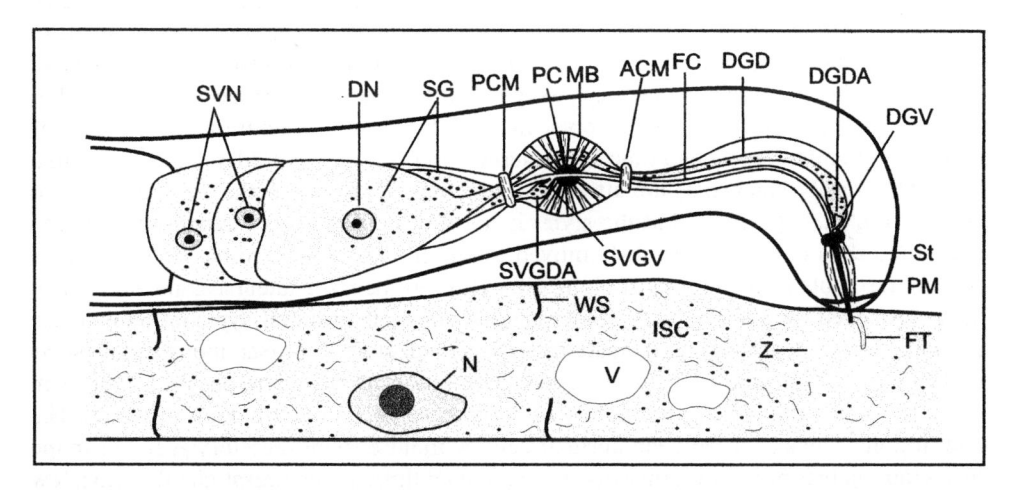

Figure 3. Schematic representation of a J_2 *H. schachtii,* feeding from an initial syncytial cell (ISC), about 36 h after its induction, here during feeding phase I. Abbreviations: FT, feeding tube; N, hypertrophied nucleus; V, vacuole; WS, wall stubs; Z, zone of modified cytoplasm. All other abbreviations as in *Figure 2*.

The ultrastructure of feeding tubes, produced by *H. schachtii* juveniles in *A. thaliana* was recently examined in detail [53]. They consist of an osmiophilic wall and an electron translucent lumen, containing membraneous structures, most probably derived from the tubular endoplasmic reticulum, which is frequently attached to the tubes. The feeding tubes are not surrounded by the plasmalemma and obviously function as a filter for a selective uptake of syncytial substances. Microinjection of fluorescence-labelled dextrans of different molecular weights into the syncytia, showed that only dextrans of 3, 10 and 20 kDa but not of 40 and 70 kDa are ingested by the nematodes. These results suggest that only molecules of up to maximum Stokes radius of 3.2 to 4.4 nm pass from the syncytium into the feeding tubes ([8, 9] and Grundler and Böckenhoff, this volume).

Female cyst nematodes have three feeding juvenile stages (J_2, J_3, J_4), whereas male nematodes have only two (J_2, J_3). The latter become vermiform again during the moult to the J_4 stage and emerge after the last moult through the juvenile cuticles in order to copulate with females. The total time of feeding activity of J_3 males exceeds that of J_3 females by at least one third [65], but altogether males require considerably less food than female nematodes. Sexual differentiation, already observed before the J_2 moult [23, 65], is obviously affected by the age of the host plant, its physiological stage as well as the diameter of invaded roots. In *A. thaliana* nearly all J$_2$ develop into males when they invade the thin lateral roots of 3 weeks old plants, whereas females develop mainly in younger primary roots [53].

4.2. ROOT-KNOT NEMATODES

It has generally been claimed that the J$_2$ of root-knot nematodes migrate intercellularly within roots of their host plants before they reach their permanent feeding site within the vascular cylinder [e.g. 17]. However, with the exception of an early study, conducted more than 50 years ago with a rather simple light microscopic equipment [36], a detailed documentation on the parasitic behaviour of the infective juveniles from root invasion to giant cell induction was lacking. Gaps were recently closed by studying the behaviour of *Meloidogyne incognita* J$_2$ inside roots of *A. thaliana* with the aid of video-enhanced contrast light microscopy and time lapse studies [67] (see also von Mende, this volume).

The freshly hatched J$_2$ are attracted to the tip of growing roots. Local exploration differs from *H.schachtii* in that the stylet is not thrusted vigorously against the wall of epidermal cells but is continuously moved to and fro within the stoma. The J$_2$ usually invade the roots in the region of elongation close to the meristematic zone. After they have weakened the thin walls of epidermal and subepidermal cells by continuous head rubbings and stylet movements, including occasional stylet tip protrusions followed by metacorpal bulb pumpings of a few seconds duration, they invade between destroyed cells. Inside the roots the J$_2$ invariably orient themselves in the direction of the root tip and migrate towards it between cortical cells without causing any damage. The same behaviour, noted during root invasion is maintained throughout migration, also when the J$_2$ have reached the meristematic region close to the root's apex. No changes in the cell's protoplast are evident, when the stylet tip appears to be inserted into a meristematic cell.

Having reached the apex of the root, the J$_2$ start to turn around and destroy some of the meristematic cells, causing retardation of root growth. Then nondestructive intercellular migration is resumed, now in direction towards the region of root differentiation. From root invasion onwards, 14 -18 h usually elapse until the J$_2$ have reached the differentiating vascular cylinder. By this time the density of secretory granules in the two subventral oesophageal glands appears diminished, compared to early events of root infection, when the glands are packed with granules. The reverse is true for the dorsal gland, which now contains many more granules.

When the J$_2$ have reached the vascular cylinder, they continue to migrate between the cambial cells for several hours without any changes in the behavioral pattern described for root invasion and intercellular migration. Eventually forward migration comes to an end when initial giant cells are induced. These cells become multinucleate within a few hours. About one day later, they are packed with nuclei. Even then and also during later stages, the head of the sedentary J$_2$ still moves in all directions, performing the same behaviour as described for the early events of root infection. Now, however, periods of stylet tip protrusion and metacorpal pumping increase with time. In contrast to earlier phases, when it is difficult to determine whether the short periods (few seconds) of metacorpal pumping are involved in food ingestion, the J$_2$ now withdraw nutrients from the initial giant cells. The growing juveniles become saccate, and having completed the J_2 stage, they enter an expanded moulting cycle, in which they moult three times in quick succession without any feeding activity. During the following adult stage, females continue feeding from the giant cells and start producing eggs, which are deposited into a gelatinous matrix, the so called egg sac. Reproduction of root-knot nematodes is almost exclusively parthenogenetic. Nevertheless, males occur regularly, especially under unfavourable conditions.

4.3. CELL RESPONSES TO SEDENTARY ENDOPARASITES

Nurse cells and nurse cell systems of sedentary nematodes (Fig. 1; 6A-G) are metabolically highly active, showing a pronounced increase in cytoplasmic density, accompanied by an increase in ribosomes, polyribosomes and cell organelles. The large central vacuole is replaced by numerous small secondary vacuoles and nuclei as well as nucleoli enlarge considerably. At later stages nearly all nuclei become amoeboid in profile due to an increase in nuclear-cytoplasmic exchange. The different types of cell responses induced are nematode-specific, regardless of the tissue and host in which they develop and can be classified into four groups [14,51, 56, 64].

4.3.1. *Discrete nonhypertrophied uninucleate nurse cells*
This nurse cell type appears to be common to some members of the subfamily Tylenchulinae (family Tylenchulidae). The nurse cells are usually induced in the cortex of the root and may consist of a single cell (Fig.1; 6A), such as in *Trophotylenchulus obscurus* [59], or of a cluster of cells (Fig. 1; 6B) as in *Tylenchulus* species [e.g. 2, 32]. *Trophotylenchulus floridensis* was reported to induce such cells in the vascular cylinder of the root [10].

4.3.2. *Single uninucleate giant cells*
This cell type is commonly found in hosts responding to non-cyst-forming genera in the subfamily Heteroderinae (family Heteroderidae). These nematodes, considered to be the ancestors of cyst nematodes, usually live in the roots of woody plants and have received attention mainly for phylogenetic studies (e.g. 3). The first description [57] of this type of nurse cell in a non-cyst-forming member (*Hylonema ivorense*) was followed by reports for other genera [3]. The invaded J_2 induce the single giant cell primarily in the pericycle, from where it expands into the stele (Fig. 1; 6D). The single nucleus is considerably enlarged and deeply invaginated to accommodate an increased rate of nuclear-cytoplasmic exchange. The wall of the giant cell is extremely thickened only at the head end of the nematode (Fig.1; 6D); elsewhere it contains numerous pit fields with many plasmodesmata for bulk transport of solutes from adjacent cells. In spite of several ultrastructural studies [e.g. 41], feeding tubes were not yet recorded in these cells. However, in the first single uninucleate giant cell described, in this case for a non-heteroderid nematode, *Rotylenchulus macrodoratus* [12], a feeding tube was shown to be attached to the nematode's stylet.

4.3.3. *Syncytia*

The syncytium (Fig. 1; 6C, 6E, 6F) is obviously the most common type of nurse cell system induced by sedentary endoparasites. It is formed by the expansion of several cells whose protoplasts become fused after partial cell wall dissolution. Mitosis is not stimulated but nuclei and nucleoli become considerably enlarged, concurrent with an increase in syncytial metabolism. Syncytia are most commonly formed within the vascular cylinder (Fig.1, 6F) as described for all cyst nematodes of the economically important genera *Globodera* [e.g. 40] and *Heterodera* [e.g. 6, 7, 19, 39], as well as in the economically less important genera *Cactodera* [4] and *Punctodera* [55]. Sometimes syncytia develop specifically in the pericycle, for instance in the case of *Rotylenchulus reniformis* (Fig. 1, 6E) in the subfamily Rotylenchulinae (family Hoplolaimidae) and of *Meloidoderita* sp. [13] and *Sphaeronema rumicis* [60], both belonging to the subfamily Tylenchulinae (family Tylenchulidae). Only occasionally syncytia may also develop in the cortex (Fig. 1; 6C) as reported for the non-cyst-forming heteroderid nematode *Verutus volvingentis* [11].

Increasing nutritional demands imposed by growing cyst nematodes stimulate polarized syncytial wall ingrowths adjacent to conductive tissues, especially xylem tracheary elements. These ingrowths are lined with plasmalemma and thus enhance short distance solute transport between the apoplast and symplast. Such modifications, typical for transfer cells, have not been detected in the syncytia of other nematodes, e.g. *Nacobbus aberrans*, in which a high plasmodesmatal frequency ensures a symplastic solute influx [34].

Syncytium development has recently been studied in roots of *Arabidopsis thaliana*, infected by *H. schachtii* [20, 53] (see also Golinowski *et al.*, this volume). In three weeks old lateral roots invaded J_2 develop into males [53]. Syncytia of male nematodes are apparently only successfully induced in a pericyclic cell, from where they extend in both axial directions within the pericycle, gradually incorporating more and more procambial and cambial cells, which, however, become not strongly hypertrophied. On the other hand syncytia of female nematodes are induced in a procambial cell, from where they extend within the procambial and cambial cells of the vascular cylinder. Incorporated cells are highly hypertrophied and connected to each other by extensive cell wall dissolutions. In both sexes the syncytial walls become rapidly thickened by wall depositions, which close the few plasmodesmata [53]. The syncytium thus becomes symplastically isolated from adjacent cells. Even if plasmodesmata should be preserved, they could no longer function, as the turgor pressure inside syncytia reaches high values of up to 10 000 hPa [8, 9]. Food supply of the syncytia is hence via the apoplast, i.e. through cell wall ingrowths adjacent to xylem tracheary elements. Recently it could convincingly be demonstrated that syncytia are also strong metabolic sinks for assimilates. New phloem sieve tubes are formed around the developing syncytium [53], and a specific phloem unloading into the syncytium could be proven with the fluorochrome 5(6)-carboxyfluorescein as well as with radioactively labelled photosynthates (^{14}C -sucrose) which was withdrawn by the feeding juveniles ([8] and Grundler and Böckenhoff, this volume).

4.3.4. Multinucleate giant cells

Root-knot nematodes (*Meloidogyne* spp.) infect a large number of host plants, and in all of them they induce a specific type of nurse cell system, classified as multinucleate giant cells ([27, 33] and Bleve-Zacheo and Melillo, this volume). These cells (Fig.1; 6G) develop by the expansion of about half a dozen cambial cells within the differentiating vascular cylinder. Each of the incorporated cells becomes multinucleate by repeated synchronous mitoses in the absence of cytokinesis. Mature giant cells function as transfer cells and are metabolically highly active, as revealed by the presence of aneuploid nuclei with 14-16 times more DNA than in root tip nuclei of uninfected plants [61]. The adult females produce feeding tubes in each giant cell, which are sealed at their distal ends and consist of an electron-dense crystalline wall. The tubes become surrounded by an elaborate membrane system [29], composed of smooth endoplasmic reticulum in contact with the tube and of rough endoplasmic reticulum at its periphery. The system probably synthesizes and/or transports solutes to the feeding tube, when it is attached to the stylet orifice of feeding females. Feeding tubes produced by *Meloidogyne* spp. differ in their structure from those produced by other sedentary endoparasites [e.g. 19, 42, 47, 53, 73] but as a common feature, they are always surrounded by an extensive membrane system.

References

1. Atkinson, H J., and Harris, P.D. (1989) Changes in nematode antigens recognized by monoclonal antibodies during early infections of soya beans with the cyst nematode *Heterodera glycines*, Parasitology 98, 479-487.
2. B'Chir, M.M. (1988) Organisation ultrastructurale du site trophique induit par *Tylenchulus semipenetrans* dans les racines de citrus, Revue Nématol. **11**, 213-222.
3. Baldwin, J.G. (1986) Testing hypotheses of phylogeny of Heteroderidae, in F. Lamberti and C. E. Taylor (eds.), *Cyst nematodes*, NATO ASI Series, Plenum, New York, pp. 75-100.
4. Baldwin, J.G., and Bell, A.H. (1985) *Cactodera eremica* n.sp., *Afenestrata africana* (Luc *et al.*, 1973) n. gen., n. comb., and an emended diagnosis of *Sarisodera* Wouts and Sher, 1971 (Heteroderidae), J. Nematol. **17**, 187-201.
5. Bleve-Zacheo, T., Lamberti, F., and Chinappen, M. (1987) Root cell response in rice attacked by *Hemicycliophora typica*, Nematol. medit, **5**, 129-138.
6. Bleve-Zacheo, T., Melillo, M.T., Andres, M., Zacheo G., and Romero, M.D. (1995) Ultrastructure of initial response of graminaceous roots to infection by *Heterodera avenae*, Nematologica **41**, 80-97.
7. Bleve-Zacheo, T., and Zacheo, G. (1987) Cytological studies of the susceptible reaction of sugarbeet roots to *Heterodera schachtii*, Physiol. Mol. Plant Pathol. **30**, 13-25.
8. Böckenhoff, A (1995) Untersuchungen zur Physiologie der Nährstoffversorgung des Rübenzystennematoden *Heterodera schachtii* und der von ihm induzierten Nährzellen in Wurzeln von *Arabidopsis thaliana* unter Verwendung einer speziell adaptierten *in situ* Mikroinjektionstechnik, Ph.D. thesis, University of Kiel, 135 pp.
9. Böckenhoff, A., and Grundler, F.M.W. (1994) Studies on the nutrient uptake by the beet cyst nematode *Heterodera schachtii* by *in situ* microinjection of fluorescent probes into the feeding structures in *Arabidopsis thaliana*, Parasitology **109**, 249-254.
10. Cohn, E., and Kaplan, D.T. (1983) Parasitic habits of *Trophotylenchulus floridensis* (Tylenchulidae) and its taxonomic relationship to *Tylenchulus semipenetrans* and allied species, J. Nematol. **15**, 514-523.
11. Cohn, E., Kaplan, D.T., and Esser, R.P. (1984) Observations on the mode of parasitism and histopathology of *Meloidodera floridensis* and *Verutus volvingentis* (Heteroderidae), J. Nematol. **16**, 256-264.
12. Cohn, E., and Mordechai, M. (1977) Uninucleate giant cell, induced in soybean by the nematode

Rotylenchulus macrodoratus, Phytoparasitica **5**, 85-93.
13. Cohn, E., and Mordechai, M. (1982) Biology and host-parasite relations of a species of *Meloidoderita* (Nematoda: Criconematoidea), *Revue Nématol.* **5**, 247-256.
14. Cohn, E.,and Spiegel, Y. (1991) Root-nematode interactions, in Y. Waisel, A. Eshel and U. Kafkafi (eds.), *Plant Roots; The Hidden Half,* Marcel Dekker, Dordrecht, pp. 789-805.
15. Davis, E.L., Allen, R., and Hussey, R.S. (1994) Developmental expression of esophageal gland antigens and their detection in stylet secretions of *Meloidogyne incognita, Fundam. appl. Nematol.* **17**, 255-262.
16. Endo, B.Y. (1984) Ultrastructure of the esophagus of larvae of the soybean cyst nematode, *Heterodera glycines, Proc. Helminthol. Soc. Wash.* **51**, 1-24.
17. Endo, B.Y. (1987) Histopathology and ultrastructure of crops invaded by certain sedentary endoparasitic nematodes, in J. A, Veech and D.W. Dickson (eds.), *Vistas on Nematology,* , Soc. Nematol., Hyattsville, MD, pp. 196-210.
18. Endo, B.Y.(1990) Ultrastructure of nematode-plant interactions, in K. Mendgen and D.E. Lesemann (eds.), *Electron Microscopy of Plant Pathogens*, Springer, Berlin, pp. 291-305.
19. Endo, B.Y. (1991) Ultrastructure of initial responses of susceptible and resistant soybean roots to infection by *Heterodera glycines, Revue Nématol.* **14**, 73-94.
20. Golinowski, W., Grundler, F.M.W., and Sobczak, M. (1996) Changes in the structure of *Arabidopsis thaliana* during female development of the plant-parasitic nematode *Heterodera schachtii, Protoplasma,* **194**, 103-116.
21. Goverse, A., Davis, E.L., and Hussey, R..S. (1994) Monoclonal antibodies to the esophageal glands and stylet secretions of *Heterodera glycines, J. Nematol.* **26**, 251-259.
22. Griffiths, B.S., and Robertson, W.M. (1984) Morphological and histochemical changes occurring during the life-span of root-tip galls on *Lolium perenne* induced by *Longidorus elongatus. J. Nematol.* **16**, 223-229.
23. Grundler, F. (1989) Untersuchungen zur Geschlechtsdetermination des Rübenzystennematoden *Heterodera schachtii* Schmidt, Ph.D. thesis, University of Kiel , 114 pp.
24. Grundler, F.M.W., Schnibbe, L., and Wyss, U. (1991) *In vitro* studies on the behaviour of second-stage juveniles of *Heterodera schachtii* (Nematoda: Heteroderidae) in response to host plant root exudates, *Parasitology* **103**, 149-155.
25. Grundler, F.M.W., Böckenhoff, A., Schmidt, K-P., Sobczak, M., Golinowski, W., and Wyss, U. (1994) *Arabidopsis thaliana* and *Heterodera schachtii* : a versatile model to characterize the interaction between host plants and cyst nematodes, in *Advances in molecular Plant* Nematology. NATO ASI Series, eds. F. Lamberti, C. De Giorgi, and D. McK Bird, pp. 171-80. New York, Plenum, 309 pp.
26. Grundler, F.M.W., and Wyss, U. (1995) Strategies of root parasitism by sedentary plant parasitic nematodes, in K. Kohmoto, U.S. Singh and R.P. Singh (eds), *Pathogenesis and Host Specificity in Plant Diseases, Histopathological, Biochemical, Genetic and Molecular Bases. Vol II* Eukaryotes, Pergamon, Oxford, pp. 309-319.
27. Huang. C.S. (1985) Formation, anatomy and physiology of giant cells induced by root-knot nematodes, in J.N. Sasser and C.C. Carter (eds.), *An Advanced Treatise on Meloidogyne, Vol. I, Biology and Control,* NC State Univ. Graphics, Raleigh, NC, pp. 155-164.
28. Hussey, R.S. (1989) Disease-inducing secretions of plant-parasitic nematodes, *Annu. Rev. Phytopathol.* **27**, 123-141.
29. Hussey, R.S., and Mims, C.W. (1991) Ultrastructure of feeding tubes formed in giant cells induced in plants by the root-knot nematode *Meloidogyne incognita, Protoplasma* **162**, 99-107.
30. Hussey, R.S., Mims, C.W. ,and Westcott, S.W. (1992) Ultrastructure of root cortical cells parasitized by the ring nematode *Criconemella xenoplax, Protoplasma* **167**, 55-65.
31. Hussey, R.S., Mims, C.W., and Westcott, S.W. (1992) Immunocytochemical localization of callose in root cortical cells parasitized by the ring nematode *Criconemella xenoplax, Protoplasma* **171**, 1-6.
32. Inserra, R.N., Vovlas, N., and O'Bannon, J.H. (1988) Morphological and biological characters of diagnostic significance in *Tylenchulus* and *Trophotylenchulus* species, *Nematologica* **34**, 412-421.
33. Jones, M.G.K. (1981) Host cell responses to endoparasitic nematode attack: structure and function of giant cells and syncytia, *Ann. appl. Biol.* **97**, 353-372.
34. Jones, M.G.K., and Payne, H.L. (1977) The structure of syncytia induced by the phytoparasitic nematode *Nacobbus aberrans* in tomato roots, and the possible role of plasmodesmata in their nutrition, *J. Cell Sci.* **23**, 299-313.

35. Jones, R.K. (1978) Histological and ultrastructural changes in cereal roots caused by feeding of *Helicotylenchus* spp., *Nematologica* **24**, 393-397.
36. Linford, M.B. (1942) The transient feeding of root-knot nematode larvae, *Phytopathology* **32**, 580-89.
37. Maggenti, A.R. (1981) *General Nematology*. Springer, New York, 372 pp.
38. Maggenti, A.R. (1991) Nemata: higher classification, in W.R. Nickle (ed.), *Manual of Agricultural Nematology*, M. Dekker, New York, pp. 147-187.
39. Magnusson, C., and Golinowski, W. (1991) Ultrastructural relationships of the developing syncytium induced by *Heterodera schachtii* (Nematoda) in root tissues of rape, *Can. J. Bot.* **69**, 44-52.
40. Melillo, M.T., Bleve-Zacheo, T., and Zacheo, G. (1990) Ultrastructural response of potato roots susceptible to cyst nematode *Globodera pallida* pathotype Pa 3, *Revue Nématol.* **13**, 17-28.
41. Mundo-Ocampo, M., and Baldwin, J.G. (1984) Comparison of host response of *Cryphodera utahensis* with other Heteroderidae, and a discussion of phylogeny, *Proc. Helminthol. Soc. Wash.* **51**, 25-31.
42. Rebois, R.V. (1980) Ultrastructure of a feeding peg and tube associated with *Rotylenchulus reniformis* in cotton, *Nematologica* **26**, 396-405.
43. Rhoades, H.L., and Linford, M.B. (1961) A study of the parasitic habit of *Paratylenchus projectus* and *P. dianthus*, *Proc. Helminthol. Soc. Wash.* **28**, 185-190.
44. Robertson, W.M., Trudgill, D.L., and Griffiths, B.S. (1984) Feeding of *Longidorus elongatus* and *L. leptocephalus* on root-tip galls of perennial ryegrass (*Lolium perenne*), *Nematologica* **30**, 222-229.
45. Robertson, W.M., and Wyss, U. (1979) Observations on the ultrastructure and function of the dorsal oesophageal gland cell in *Xiphinema index*, *Nematologica* **25**, 391-396.
46. Robertson, W.M., and Wyss, U. (1983) Feeding processes of virus-transmitting nematodes, in K.F. Harris (ed.), *Current Topics in Vector Research Vol 1*, Praeger, New York, 325 pp.
47. Rumpenhorst, H.J. (1984) Intracellular feeding tubes associated with sedentary plant parasitic nematodes, *Nematologica* **30**, 77-85.
48. Rumpenhorst, H.J., and Weischer, B. (1978) Histopathological and histochemical studies on grapevine roots damaged by *Xiphinema index*, *Revue Nématol.* **1**, 217-225.
49. Schuerger, A.C., and McClure, M.A. (1983) Ultrastructural changes induced by *Scutellonema brachyurum* in potato roots, *Phytopathology* **73**, 70-81.
50. Sijmons, P. C. (1993) Plant nematode interactions, *Plant Mol. Biol.* **23**, 917-931.
51. Sijmons, P.C., Atkinson, H.J., and Wyss, U. (1994) Parasitic strategies of root nematodes and associated host cell responses, *Annu. Rev. Phytopathol.* **32**, 235-259.
52. Sijmons, P.C., Grundler, F.M.W., von Mende, N., Burrows, P.R., and Wyss, U. (1991) *Arabidopsis thaliana* as a new model host for plant-parasitic nematodes, *Plant J.* **1**, 245-254.
53. Sobczak, M. (1996) Investigations on the structure of syncytia in roots of *Arabidopsis thaliana* induced by the beet cyst nematode *Heterodera schachtii* and its relevance to the sex of the nematode, Ph.D. thesis, University of Kiel.
54. Steinbach, P. (1973) Untersuchungen über das Verhalten von Larven des Kartoffelzystenälchens (*Heterodera rostochiensis* Wollenweber, 1923) an und in Wurzeln der Wirtspflanze *Lycopersicon esculentum* Mill.). III. Die Nahrungsaufnahme von Kartoffelnematodenlarven, *Biol. Zbl.* **92**, 563-582.
55. Suarez, Z., Sosa Moss, C., and Inserra, R.N. (1985) Anatomical changes induced by *Punctodera chalcoensis* in corn roots, *J. Nematol.* **17**, 242-244.
56. Subbotin, S.A. (1993) Evolution of modified food cells induced by sedentary nematodes in plant roots, *Russian J. Nematol.* **1**, 17-26.
57. Taylor, D.P., Cadet, P., and Luc, M. (1978) An unique host-parasite relationship between *Hylonema ivorense* (Nematoda: Heteroderidae) and the roots of a tropical rainforest tree, *Revue Nématol.* **1**, 99-108.
58. Townshend, J.L., Stobbs, L., and Carter, R. (1989) Ultrastructural pathology of cells affected by *Pratylenchus penetrans* in alfalfa roots, *J. Nematol.* **21**, 530-539.
59. Vovlas, N. (1987) Parasitism of *Trophotylenchulus obscurus* on coffee roots, *Revue Nématol.* **10**, 337-342.
60. Vovlas, N., and Inserra, R.N. (1986) Morphometrics, illustration, and histopathology of *Sphaeronema rumicis* on cottonwood in Utah, *J. Nematol.* **18**, 239-46.
61. Wiggers, R.J., Starr, J.L., and Price, H.J. (1990) DNA content and variation in chromosome number in plant cells affected by *Meloidogyne incognita* and *M.arenaria*, *Phytopathology* **80**, 1391-1395.
62. Wyss, U. (1981) Ectoparasitc root nematodes: Feeding behavior and plant cell responses, in B.M. Zuckerman and R.A. Rhode (eds.), *Plant Parasitic Nematodes. Vol. III*, Academic, New York, pp.

325-351.
63. Wyss, U. (1987) Video assessment of root cell responses to dorylaimid and tylenchid nematodes, in J. A, Veech and D.W. Dickson (eds.), *Vistas on Nematology.*, Soc. Nematol., Hyattsville, MD, pp. 211-220.
64. Wyss, U. (1988) Pathogenesis and host-parasite specificity in nematodes, in R.S. Singh, U.S. Singh W.M. Hess and D.J. Weber (eds.), *Experimental and Conceptual Plant Pathology, Vol. 2. Pathogenesis and Host-Parasite Specificity*, Gordon and Breach, New York, pp. 417-32.
65. Wyss, U. (1992) Observations on the feeding behaviour of *Heterodera schachtii* throughout development,including events during moulting, *Fundam. appl. Nematol.* **15**, 75-89.
66. Wyss, U., and Grundler, F.M.W. (1992) Seminar: *Heterodera schachtii* and *Arabidopsis thaliana*, a model host-parasite interaction, *Nematologica* **38**, 488-493.
67. Wyss, U., Grundler, F.M.W. and Münch, A. (1992) The parasitic behaviour of second-stage juveniles of *Meloidogyne incognita* in roots of *Arabidopsis thaliana, Nematologica* **38**, 98-111.
68. Wyss, U., and Inst. Wiss. Film (1974) *Trichodorus similis* (Nematoda) - Response of protoplasts in root hairs (*Nicotiana tabacum*) to feeding, Film E 2045, IWF Göttingen, 19 pp.
69. Wyss, U., and Inst. Wiss. Film (1988) Responses of root-tip cells (*Ficus carica*)) to feeding of the nematode *Xiphinema index*, Film D 1657, IWF Göttingen, 22 pp.
70. Wyss, U., Jank-Ladwig, R., and Lehmann, H. (1979) On the formation and ultrastructure of feeding tubes produced by trichodorid nematodes, *Nematologica* **25**, 385-390.
71. Wyss, U., Lehmann, H., and Jank-Ladwig, R. (1980) Ultrastructure of modified root-tip cells in *Ficus carica*, induced by the ectoparasitic nematode *Xiphinema index, J. Cell Sci.* **41**, 193- 208.
72. Wyss, U., Robertson, W.M., and Trudgill, D.L. (1988) Oesophageal bulb function of *Xiphinema index* and associated root cell responses, assessed by video-enhanced contrast light microscopy, *Revue Nématol.* **11**, 253-261.
73. Wyss, U., Stender, C., and Lehmann, H. (1984) Ultrastructure of feeding sites of the cyst nematode *Heterodera schachtii* Schmidt in roots of susceptible and resistant *Raphanus sativus* L. var. *oleiformis* Pers. cultivars, *Physiol. Plant Pathol.* **25**, 21-37.
74. Wyss, U., and Zunke, U. (1986) Observations on the behaviour of second stage juveniles of *Heterodera schachtii* inside host roots, *Revue Nématol.* **9**, 153-165.
75. Wyss, U., Zunke, U., and Inst. Wiss. Film. (1986) *Heterodera schachtii* (Nematoda) - Behaviour inside the root (rape).,Film E 2904, IWF Göttingen, 21 pp.
76. Zacheo, G., and Bleve-Zacheo, T. (1995) Plant-nematode interactions: histological, physiological and biochemical interactions, in K. Kohmoto, U.S. Singh and R.P. Singh (eds.), *Pathogenesis and Host Specificity in Plant Diseases. Histo-pathological, Biochemical, Genetic and Molecular Bases,Vol.II Eukaryotes*, Pergamon, Oxford, pp.321-353.
77. Zunke, U. (1990) Ectoparasitic feeding behaviour of the root lesion nematode, *Pratylenchus penetrans*, on root hairs of different host plants, *Revue Nématol.* **13**, 331-337.
78. Zunke, U. (1990) Observations on the invasion and endoparasitic behavior of the root lesion nematode *Pratylenchus penetrans, J. Nematol.* **22**, 309-320.

A: CELLULAR AND MOLECULAR ASPECTS OF THE INTERACTION

ROOT ANATOMY AND DEVELOPMENT, THE BASIS FOR NEMATODE PARASITISM

Ben SCHERES, Peter C. SIJMONS[1], Claudia van den BERG, Heather McKHANN, Geert de VRIEZE, Viola WILLEMSEN and Harald WOLKENFELT
Dpt. of Molecular Cell Biology, Padualaan 8, NL-3584CH Utrecht, The Netherlands
[1]*ATO-DLO, Bornsesteeg 59, NL-6708 PM Wageningen, The Netherlands*

Abstract

Plant parasitic nematodes appear to rely on very specific interactions with root cells to establish a feeding site. To understand these interactions in detail, it is of advantage to achieve a basic understanding of root development. *Arabidopsis thaliana* is a suitable plant to investigate root development genetically and molecularly, and it can act as a host plant for plant parasitic nematodes. The anatomy and the ontogeny of the *Arabidopsis* root can be described in considerable detail. Despite the rigorous lineage relationships in the root, laser ablation experiments demonstrate the presence of continuous information in the root meristem. This information guides cells to differentiate appropriately, according to position. A large spectrum of promoter traps that are specifically expressed in roots are examined in detail, and put into four categories. These expression patterns can be complex, and a relation between the tagged gene and cell type is not always obvious. As a complementary approach, genetic analysis, using specific mutants, is now beginning to unravel key genes that are involved in setting up the pattern of cell differentiation in the root. Combining promoter trap analyses with mutant analysis may create novel strategies for nematode control.

1. Introduction

Controlling an organism by manipulating its favorite food is an ancient strategy. For this control, a thorough knowledge on the organism's gastronomic preferences is required. In this context, this chapter focuses on the obligate partner of parasitic nematodes, the plant. A biological interaction between two different organsims, in this case nematodes and plants, can only be understood in molecular detail if both partners are investigated. For example, study of the *Rhizobium*-legume symbiosis, initially facilitated by the application of bacterial genetics, increasingly relies on analysing the plant response [14]. In a similar vein, knowledge of the genetic program by which the root normally develops is a prerequisite for understanding how a nematode selects particular cells and modifies them to serve as a feeding site. Much effort is currently aimed at the establishment of "catalogues" of genes that are up- or down- regulated in feeding sites, using promoter trapping and differential display techniques (see chapters by Fenoll *et al.* and Ohl *et al.*, this volume). Basic knowledge of plant structure and development may help to put this information into developmental perspective. In this chapter we will review recent studies on root development in *Arabidopsis thaliana*, a weed ideally suited for genetic analysis which can be infected by various nematodes [13]. Data on root anatomy and ontogeny, promoter/enhancer trap expression patterns and developmental mutants will be presented. We will conclude with the potential relevance of these data for understanding a fundamental aspect of plant-nematode interactions: the establishment of a feeding site.

2. Anatomy of Arabidopsis Roots

Plant roots have a simple tissue organisation, in which outer rings of epidermal and cortical cells surround a central vascular bundle. *Arabidopsis* roots are a paragon of this regularity. They contain a surprisingly constant number and arrangement of cells in cross-section (Fig. 1A). In longitudinal view, files of each cell type terminate in so-called initial cells (Fig. 1B). A small set of initials for all tissues surround four quiescent cells [6]. This quiescent centre contacts all the initials, an observation that suggests regulatory functions. Quiescent centre and initials together are termed the promeristem, the minimal construction centre of the root [4]. All cells within the promeristem are laid down during embryogenesis, and exhibit the division pattern typical for the root meristem from the heart stage of embryogenesis onward. Cells that leave the meristem as a result of this division pattern progressively differentiate into the various mature cell types as predicted by their position.

3. Ontogeny of Arabidopsis Roots

The ontogeny of an organ describes the way in which it arises from its precursor cells. Ontogenetic studies can be performed by analysing anatomical changes that define a series of developmental stages. This is particularly easy if cells can be readily distinguished either by positional or by morphological criteria. If the latter is not the case, genetic markers can be used to recognise the progeny of single cells ("clonal analysis"). We have performed both anatomical and clonal analysis to study how the primary root meristem is laid down during embryo development. We analysed blue sectors which arose by transposon excision from the *uidA (GUS)* marker gene in transgenic plants. Large sectors mark the progeny of a single embryonic cell. The end points and width of these sectors allowed us to deduce a complete fate map for the *Arabidopsis* root [11]. The root promeristem arises from daughters of both the basal and apical cell which are separated at the first zygotic division: the quiescent centre and

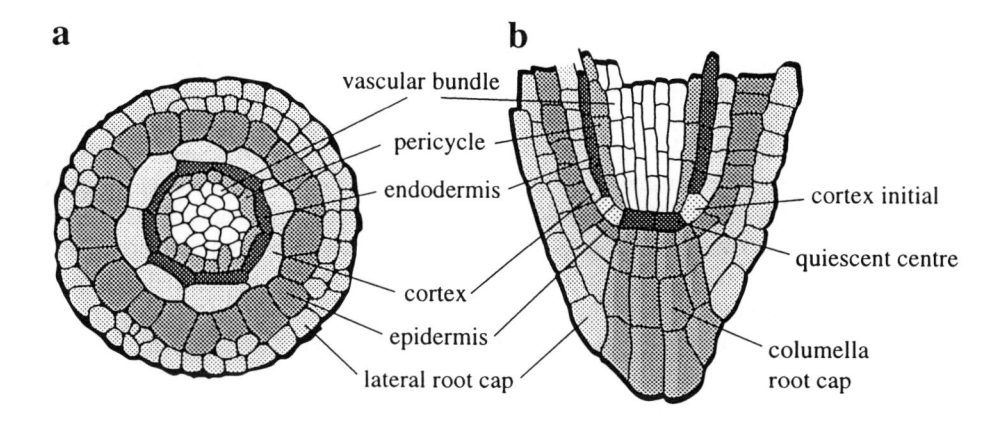

Figure 1. Anatomy of the *Arabidopsis* root. (A) Transverse view. (B) Longitudinal view.

columella root cap arise from the hypophyseal cell that is, in turn, derived from the basal cell, while the proximal initials arise from the apical cell (Fig. 2). Apparently, the daughters of the hypophyseal cell come to cooperate with the proximal initials to give rise to the functionally integrated root meristem.

The separation of the main tissue types: protoderm, ground tissue and vascular cambium, also occurs early during embryogenesis (cf. protoderm in Fig. 2). These divisions, like the first zygotic division, act as clonal boundaries that separate different cell fates. Root meristem initials that continue to produce protoderm, ground tissue and vascular cells, respectively, are set apart from these tissues at a later stage.

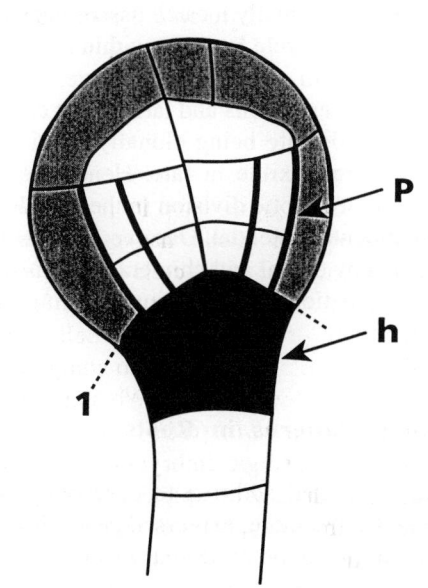

Figure 2. Fate map of early globular embryo. P (light shading): protoderm. h (dark shading): hypophyseal cell. 1: plane of first zygotic division.

4. Flexibility of Cell Fate in Arabidopsis Roots

The strong correlation between cell type and embryonic lineage in the *Arabidopsis* promeristem could be explained in terms of lineage-dependent development after early instructions on cell fate during embryogenesis. On the other hand, positional signalling might continuously determine cell fate. Such signals would have to act at a single-cell resolution, since many layers in the root meristem comprise only one cell. Being superimposed on a rigid cell lineage, such a position-dependency would go undetected. To demonstrate positional signalling in the *Arabidopsis* root, we performed laser ablation experiments. In such experiments, one meristem cell is killed and daughters of neighbouring cells can occupy its position. If the neighbouring cell ends up in a different tissue it crosses a clonal boundary. If commitment of cells were irreversible, the cells would develop into their "old" tissue type. On the other hand, if positional information were continuously operative in the *Arabidopsis* root meristem, the fate of incoming cells after laser ablation would change.

We investigated the ability of root meristem cells to switch fate when they cross the first zygotic division plane (*i.e.* the separation between the quiescent centre and columella on the one hand, and the proximal meristem on the other hand). Upon ablation of quiescent centre cells, the underlying (more distal) columella cells ceased to divide. As a result of this, the dead quiescent centre cells were carried off distally and cell files, continuous with the vascular bundle, were displaced toward the root tip. By using promoter-marker gene fusions, we have shown that these displaced cells switch fate, and display columella-specific gene expression instead of the former vascular- specific expression [15]. Therefore, the boundary set by the first zygotic division does not restrict the developmental potential of the resulting daughter cells.

Upon ablation of cortical and epidermal initial cells, the dead cells are compressed toward the periphery of the root and cells from more internally located tissues take up the position of the ablated cell. Cells derived from the pericycle but now within the cortical cell layer were capable of switching fate and formed both endodermis and cortex. Cortical initials invading the epidermal cell layer formed both epidermis and lateral root cap [15]. We concluded that proximal root meristem cells, despite being clonally restricted to tissue layers at early stages of embryogenesis, are flexible in fate. Hence the early embryonic divisions in the radial plane, like the first zygotic division in the apical-basal plane, are not instrumental in restricting developmental potential. Our experiments are in line with indirect evidence that has been obtained by clonal and chimera analysis of the shoot meristem [*e.g.* 10], and demonstrate that positional information is acting in the regularly patterned root meristem as it is in the shoot.

5. Promoter/Enhancer Trap Expression Patterns in Roots

Patterns of gene expression in an organ form a useful addition to describing organ formation with the aid of morphological criteria. Furthermore, expressed genes that mark a particular cell type or region provide an entrance into the molecular mechanisms that are adopted to specify cellular or regional identity. An easy way to visualise patterns of gene expression is to transform plants with a promoterless marker gene ("promoter trap"), or with a marker gene containing a minimal promoter that can be activated under the control of nearby enhancer elements ("enhancer trap"). Studies in the fruit fly *Drosophila melanogaster* indicate that the majority of expression patterns found in this way correctly reflect the activity of an endogenous gene that is located nearby the marker gene insertion [*e.g.* 16]. A convenient marker gene in plants is the ß-glucuronidase (*uidA*) gene [8], which converts a colourless substrate into a blue precipitate. A joint promoter trap screen to identify nematode-controlled genes, performed as an EC-AIR concerted action, is described elsewhere in this book (see chapters by Ohl *et al.* and Fenoll *et al.*, this volume). We have investigated in some detail lines with root-specific expression patterns. In addition to these lines, we have analysed other root-expressing promoter/enhancer traps. Below we will describe a variety of expression patterns that we have observed in *Arabidopsis* roots, which can be classified broadly into four different correlation groups: tissue type; cell type; differentiation stage-dependent; complex (Table 1).

5.1. TISSUE TYPE

The protoderm, ground tissue, and provascular tissue form the three major tissue types in vascular plants. In the root, the distal root cap can be envisaged as a fourth major tissue. The *AX92* marker was identified as a ground-tissue marker in root and hypocotyl [5], and this has been confirmed in *Arabidopsis* (Claudia van den Berg; Jocelyn Malamy, pers. comm.). In addition, we have identified marker gene expression patterns in roots that coincide with the vascular tissue (pMOG553-643) and root cap (CaMV 35S::B2 subdomain). We have not yet identified lines which stain all of the root epidermis specifically.

5.2. CELL TYPE

Within each of the major plant tissues, several distinct cell types can be identified by morphological criteria. Within the root, these are from periphery to center: the root hair-carrying epidermis (trichoblast); the non-hair epidermis (atrichoblast), the cortical parenchyma; the endodermis; the pericycle; phloem and xylem vascular elements; companion cells; vascular parenchyma and root cap cells. Recently it has been shown that the *GL2* gene is expressed in non-hair epidermal cells [9]. In our analysis, we have investigated a number of cell-type specific markers. Cortex-specific, endodermis-specific as well as pericycle-specific expression patterns have been detected (Table 1). We have focused on specific markers within the root cap region. The root cap presents a complex case, since the functional cap cells slough off the root and they may undergo programmed cell death at their final stage of development (Fig.1). It is therefore possible to define the "terminally differentiated root cap cell" in at least two ways: i) the fully functional cells still attached to the root, *e.g.* the columella cells with the accumulated starch grains that are thought to be involved in the gravitropic response. ii) the cell that is lysed and detached from the root. We have detected marker genes that are expressed specifically in either of these two cell types, as well as markers specific for the lateral root cap (Tab. 1).

TABLE 1. Characteristics of marker lines. All three- and four-digit markers are from the Concerted Action screen using the pMOG533 insertion. Sources of other markers are indicated in the text.

Class	Marker line	Characteristics
Tissue-specific	643	vascular bundle
	CAMV 35S::B2	root cap
	PKU 6	root cap
	ET 271	root cap
Cell-type-specific	PKU 14	columella
	ET 283	columella
	ET244	lateral root cap
Differentiation-phase-specific	124	quiescent centre + vascular initials?
	300	columella tiers
	463	columella-tiers
	648	columella-tiers
	359	columella-tiers + lateral root cap
	649	columella-tiers + lateral root cap
	915	columella tiers + lateral root cap
	1066	columella tiers + lateral root cap
	438	meristematic zone
	826	meristematic zone
	654	elongation zone
Complex	*POLARIS*	root tip
	516	root tip
	174	columella tiers, part of embryonic hypocotyl
	134	columella tiers + vascular tissue
	959	columella tiers + vascular tissue
	1027	columella + vascular tissue
	1004	root cap + vascular tissue

5.3. DIFFERENTIATION STAGE-DEPENDENT

These markers define zones which can be grouped by common functional characteristics, which are shared by several cell types or tissues. Examples are marker genes that specify collectively all initial cells, the meristematic zone, and the elongation zone. As discussed above, markers that are expressed in specific layers of the root cap can also fall into this class, depending on the definition of the terminal root cap cell type.

It must be noted that caution must be taken when an expression pattern seems to correlate completely with a cell type. For example, the pMOG553-1027 marker is specific for the columella root cap at the seedling stage, but is procambium-specific in the embryo. When this marker is expressed in a developmental mutant, it cannot be concluded that the expressing cell is either one of these two cell types.

5.4. COMPLEX EMBRYONIC OR POST-EMBRYONIC EXPRESSION

Complex markers define zones in the embryo that cannot (yet) be grouped by common functional or cell type characteristics. For example, three marker lines show GUS expression in a small region of the root cap, but also in vascular cells immediately proximal to the quiescent centre. It remains to be established whether these gene expression patterns convey that the two cell types have something in common, or, alternatively, whether the marker correlates with postional values that are set up *e.g.* by the quiescent centre. A marker for which it is more clear that it substantiates a positional value, is the *em101 (POLARIS)* gene, which is expressed in the distal root tip in a constant region from heart stage embryogenesis onward.

6. Mutational Analysis of Root Development

The *Arabidopsis* root develops by a stereotyped scheme of cell divisions that starts during embryogenesis, superimposed on a continuously present system of signals that determine cell fate. Identification of genes that are involved in generating, perceiving, and responding to the signals that allocate cells to tissues in the root is a major strategy to unravel the molecular details of this process. Below we will summarize the results of genetic screens aimed at identifying genes that are required for the formation of specific elements of the cellular pattern that comprises the root.

6.1. RADIAL ORGANISATION

A number of mutants have been described that alter the radial organisation of the root [1; 12]. Given the allele frequencies of the identified loci, probably most of these genes are as yet unidentified. The three mutants *shortroot, scarecrow* and *pinocchio* are affected in the specification of cortex and endodermis from the ground tissue. *scarecrow* and *pinocchio* have recently been shown to represent two different mutations in the same gene (P.N.Benfey, pers. comm.). *wooden leg* and *gollum* interfere with the specification of the vascular tissue. Noteworthy, the layer-specific phenotypes persist in the hypocotyl, and all five mutants have an embryonic phenotype throughout the embryonic axis. Hence all genes identified so far that influence the pattern affect the complete seedling axis from embryogenesis onward. The tentative conclusion is that the information in the root meristem originates as the result of gene activities during embryonic radial pattern formation. Based on these results and those from laser ablation experiments, it becomes attractive to envisage the root meristem as a group of dividing cells that is competent to react to signals from more mature cells. These signals, in turn, depend on the correct activity of pattern formation genes that are first active during embryogenesis. Since all the radial mutants display similar phenotypes in lateral roots and in roots derived from callus, the corresponding gene activities are not restricted to embryogenesis, but appear to be employed again when secondary roots are formed.

Cells in the distal region of the root, containing the quiescent centre and the root cap, are programmed differently since they do not form elements of the radial tissue pattern. Yet the laser ablation experiments show that this region also is programmed by positional information. In the next section we will present genetic data that suggest the great importance of this distal region in establishing a root meristem.

6.2. MERISTEM SPECIFICATION

In addition to radial specification of cell layers, the root and root meristem are specified as elements of the apical-basal embryonic pattern. Clonal analyses showed that the boundary between the root and the hypocotyl does correlate, but not with cellular precision, to early embryonic divisions [11]. This indicates that, during early pattern formation, root and hypocotyl fate are connected. A few *Arabidopsis* genes have been described which are required for the formation of a root. Among these, the *MONOPTEROS* gen [2] has been analysed in detail. This gene appears to be required for the specification of both root and hypocotyl in the embryonic context. The *mp* phenotype also indicates the intimate relation between root and hypocotyl specification.

Mutations in the *ROOT MERISTEMLESS* loci display no embryonic phenotype but lack post-embryonic cell division in the root meristem (3; Viola Willemsen and Ben Scheres, unpublished). These *rml* mutants appear to be disturbed in the re-initiation of cell division within the root meristem upon germination.

We have concentrated on defining loci which are involved in the programming of a correctly patterned root meristem. The fate map of the *Arabidopsis* seedling shows that the majority of the primary root cells arise from the root meristem. However, cells within a small region covered with root hairs that connects root and hypocotyl originate from a different region in the embryo and are referred to as the "embryonic root". We performed a genetic screen for mutant seedlings that contained the embryonic root, but lacked an organised promeristem. In this chapter we will discuss in some detail four genetic loci that upon mutation confer embryonic defects in the hypophyseal cell region (*e.g.* the prospective quiescent centre and columella):*HOBBIT, BOMBADIL, ORC,* and *GREMLIN* (Figures. 3,4). A number of other mutants were identified, which have similar phenotypes to the ones stated above but which are not allelic. All the mutants are fully recessive and, with the exception of *orc*, seedling-lethal or sterile.

6.3. THE HOBBIT PHENOTYPE GROUP

The *HOBBIT* locus is defined by a series of independent allelic mutations which lead to a very similar, "root meristemless" appearance (Fig. 4; Willemsen *et al.*; in prep.). Seven independent alleles were identified in our screen, and four more alleles were kindly provided by Prof. G. Jürgens (Univ. Tübiningen, Germany) and Drs. H. Höfte and C. Bellini (INRA Versailles, France). Seedlings homozygous for strong *hbt* alleles display no root meristem activity, while seedlings homozygous for weak alleles allow some residual activity. All seedlings homozygous for *hbt* alleles have abnormal root meristem anatomy. The most conspicuous anatomical defects are the irregularities in cell shape, number, and arrangement of the columella and quiescent centre region. Mutants homozygous for strong *hbt* alleles contain no differentiated columella root cap, based on the absence of starch granules. *hbt* mutants carrying either strong or weak alleles show

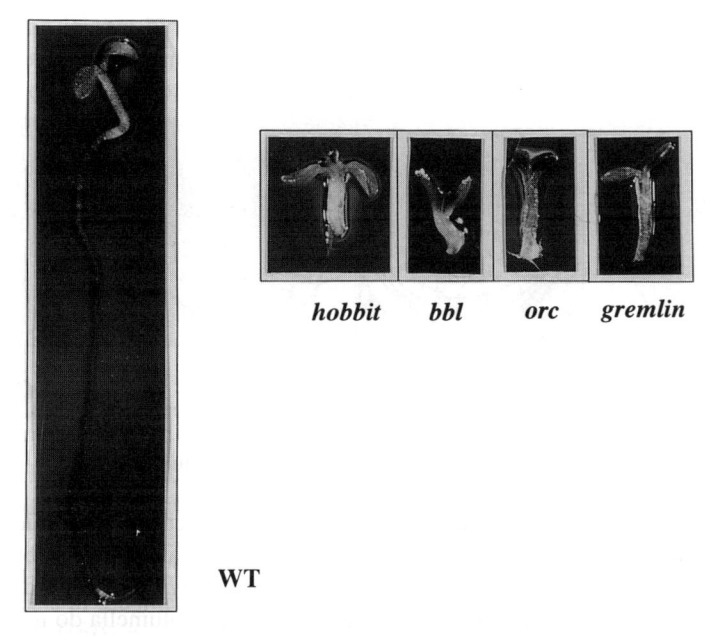

Figure 3. Seedling appearance of wildtype and "hypophyseal cell group" mutants.

abnormalities in division pattern within the hypophyseal cell from early globular stage embryo onward. These abbreviations seem restricted to the hypophyseal cell at early stages of embryogenesis, but the proximal initials, most notably the epidermal initials that should form a lateral root cap, can also become abnormal (Willemsen *et al.;* in prep). This mutant phenotype suggests three functions of the *HBT* gene: i) specifying hypophyseal cell descendants; ii) triggering of activity in the proximal meristem, and iii) proper formation of a lateral root cap. It remains to be clarified whether the second and third function are a direct downstream result of the presence of a correctly programmed hypophyseal cell.

The root phenotype in *hbt* mutants is not embryo-specific. Adventitious roots from *hbt* mutants, generated from the hypocotyl of seedlings or via tissue culture, have the characteristic mutant phenotype and arrest development. Therefore the *HBT* gene, unlike the *MP* gene, is not required just for embryonic root formation but for root formation in all developmental contexts [2; Willemsen *et al.*, in prep.]. However, only the root meristem forms abnormally. *Hobbit* seedlings contain basal root hairs and express the basal embryo marker gene *POLARIS* .

6.4. THE BOMBADIL PHENOTYPE GROUP

Further evidence for more than one gene acting in a similar specification pathway comes from the phenotype of the *bbl* mutants. *bbl* seedlings are distinguishable from *hbt* seedlings (Fig. 3), and they form a different complementation group with different map patterns in the root mutants will identify the promoter trap lines that are correlated to cell differentiation. Careful analysis of such selected tags during nematode infection may provide valuable information on cellular differentiation characteristics that are essential for feeding cell induction. Furthermore, understanding how cell differentiation pathways are connected with each other seems pivotal if one wants to achieve effective nematode control using plant genes without unwanted secondary effects on plant development. Cell cycle regulation is also likely to be of importance for feeding site establishment, as judged from the expression pattern of *CDC2* and cyclin genes (see Gheysen et al., this volume). In this case also, the combination of cell cycle marker genes with genetic approaches can be useful. It will for example be interesting to determine whether parasitic nematodes are capable of regulating the *RML* loci, which have been shown to be involved in the stimulation of cell division in the root.

Whether any of the expectations stated above will turn out to have practical value remains to be seen, but it is becoming clear that continuing studies on root development will further enhance our knowledge of the dinner table preferences of parasitic nematodes. We hope that such knowledge will enable the birth of novel concepts in pest control.

References

1. Benfey, P.N., Linstead, P.J., Roberts, K., Schiefelbein, J.W., Hauser, M-T. and Aeschbacher, R.A. (1993) Root development in *Arabidopsis*: four mutants with dramatically altered root morphogenesis, *Development* **119**, 57-70.
2. Berleth, T. and Jürgens, G. (1993) The role of the monopteros gene in organising the basal body region of the Arabidopsis embryo, *Development* **118**, 575-587.
3. Cheng, J-C, Seeley, K.A. and Sung, Z.R., (1995) RML1 and RML2, Arabidopsis genes required for cell proliferation at the root tip, *Plant Physiol.* **107**, 365-376.
4. Clowes, F.A.L., (1961) Apical Meristems. Oxford: Blackwell.
5. Dietrich, R.A., Radke, S.E, and Harada, J.J., (1992) Downstream DNA sequences are required to activate a gene expressed in the root cortex of embryos and seedlings, *Plant Cell* **4**, 1371-1382.
6. Dolan, L., Janmaat, K., Willemsen, V., Linstead, P., Poethig, S., Roberts, K. and Scheres, B., (1993) Cellular organisation of the Arabidopsis root, *Development* **119**, 71-84.
7. Golinowski, W., Grundler, F., Sobczak, M., 1996 Changes in the structure of *Arabidopsis thaliana* induced during development of females of the plant parasitic nematode *Heterodera schachtii*, *Protoplasma*, **194**, 103-116.
8. Jefferson, R,A., Kavanagh, T.A. and Bevan, M.W., (1987) GUS fusions: glucoronidase as a sensitive and versatile gene fusion marker in higher plants, *EMBO J.* **6**, 3901-3907.
9. Masucci, J.D., Rerie, W.G., Foreman, D.R., Zhang, M., Galway, M.E., Marks, M.D. and Schiefelbein, J.W., (1996) The homeobox gene *GLABRA 2* is required for position-dependent cell differentiation in the root epidermis of *Arabidopsis thaliana*, *Development* **122**, 1253-1260.
10. Poethig, R.S., (1987) Clonal analysis of cell lineage patterns in plant development, *Am. J. Bot.* **74**, 581-594.
11. Scheres, B., Wolkenfelt, H., Willemsen, V., Terlouw, M., Lawson, E., Dean, C., Weisbeek, P., (1994) Embryonic origin of the Arabidopsis primary root and root meristem initials, *Development* **120**, 2475-2487.
12. Scheres, B., Di Laurenzio, L., Willemsen, V., Hauser, M-T., Janmaat, K., Weisbeek, P., Benfey, P.N., (1995) Mutations affecting the radial organisation of the Arabidopsis root display specific defects throughout the embryonic axis, *Development* 12153-62.
13. Sijmons, P.C., Grundler, F.M.W., von Mende, N., Burrows, P.R., Wyss, U., (1991) *Arabidopsis thaliana* as a new model host for plant-parasitic nematodes, *Plant J.* **1**, 245-254.
14. Spaink, H.P. (1995) The molecular basis of infection and nodulation by Rhizobia: the ins and outs of

sympathogenesis, *Ann. Rev. Phytopathol.* **33**, 345-368.
15. Van den Berg, C., Willemsen, V., Hage, W., Weisbeek, P. and Scheres, B. (1995) Cell fate in the Arabidopsis root meristem determined by directional signalling, *Nature* **378**, 62-65.
16. Wilson, C., Pearson, R.K., Bellen, H.J., O'Kane, C.J., Grossniklaus, U. and Gehring, W.J., (1989) P-element-mediated enhancer detection: an efficient method for isolating and characterizing developmentally regulated genes in *Drosophila, Genes Devel.* **3**, 1201-1213.

PLANT SIGNALS IN NEMATODE HATCHING AND ATTRACTION

Roland N. PERRY
Entomology and Nematology Department, IACR- Rothamsted, Harpenden, Hertfordshire, AL5 2JQ, UK

Abstract

Plant parasitic nematodes use plant signals (allelochemicals) to ensure close synchrony between host and parasite life cycles. The use by nematodes of plant signals is most extensively developed in the sedentary species, especially in relation to hatching and host location. There is also evidence that plant signals, acting via the feeding female, may influence the physiology of the developing juveniles. Detailed analysis of nematode sensory perception and characterization of the chemicals involved are required to extend understanding of the complex host-parasite interactions of sedentary plant parasitic nematodes.

1. Introduction

Species of plant parasitic nematodes have stages in their life cycle in soil which require for movement at least a film of water on the surface of soil particles. Signal molecules eliciting nematode responses in soil are likely to be water-soluble to move through the aqueous environment that nematodes inhabit. Chemicals that mediate inter- and intraspecific interactions between organisms are termed semiochemicals. The term allelochemical describes a semiochemical that mediates interspecific interaction between organisms; thus, allelochemicals include chemical signals from plants which elicit nematode responses.

Plant allelochemicals are clearly an essential component of the host-parasite interaction and much work has centred on their involvement in certain soil-based phases of the nematode life cycle, such as hatching and location of host roots. After invasion, plant signals are probably also involved in orientated movement of the nematode through host tissue to locate a cell for feeding site initiation. Chemoreception during these phases may be particularly sensitive to disruption by using naturally occurring compounds to give an environmentally benign control strategy and this has generated increasing interest in nematode sensory perception.

There is considerable evidence of chemotaxis in nematodes but little is known about the functional physiology of nematode sensilla (=sense organs). The amphids are considered to be the primary chemosensilla and their structure is conserved in a range of sedentary nematodes including second stage juveniles (J2) and adult males of *Meloidogyne incognita* [5, 87], *Heterodera glycines* [6, 23] and *Globodera rostochiensis* [41]. The sensory and secretory roles of the amphids have been investigated with the aim of perturbing sensory perception, particularly in relation to host finding.

This chapter examines the involvement of plant signals in three major aspects of the life cycles of sedentary nematodes, hatch stimulation, host location and dormancy, and includes brief mention of recent work on the functioning of the amphids.

2. Hatch Stimulation

The reliance on host root diffusates to stimulate hatch varies between different species of sedentary nematodes. In general, given favourable environmental conditions, such as temperature, oxygen availability and soil moisture levels, and an absence of physiological barriers, such as diapause (see section 3.2), hatch of most species occurs once development to J2 has been completed. In many species root diffusates enhance the rate of hatch but the almost complete dependence on diffusates for hatch, as shown by *G. rostochiensis* for example, is unusual. Despite this, the hatching mechanism of this species has been the most extensively studied and has frequently been used as the basis for comparison with other species. It provides one of the clearest examples of the role of plant signals in synchronising host and parasite life cycles. Identification of specific plant allelochemicals, or hatching factors, in diffusates [54] still remains a possible precursor to novel control strategies for this species and for *G. pallida* and *H. glycines*. Host diffusates also stimulate movement of hatched J2 and may aid in host location, so analysis of these allelochemicals may have broader control implications.

2.1. HATCHING MECHANISMS

The sequence of events in the hatching process of *G. rostochiensis* has been reviewed in detail [52, 53]. Potato root diffusate (PRD) has a bimodal action, altering eggshell permeability and directly affecting the J2 by stimulating movement [15, 86] and the dorsal oesophageal glands [4, 63]. Eggshell permeability is altered through a Ca^{2+}-mediated change of the inner lipoprotein membranes. This results in the escape of trehalose from the perivitelline fluid through the eggshell thus reducing the osmotic pressure on the unhatched J2. The water content and metabolic activity of the J2 increase enabling it to move and cut its way out of the rigid eggshell. The metabolic responses include increased O_2 consumption [1] and lipid utilisation [71], a fall in the adenylate energy charge [2] and a change in cAMP levels [4]. J2 of *H. goettingiana* also show an increase in water content after exposure to host root diffusates and before hatching but this occurs over a longer period and was correlated with a much slower rate of hatching compared with *G. rostochiensis* [59]. The oesophageal glands of *G. rostochiensis* appear not to be involved in eclosion but accumulate secretions for the subsequent invasion and feeding phases [63]. Initiation of hatching is rapid as exposure to PRD for as short as 5 min triggers the hatching sequence of *G. rostochiensis* [56] and *G. pallida* [27].

There are differences between the two species of potato cyst nematodes in their hatching response [73] and Den Nijs and Lock [18] considered that *G. pallida* is more dependent on PRD whereas *G. rostochiensis* responds more readily to non-specific hatching triggers. However, difference in the hatching response may be related, in part at least, to difference in hatching factor preference (see section 2.3).

Changes were observed in the structure of the amphids of *G. rostochiensis* J2 during the hatching process [41]. The absence of secretory material and the shrunken state of the sheath cell in unhatched nematodes indicated that the amphids may not be functional before hatching and, thus, have no role in the detection of the hatching factors. The change to a functional appearance is not associated specifically with exposure to PRD but is a more general characteristic of naturally hatched J2; the structure of the amphids of *G. rostochiensis* altered very little during subsequent phases of the life cycle [41]. Few changes in gene expression appear to be induced directly by PRD but they seem to occur during or immediately after the hatching process [42].

Hatching of *H. schachtii* is not dependent on stimulation by host root diffusates. The perivitelline fluid surrounding the unhatched J2 has a lower osmotic pressure than that of *G. rostochiensis*, and the unhatched juvenile contains sufficient water to move and is less affected by osmotic stress [58]. Although this helps to explain the large water hatch of *H. schachtii*, host root diffusates stimulate an additional hatch, indicating that a change in eggshell permeability may be required for these eggs to hatch. Lipoprotein membranes have only been detected in eggs from cysts with no fungal contamination [62] and it is possible that fungal enzymes, such as lipases, could disrupt the lipoprotein membranes thus resulting in juvenile hatch; eggs or cysts not affected by fungal contamination would still require root diffusate to alter membrane permeability.

Enzymes involved in the hatching process may originate from the J2 and/or may be present in the perivitelline fluid and kept inactive either by separation from their substrates by the lipid membrane or by an inhibitor, such as trehalose. The involvement of enzymes in hatching of *Meloidogyne* was postulated several years ago [10]. The cumulative percentage hatch of *M. incognita* was positively correlated with lipase activity; proteinase activity, including chitinase and collagenase, was also detected and these enzymes are likely to be associated with increased flexibility of the eggshell before eclosion [60]. However, enzyme activity appears not to be dependent on plant signals; J2 of species of *Meloidogyne* hatch when environmental conditions are favourable and root diffusates are not required for hatching of most species although they can enhance the rate of hatching. Similarly, leucine aminopeptidase (LAP) activity has been implicated in hatching of *H. glycines* [77] but host root diffusates had no direct effect on LAP activity [78].

In some species of cyst nematodes a number of eggs are deposited into an egg sac attached to the cyst but there is no evidence that these eggs are dependent on plant signals for hatch [31, 39]. Many species complete several generations during the host growing season and with some species there is a change between generations in the dependence of encysted eggs on root diffusates [31, 32, 38]. For the first four generations, J2 in cysts of *H. cajani* hatched well in water with no enhancement of hatch by root diffusates but, in the fifth and sixth generations, 18-22% of the eggs required host root diffusate to stimulate hatch [32]. With the onset of plant senescence, females of *H. sacchari* developed into cysts which contained approximately 20% more eggs which were refractory to hatching stimuli and an additional 10-15% which depended on host root diffusate for hatch stimulation, compared to cysts produced on younger plants [38]; this 30-35% of viable, dormant J2 provides a 'carry-over' between crops. Similarly, there are at least three kinds of eggs in cysts of *H. sorghi*: ones that hatch freely in soil leachates, those that require stimulation from host root diffusates to hatch and a large percentage which do not hatch immediately; the proportions of these three types of eggs changed with successive generations [31]. The change in response to plant signals by *H. cajani*, *H. sacchari* and *H. sorghi* ensures that a large proportion of J2 from the later generations do not hatch and remain protected by the egg and cyst during the intercrop period; this aspect is discussed further in relation to dormancy (section 3).

2.2. ROOT DIFFUSATE ACTIVITY

In the research reported in the preceding section, cysts were exposed to optimum *in vitro* conditions for hatch including root diffusate with maximum hatching activity. Activity of root diffusates is not constant throughout the life of some host plants and the decline in activity as plants age is an additional factor ensuring that J2 do not hatch immediately before the intercrop period. For example, substantial hatch of *H. goettingiana*, occurred in diffusates from 4- and 6-week-old pea plants but diffusates from 2- or 10-week-old plants elicited hatches of less than 10% [57]. Most J2 of *H. carotae* hatch in diffusate from 5-7 week-old carrots [33].

Evolution of the host-parasite interaction has resulted in host specific responses. In general, where diffusates induce or enhance hatch they are from good hosts; non-hosts or poor hosts tend not to produce active diffusate. For example, non-hosts or poor hosts do not stimulate hatch of *H. glycines* [74]. Nightshade stimulated more hatch of *G. tabacum tabacum* than tobacco or tomato [45] which may be correlated with reports that nightshade is the more efficient host. Similarly, although J2 of *Rotylenchulus reniformis* hatch freely in water, an enhancement of hatch was found in root diffusates from good hosts and diffusates from poor or non-hosts suppressed hatch [43]. By contrast, activity of diffusates does not appear to be linked to host resistance or susceptibility. Cultivars of wheat, oats and barley resistant to *H. avenae* stimulated hatch [88] and resistant potato cultivars produced active diffusates [22, 67, 80].

However, there is great difficulty in evaluating and comparing the numerous reports of root diffusate activity. Variations in PRD activity are probably a reflection of plant growth rather than intrinsic differences [25, 79] but some potato hybrids may produce diffusate of reduced hatching activity [26, 28]. Apart from plant age, diffusate dilution and possible restriction of roots in pots, there are several other influential factors. In the non-sterile soil environment many components will modify diffusate production and activity. When diffusate was collected from aseptically grown potato plants it lacked several hatching factors compared with diffusate collected by standard methods [19]. Rhizobacteria that reduce the hatch of PCN have been isolated [66] and their antagonistic action may be the result of bacterial alteration of plant signals that influence nematode hatch. There is also no accepted standard way of demonstrating diffusate activity. For effective comparisons, results may need to be expressed as number (or percentage) of J2 hatched per gram of root or per unit root length; the latter more accurately reflects the physiological contribution of the root. Rawsthorne and Brodie [67] found that hatching of *G. rostochiensis* was positively correlated with increased root weight only during the first 3 weeks after plant emergence.

Hatching factors are present in very low concentrations and Masamune [49] reported that those causing hatch of *H. glycines* were active at dilutions as low as 10^{-14}g ml^{-1}. Diluting PRD enhances its hatching activity, probably because the influence of hatching inhibitors is reduced. Hatching inhibitors may be important in novel control approaches but they have been investigated only to a limited extent. Recent research has shown that hatch of *G. rostochiensis* in soil under host and non-host solanaceous and non solanaceous species was positively correlated with the hatching factors : hatching inhibitors ratio [81]. Inhibitors are not always present in diffusates: hatching tests with *H. goettingiana* indicated they are absent in bean and pea root diffusates, for example [57].

2.3. CHEMISTRY OF THE HATCHING FACTORS

Chemical signals involved in the hatching of *G. rostochiensis, G. pallida* and *H. glycines* have been the subject of detailed analysis. Research on the potato cyst nematodes has focussed on the hatching factors present in host root diffusates whereas, for *H. glycines*, extracts from macerated kidney bean (*Phaseolus vulgaris*) roots have been analysed. Several empirical formulae have been given for hatching factors for potato cyst nematodes: $C_{18}H_{24}O_8$ [48], $C_{19}H_{28}O_8$ [40], $C_{13}H_{12}O_3$ [34], $C_{11}H_{16}O_4$ [12] and $C_{27}H_{30}O_9$ [51]. Initial studies indicated that at least four different hatching factors were present in potato root diffusate [13, 14] and different factors affected different events during the hatching process [3, 54]. More recent work has revealed the presence of at least twenty-five hatching factors [11]. Both *G. rostochiensis* and *G. pallida* responded to each of these hatching factors but differences in hatch between species were found, with a greater hatch of *G. pallida* induced by the less polar hatching factors which were produced principally at the later stages of host growth; *G. pallida* hatched at the same time as *G. rostochiensis* when the diffusate solution applied to infested soil included different ratios of species-preferred hatching factors. Thus, the difference in hatch between the two species may be due primarily to the plant signals involved in inducing the hatching response. Two hatching factors for *G. rostochiensis* have been identified as the potato glycoalkaloids, solanine and α-chaconine [19].

The molecular mass of hatching factors in PRD was reported to be 358-437 Da [4] and 498 Da [51]; these values are similar to the value of 446 Da for the hatching factor for *H. glycines* isolated from kidney bean roots [50]. The isolation [50] and structural determination [29] of this hatching factor, termed glycinoeclepin A with the formula $C_{25}H_{34}O_7$, was followed by the isolation of two new nortriterpins, designated glycinoeclepins B and C [30]. Research by several groups has focussed on synthesizing glycinoeclepin A because it is not present in kidney bean root tissue in quantities sufficient for agricultural use; total synthesis of the compound was described by Watanabe and Mori [85]. Analogues and precursors of glycinoeclepin A have been synthesized and screened for *H. glycines* hatching activity; one compound enhanced hatch

[82] and a second compound, a precursor, inhibited hatch [83]. Subsequent work revealed several analogues which inhibited hatch and the minimum functionality for inhibition appeared to be a keto diacid.

3. Plant Signals and the Induction of Dormancy

Dormancy has been separated into quiescence and diapause: quiescence is an arrest in development induced in response to unfavourable conditions and development is resumed soon after the return of favourable conditions whilst diapause is a condition in which development has been arrested and cannot be resumed until specific requirements have been recognised, even if favourable conditions return [24].

3.1. INDUCTION OF DORMANCY

The hatching pattern of *H. avenae* in areas such as the Mediteranean and Australia is related to a well defined diapause mediated by temperature but without the influence of plant signals; similarly, diapause in species of *Meloidogyne* is initiated by environmental cues [24]. By contrast, diapause in *G. rostochiensis* appears to be initiated by signals from the plant during the growing season [36]. Diapause is not an intrinsic property of the eggs of *G. rostochiensis* and Hominick [35] considered that photoperiod, acting on the potato plant, affected the developing females of *G. rostochiensis* and influenced the hatching mechanism of the developing juveniles.

In an analysis of the hatching from cysts of *H. schachtii* harvested at the end of the growing season, Zheng and Ferris [89] identified four types of egg hatch and linked them to forms of dormancy. About 40-50% of J2 hatch rapidly in water (non-dormant), an additional 10% hatch rapidly in root diffusate (obligate quiescence), about 10% which do not hatch immediately in water will hatch in water over a longer period (time-mediated obligate diapause) and the remaining eggs hatch over a long period of exposure to host root diffusates (host-mediated obligate diapause). In California, *H. schachtii* can complete six generations during a host growing season. It would be interesting to determine whether the proportion of the four types of egg hatch remained constant in successive generations.

The J2 of *H. sacchari* and *H. cajani* (see section 2.1) which hatched immediately on stimulation with host root diffusate were quiescent and those J2 of *H. sacchari* which were refractory to host diffusates, even under favourable conditions, were likely to be in a state of diapause. It is probable that changes in plant signals from the syncytia of different age plants relate indirectly to changes in juvenile biochemistry and physiology. The physiological state of the host plant is known to influence egg deposition and hatching of cyst nematodes and this may be related directly to the availability of nutrients. The presence or absence of certain plant signals or nutritional elements in the syncytium associated with plant senescence, may be the trigger for induction, in the J2, of diapause and concomitant preparation for a period of survival in the absence of a suitable host. The increased lipid reserves in encysted J2 of *H. cajani* in the final generation, combined with the lowered metabolism during quiescence [24], are likely to enhance considerably the period that encysted J2 remain viable.

3.2. BIOCHEMICAL CHANGES

Elements present in syncytia induced by *H. trifolii* have been investigated by scanning electron microscopy and energy dispersive X-ray microanalysis and changes were found in the elemental composition of syncytia compared to unmodified tissue [16]. There were variations in elemental composition with age of the syncytia and the greatest changes were associated with maturing nematodes during egg production. It may be feasible to use this technique to compare the elemental composition of syncytia at a comparable developmental phase but from host plants of different ages. Amino acids influence the development of nematodes and Betka *et al.* [8] demonstrated that glutamine enhanced development of *H. schachtii* whereas methionine, phenylaline, lysine and tryptophan inhibited development (see also chapter by Grundler and Böckemhoff, this volume).

4. Host Location

Active, infective stages of sedentary nematodes are vulnerable to environmental extremes and while in the soil they are dependent on their food reserves. For example, the J2 of *G. rostochiensis* need to locate host roots soon after hatching as their infective life is only 6-11 days under optimal conditions for motility [72]. The ability to orientate towards stimuli from host roots increases the likelihood that food reserves will not be utilised before location of a host root and, thus, enhances the chance of successful invasion.

4.1. ROLE OF ROOT DIFFUSATES IN HOST LOCATION

In the soil, nematodes move through a three dimensional matrix and respond to gradients of a variety of stimuli. Ideally, *in vitro* bioassays should allow the nematode to move as naturally as possible but, although some researchers have attempted this, most bioassays are based on radial two-dimensional attraction gradients in agar. Studies on the ability of plant parasitic nematodes to locate host roots or to respond to a chemical stimulus have used a variety of experimental conditions and it is frequently difficult to compare results. It is generally accepted that host roots are attractive to infective stages of sedentary nematodes but repulsion or a total absence of any effect has been reported [65]. Although, in general, the attraction is non-specific, there are indications that the attractiveness of a host to the pest species is correlated with its efficiency as a host [46, 84]. Other factors, such as age of the root or presence of microorganisms, also condition the root's attractiveness; for example, attraction is lost when the root's growth is stopped or limited [47] but the reason for this is unknown.

Although experimental results are not consistent, it appears that many species use the carbon dioxide gradient established around roots to orientate towards the host [37, 64]. Several other gradients exist around physiologically active roots including amino acid, ion, pH and sugar gradients. Many of these gradients constitute general, non-specific attractants for long distance migration. A "kairomone" action of sucrose on the orientation behaviour of *M. naasi* has been suggested [7] but this may be due to the stimulatory effect of sucrose on root tip metabolism. Cells at the root tip of potato plants produced a more active diffusate than cell located elsewhere but diffusate appeared to be produced along the entire root [67]. Specific plant signals may be used to attract infective stages to host plants and other allelochemicals may be responsible for orientation to preferred invasion sites such as the root tip; these attractants remain to be identified.

Bird [9] was the first to suggest that nematodes orientate along a potential gradient created by a lower redox potential at the root's surface. Subsequent studies by several authors have demonstrated that nematode movement can be orientated by redox potential or an electric field created by the roots [70] but the relative importance of chemical and electrical attractants has not been assessed. Electrical potential gradients at the region of the elongation zone may be involved in attracting J2 of cyst and root-knot nematodes to the preferred invasion site at the root tips [7].

4.2. MECHANISMS OF ATTRACTION

Molecules of chemical substances initially come into contact with the amphidial secretions filling the amphidial duct and, thus, information on the specific nature of these secretions may enhance understanding of nematode sensory perception. In the amphidial ducts, plant signals may reach the receptors exclusively by diffusion. Phytohormones may be triggers for the induction of secretions [20] and the possibility that auxin was acting as a host cue has been supported by evidence of a putative auxin binding protein from *G. pallida*, possibly associated with the amphids [21]. The involvement of nematode secretions in host-parasite interactions is reviewed by Jones and Robertson in this volume.

Host location could be affected either by blocking plant signals or by blocking the chemoreceptors [90]. Sensory perception may be inhibited by lectins but the penetration of sensilla secretions by putative blocking agents needs to be investigated in more detail using various species of nematodes on a comparative basis, especially as there are indications of fundamental differences in the composition of sensilla secretions between species of nematodes [55]. A 32 kDa glycoprotein (termed gp32), associated with the secretory material filling the amphids, was found in six species of *Meloidogyne* but appears to be genus-specific as it was not found in representatives from eight other genera including *Globodera* and *Heterodera* [76]. Gp32 was expressed in all stages of the *Meloidogyne* life cycle, including males of *M. javanica*, but not in the sedentary adult female, where the amphids appear to be non-functional; differences in the composition of amphidial secretions between J2 and females of *M. incognita* have also been demonstrated by Davis *et al.* [17]. Thus, there are indications that at certain stages of the life cycle the amphids of some species of sedentary nematodes may not be involved in the perception of plant signals. Incubation of J2 of *M. javanica* in the polyclonal antiserum for gp32 significantly retarded orientation to host roots [75]. The orientation of plant parasitic nematodes to known stimuli is also impaired by low concentrations of nematicides, although motility is not inhibited; the sensilla may be the primary sites of action of these nematicides.

In electrophysiological studies, exposure of intact males of both *G. rostochiensis* and *G. pallida* to PRD elicited no significant response and it appears that PRD plays no part in orientating males [68]. Males exit from the roots into the soil and probably remain close to the roots, requiring only sex pheromones to attract them to the sessile females. Some interesting contrasts were found in the responses of males to various chemicals [69]. When males of *G. pallida* and *G. rostochiensis* were exposed to L-glutamic acid the frequency of spike activity increased significantly yet exposure to D-glutamic acid did not elicit a similar response. In insects, the D-isomer of many amino acids usually elicts a phagostimulatory response whilst many L-amino acids are feeding deterrents. Neither glycine nor citric acid induced any marked electrophysiological responses and only males of *G. pallida* showed significant behavioural response, moving towards glycine and away from citric acid. The electrophysiological and behavioural responses to γ-GABA and α-GABA were complimentary, with *G. rostochiensis* showing significant response only to the latter and *G. pallida* responding significantly only to the former.

Chemoreception is essential for suitable food selection and the electrophysiological analysis of responses of various species of nematodes to specific plant compounds will provide information about feeding deterrents and stimulants which may relate to plant host suitability and the biochemical nature of the resistance response [61]. The technique will also enable detailed evaluation of allelochemicals responsible for attracting nematodes to plant roots.

5. Conclusions

Information on the plant signals responsible for hatching and attracting nematodes to roots is limited. Research on attractants needs to discriminate between non-specific, general plant signals which may serve to attract the nematode from a distance towards the root and the specific signals required for the nematode to orientate to the site of invasion. Given the well developed host-parasite relationship of certain sedentary nematodes with restricted host ranges and also that some of the hatching factors preferentially hatch only one species of *Globodera*, it would be expected that some signals responsible for short distance orientation would be host specific. The recent advances in fractionating and characterising active hatching factors for potato cyst nematodes in host root diffusates could now provide the basis for evaluating further the bimodal role of diffusates [54] in initiating the hatching sequence and directly stimulating juvenile metabolism. It is not difficult to envisage at least some of the active hatching factors having a role in attracting nematodes to roots. The possible role of inhibitors and repellants should not be ignored. In certain plant species, allelochemicals which may be attractants could have their effect nullified by stronger repellant effects of other compounds in root diffusates. Research on isolated plant signals is of value but additional work on synergistic or antagonistic effects will considerably enhance understanding of the complex role of plant signals in hatching and attraction.

ACKNOWLEDGEMENTS

IACR receives grant-aided support from the Biotechnology and Biological Sciences Research Council of the United Kingdom.

References

1. Atkinson, H.J., and Ballantyne, A.J. (1977) Changes in the oxygen consumption of cysts of *Globodera rostochiensis* associated with the hatching of juveniles, *Ann. appl. Biol.* **87**, 159-166.
2. Atkinson, H.J., and Ballantyne, A.J. (1977) Changes in the adenine nucleotide content of cysts of *Globodera rostochiensis* associated with the hatching of juveniles, *Ann. appl. Biol.* **87**, 167-174.
3. Atkinson, H.J., Fowler, M., and Isaac, R.E. (1987) Partial purification of hatching activity for *Globodera rostochiensis* from potato root diffusate, *Ann. appl. Biol.* **110**, 115-125.
4. Atkinson, H.J., Taylor, J.D., and Fowler, M. (1987) Changes in the second stage juveniles of *Globodera rostochiensis* prior to hatching in response to potato root diffusate, *Ann. appl. Biol.* **110**, 105-114.
5. Baldwin, J.G., and Hirschmann, H. (1973) Fine structure of the cephalic sense organs in *Meloidogyne incognita* males, *J. Nematol.* **5**, 285-302.
6. Baldwin, J.G., and Hirschmann, H. (1975) Fine structure of cephalic sense organs in *Heterodera glycines* males, *J. Nematol.* **7**, 40-53.
7. Balhadère, P., and Evans, A.A.F. (1994) Characterization of attractiveness of excised root tips of resistant and susceptible plants for *Meloidogyne naasi*, *Fundam. appl. Nematol.* **17**, 527-536.
8. Betka, M., Grundler, F., and Wyss, U. (1991) Influence of changes in the nurse cell system (syncytium) on the development of the cyst nematode *Heterodera schachtii*: single amino acids, *Phytopathology* **81**, 75-79.
9. Bird, A.F. (1959) The attractiveness of roots to the plant parasitic nematodes *Meloidogyne javanica* and *M. hapla*, *Nematologica* **4**, 322-35.
10. Bird, A.F. (1968) Changes associated with parasitism in nematodes. III. Ultrastructure of the egg shell, arval cuticle, and contents of the subventral esophageal glands in *Meloidogyne javanica*, with some observations on hatching, *J. Parasitol.* **54**, 475-489.
11. Byrne, J., Walsh, D., Devine, K., and Jones, P. (1996) Investigations into the cause of the delayed hatch of *Globodera pallida* compared to *G. rostochiensis*. *Abstract: Third International Nematology Congress, Guadeloupe*, 161.
12. Clarke, A.J. (1960) Eelworm hatching factors, *Report of Rothamsted Experimental Station for 1959*, 112-113.
13. Clarke, A.J. (1970) Hatching factors and sex attractants of cyst nematodes, *Report of Rothamsted Experimental Station for 1969*, 179.
14. Clarke, A.J. (1971) Hatching factors and sex attractants of potato cyst nematode, *Report of Rothamsted Experimental Station for 1970*, 147-148.
15. Clarke, A.J., and Hennessy, J. (1984) Movement of *Globodera rostochiensis* (Wollenweber) juveniles stimulated by potato root exudate, *Nematologica* **30**, 206-212.
16. Cook, R., Thomas, B.J., and Mizen, K.A. (1992) X-ray microanalysis of feeding syncytia induced in plants by cyst nematodes, *Nematologica* **38**, 36-49.
17. Davis, E.L., Aron, L.M., Pratt, L.H., and Hussey, R.S. (1992) Novel immunization procedures used to develop antibodies that bind to specific structures in *Meloidogyne* spp., *Phytopathology* **82**, 1244-50.
18. Den Nijs, L.J.M.F., and Lock, C.A.M. (1992) Differential hatching of the potato cyst nematodes *Globodera rostochiensis* and *G. pallida* in root diffusates and water of differing ionic composition, *Netherlands J. Plant Path.* **98**, 117-128.
19. Devine, K.J., Byrne, J., Maher, N., and Jones, P.W.: Resolution of natural hatching factors for the golden potato cyst nematode, *Globodera rostochiensis*. *Ann. appl. Biol.* (in press).
20. Duncan L.H., Robertson, W.M., and Kusel, J.R. (1995) Induction of secretions in *Globodera pallida*. *Abstract: British Society for Parasitology Meeting, Edinburgh.*, 13.
21. Duncan L.H., Robertson, W.M., Kusel, J.R., and Phillips, M.S. (1996) A putative nematode auxin binding protein from the potato cyst nematode *Globodera pallida*. *Abstract: Third International*

Nematology Congress, Guadeloupe, 116.
22. Ellenby, C. (1954) Tuber forming species and varieties of the genus *Solanum* tested for resistance to the potato root eelworm *Heterodera rostochiensis* Wollenweber, *Euphytica* **3**, 195-202.
23. Endo, B.Y. (1980) Ultrastructure of the anterior neurosensory organs of the larvae of the soybean cyst nematode, *Heterodera glycines*, *J. Ultrastr. Res.* **72**, 349-66.
24. Evans, A.A.F., and Perry, R.N. (1976) Survival strategies in nematodes. *In:* Croll, N.A. (Ed.). *The organisation of nematodes*. London & New York, Academic Press , 383-424.
25. Evans, K. (1983) Hatching of potato cyst nematodes in root diffusates collected from twenty-five potato cultivars, *Crop Protection* **2**, 97-103.
26. Farrer, L.A., and Phillips, M.S. (1983) *In vitro* hatching of *Globodera pallida* in response to *Solanum vernei* and *S. tuberosum X S. vernei* hybrids, *Rev. Nématol.* **6**, 165-170.
27. Forrest, J.M.S., and Perry, R.N. (1980) Hatching of *Globodera pallida* eggs after brief exposure to potato root diffusate, *Nematologica* **26**, 130-132.
28. Forrest, J.M.S., and Phillips, M.S. (1984) The effect of *S. tuberosum X S. vernei* hybrids on the hatching of the potato cyst nematode *Globodera pallida*, *Ann. appl. Biol.* **104**, 521-526.
29. Fukuzawa, A., Furusaki, A., Ikura, M., and Masamune, R. (1985) Glycinoeclepin A, a natural hatching stimulus for the soybean cyst nematode, *J. Chem. Soc.* **4**, 222-224.
30. Fukuzawa, A., Matsue, H., Ikura, M., and Masamune, R. (1985) Glycinoeclepins B and C, related to glycinoeclepin A, *Tetrahedron Letters* **26**, 5539-5542.
31. Gaur, H.S., Beane, J., and Perry, R.N. (1995) Hatching of four successive generations of *Heterodera sorghi* in relation to the age of sorghum, *Sorghum vulgare. Fundam. appl. Nematol.* **18**, 599-601.
32. Gaur, H.S., Perry, R.N., and Beane, J. (1992) Hatching behaviour of six successive generations of the pigeon-pea cyst nematode, *Heterodera cajani*, in relation to growth and senescence of cowpea, *Vigna unguiculata, Nematologica* **38**, 190-202.
33. Greco, N., and Brandonisio, A. (1986) The biology of *Heterodera carotae, Nematologica* **32**, 447-460.
34. Hartwell, W.V., Dahlstrom, R.V., and Neal, A.L. (1959) Crystallisation of a natural hatching factor for larvae of the golden nematode, *Phytopathology* **49**, 540-541.
35. Hominick, W.M. (1986) Photoperiod and diapause in the potato cyst nematode, *Globodera rostochiensis, Nematologica* **32**, 408-418.
36. Hominick, W.M., Forrest, J.M.S., and Evans, A.A.F. (1985) Diapause in *Globodera rostochiensis* and variability in hatching trials, *Nematologica* **31**, 159-170.
37. Hussey, R.S. (1985) Host-parasite relationships and associated physiological changes. *In:* Sasser, J.N. and Carter, C.C. (Eds). *An advanced treatise on Meloidogyne. Volume 1 Biology and control*. Raleigh, North Carolina State University Graphics, 143-153.
38. Ibrahim, S.K., Perry, R.N., Plowright, R.A., and Rowe, J. (1993) Hatching behaviour of the rice cyst nematodes *Heterodera sacchari* and *H. oryzicola* in relation to age of host plant, *Fundam. appl. Nematol.* **16**, 23-29.
39. Ishibashi, N., Kondo, E., Muraoka, M., and Yokoo, T. (1973) Ecological significance of dormancy in plant parasitic nematodes. I. Ecological difference between eggs in gelatinous matrix and cysts of *Heterodera glycines* Ichinoe, *Appl. Ent. Zool.* **8** , 53-63.
40. Johnson, A.W. (1952) The eelworm problem: biological aspects. The potato eelworm hatching factor, *Chemistry and Industry* **40**, 998-999.
41. Jones, J.T., Perry, R.N., and Johnston, M.R.L. (1994) Changes in the ultrastructure of the amphids of the potato cyst nematode, *Globodera rostochiensis*, during development and infection, *Fundam. appl. Nematol.* **17**, 369-82.
42. Jones, J.T., Robertson, L., Perry, R.N., and Robertson, W.M.: Changes in gene expression during stimulation and hatching of the potato cyst nematode, *Globodera rostochiensis. Parasitology* (in press).
43. Khan, F.A. (1985) Hatching response of *Rotylenchulus reniformis* to root leachates of certain hosts and nonhosts, *Rev. Nématol.* **8**, 319-393.
44. Kraus, G.A., Vander Louw, S.J., Tylka, G.L., and Soh, D.H. (1996) Synthesis and testing of compounds that inhibit soybean cyst nematode egg hatch. *J. Agric. Food Chem.* **44**, 1548-1550.
45. LaMondia, J.A. (1995) Hatch and reproduction of *Globodera tabacum tabacum* in response to tobacco, tomato, or black nightshade, *J. Nematol.* **27**, 382-386.
46. Lee, Y.B., and Evans, A.A.F. (1973) Correlation between attractions and susceptibilities of rice varieties to *Aphelenchoides besseyi. Kor. J. Pl. Prot.* **12**, 147-51
47. Lownsbery, B.F., and Viglierchio, D.R. (1961) Importance of response of *Meloidogyne hapla* to an

agent from germinating tomato seeds, *Phytopathology* **51**, 219-21.
48. Marrian, D.H., Russell, P.B., Todd, A.R., and Waring, R.S. (1949) The potato eelworm hatching factor. 3. Concentration of the factor by chromatography. Observations on the nature of eclepic acid, *Biochem.J.* **45**, 524-528.
49. Masamune, T. (1976) Purification of the hatching substances of the soybean cyst nematode, *34th Meeting of the Chemical Society of Japan*, 91.
50. Masamune, T., Anetai, M., Takasugi, M., and Katsui, N. (1982) Isolation of a natural hatching stimulus, glycinoeclepin A, for the soybean cyst nematode. *Nature* **297**, 495-496.
51. Mulder, J.G., Diepenhorst, P., Plieger, P., and Brüggemann-Rotgans, I.E.M. (1992) Hatching agent for the potato cyst nematode, *Patent application No. PCT/NL92/00126*.
52. Perry, R.N. (1986) Physiology of hatching. *In:* Lamberti, F. & Taylor, C.E. (Eds). *Cyst nematodes.* New York, Plenum Press, 119-131.
53. Perry, R.N. (1987) Host induced hatching of phytoparasitic nematode eggs. *In:* Veech, J.A. and Dickson, D.W. (Eds). *Vistas on Nematology*. Hyattsville, Society of Nematologists Inc, 159-64.
54. Perry, R.N. (1989) Root diffusates and hatching factors, *Aspects appl. Biol.* **22**, 121-28.
55. Perry, R.N. (1996) Chemoreception in plant parasitic nematodes, *Annu. Rev. Phytopathol.* **34**, 181-199.
56. Perry R.N. and Beane, J. (1982) The effect of brief exposures to potato root diffusate on the hatching of *Globodera rostochiensis*. *Rev. Nématol.* **5**, 221-224.
57. Perry, R.N., Clarke, A.J., and Beane, J. (1980) Hatching of *Heterodera goettingiana in vitro. Nematologica* **26**, 493-495.
58. Perry, R.N., Clarke, A.J., and Hennessy, J. (1980) The influence of osmotic pressure on the hatching of *Heterodera schachtii. Rev. Nématol.* **3**, 3-9.
59. Perry, R.N., Clarke, A.J., Hennessy, J., and Beane, J. (1983) The role of trehalose in the hatching mechanism of *Heterodera goettingiana, Nematologica* **29**, 324-335.
60. Perry, R.N., Knox, D., and Beane, J. (1992) Enzymes released during hatching of *Globodera rostochiensis* and *Meloidogyne incognita. Fundam. appl. Nematol.* **15**, 283-288.
61. Perry, R.N., and Riga, E. (1995) Electrophysiological analysis of sensory responses of parasitic nematodes, *Jap. J. Nematol.* **25**, 61-69.
62. Perry, R.N., and Trett, M.W. (1986) Ultrastructure of the eggshell of *Heterodera schachtii* and *H. glycines* (Nematoda: Tylenchida). *Rev. Nématol.* **9**, 399-403.
63. Perry, R.N., Zunke, U., and Wyss, U. (1989) Observations on the response of the dorsal and subventral oesophageal glands of *Globodera rostochiensis* to hatching stimulation, *Revue Nématol.* **12**, 91-96.
64. Pline, M., and Dusenbery, D.B. (1987) Responses of plant parasitic nematode *Meloidogyne incognita* to carbon dioxide determined by video camera-computer tracking, *J. Chem. Ecol.* **13**, 873-888.
65. Prot, J-C. (1980) Migration of plant-parasitic nematodes towards plant roots, *Revue Nématol.* **3**, 305-18.
66. Racke, J. and Sikora, R.A. (1992) Wirkung der pflanzengesundheitsfördernden rhizobakterien *Agrobacterium radiobacter* und *Bacillus sphaericus* auf den *Globodera pallida*, *J. Phytopath.* **134**, 198-208.
67. Rawsthorne, D., and Brodie, B.B.(1986) Relationship between root growth of potato, root diffusate production, and hatching of *Globodera rostochiensis*, *J. Nematol.* **18**, 379-384.
68. Riga, E., Perry, R.N., Barrett, J. (1996) Electrophysiological analysis of the responses of males of *Globodera rostochiensis* and *G. pallida* to their female sex pheromones and to potato root diffusate, *Nematologica* **42**, 1-6.
69. Riga, E., Perry, R.N., Barrett, J., and Johnston, M.R.L.: Electrophysiological responses of males of the potato cyst nematodes, *Globodera rostochiensis* and *G. pallida*, to some chemicals, *J. Chem. Ecol.* (in press).
70. Robertson, W.M., and Forrest, J.M.S. (1989) Factors involved in host recognition by plant-parasitic nematodes. *Aspects appl.Biol.* **22**, 129-33.
71. Robinson, M.P., Atkinson, H.J., and Perry, R.N. (1985) The effect of delayed emergence on the subsequent infectivity of second stage juveniles of the potato cyst nematode *Globodera rostochiensis, Nematologica* **31**, 171-178.
72. Robinson, M.P., Atkinson, H.J., and Perry, R.N. (1987) The influence of soil moisture and storage time on the motility, infectivity and lipid utilization of second stage juveniles of the potato cyst nematodes *Globodera rostochiensis* and *G. pallida, Revue Nématol.* **10**, 343-348.
73. Robinson, M.P., Atkinson, H.J., and Perry, R.N. (1987) The influence of temperature on the hatching, activity and lipid utilization of the potato cyst nematodes *Globodera rostochiensis* and *G. pallida, Revue*

Nématol. **10**, 349-354.
74. Schmitt, D.P., and Riggs, R.D. (1991) Influence of selected plant species on hatching of eggs and development of juveniles of *Heterodera glycines, J. Nematol.* **23**, 1-6.
75. Stewart, G.R., Perry, R.N., and Wright, D.J. (1993) Studies on the amphid specific glycoprotein gp32 in different life cycle stages of *Meloidogyne* species, *Parasitology* **107**, 573-78.
76. Stewart, G.R., Perry, R.N., Alexander, J., and Wright, D.J. (1993) A glycoprotein specific to the amphids of *Meloidogyne* species, *Parasitology* **106**, 405-12.
77. Tefft, P.M., and Bone, L.W. (1985) Leucine aminopeptidase in eggs of the soybean cyst nematode *Heterodera glycines, J. Nematol.* **17**, 270-274.
78. Tefft, P.M., and Bone, L.W. (1985) Plant-induced hatching of the soybean cyst nematode *Heterodera glycines. J. Nematol.* **17**, 275-279.
79. Turner, S.J., and Stone, A.R. (1981) Hatching of potato cyst nematodes (*Globodera rostochiensis, G. pallida*) in root exudates of *Solanum vernei* hybrids, *Nematologica* **27**, 315-318.
80. Turner, S.J., and Stone, A.R. (1984) Development of potato cyst-nematodes in roots of resistant *Solanum tuberosum* spp. *andigena* and *S. vernei* hybrids, *Nematologica* **30**, 324-332.
81. Twomey, U., Devine, K., Byrne, J., and Jones, P. (1996) Roles of host-derived hatching inhibitors in controlling hatch of *Globodera rostochiensis*, *Abstract: Third International Nematology Congress*, Guadeloupe, 133.
82. Tylka, G.L., Kraus, G.A., Applegate, J.M., and Johnston, B.E. (1992) Evaluation of precursors and analogs of glycinoeclepin A, a natural hatching stimulus of *Heterodera glycines*. *Abstract: 31st Meeting of the Society of Nematologists*, 623.
83. Tylka, G.L., Kraus, G.A., Wong, A.T.S., and Vanderlouw, S.J. (1994) Inhibition of *Heterodera glycines* egg hatching with a synthetic glycinoeclepin A precursor, *Abstract: 33rd Meeting of the Society of Nematologists*, 75.
84. Viglierchio, D.R. (1961) Attraction of parasitic nematodes by plant root emanations. *Phytopathology* **51**, 136-42.
85. Watanabe, H., and Mori, K. (1991) Triterpenoid total synthesis. Part 2. Synthesis of glycinoeclepin A, a potent hatching stimulus for the soybean cyst nematode, *J. Chem. Soc. Perkin Trans.* **1**, 2919-2934.
86. Weischer, B. (1959) Experimentelle Untersuchungen über die Wanderung von Nematoden, *Nematologica* **4**, 172-186.
87. Wergin, W.P., and Endo, B.Y. (1976) Ultrastructure of a neurosensory organ in a root-knot nematode. *J. Ultrastr. Res.* **56**, 258-76.
88. Williams, T.D., and Beane, J. (1979) Temperature and root exudates on the cereal cyst nematode *Heterodera avenae, Nematologica* **25**, 397-405.
89. Zheng, L., and Ferris, H. (1991) Four types of dormancy exhibited by eggs of *Heterodera schachtii. Revue Nématol.* **14**, 419-426.
90. Zuckerman, B.M., and Esnard, J. (1994) Biological control of plant nematodes - current status and hypotheses, *Jpn. J. Nematol.* **24**, 1-13.

INVASION AND MIGRATION BEHAVIOUR OF SEDENTARY NEMATODES

Nicola von MENDE
Entomology and Nematology Department, IACR Rothamsted, Harpenden, Herts. AL5 2JQ, U.K.

Abstract

Research on the invasion and migration behaviour of sedentary nematodes is reviewed, with emphasis on the root-knot nematodes, *Meloidogyne* spp., and with particular reference to the orientation in the host, changes that occur in the nematode, the importance of nematode secretory products and the response of the plant. A comparison with cyst-forming nematodes highlights two important features during *Meloidogyne* migration which are discussed: the intercellular movement of the juveniles through plant tissues and the ability of orientation within the root tip. As the second-stage juveniles of all *Meloidogyne* species show this behaviour independent of the host, future research on these two topics may provide information leading to the development of plants with resistance to root-knot nematodes that operates during the early stages of infection.

1. Introduction

Research on host-parasite interactions of sedentary nematodes has focused mainly on the understanding of the initiation and maintenance of the feeding cells. Relatively few studies have been made on the early stages of pathogenicity, such as the invasion and migration.

About sixty years ago, Godfrey and Oliveira [31], Christie [16] and Linford [51] presented for the first time detailed observations on the migration of root-knot nematodes inside the host. They refer to Némec [59], Bessey [6] and Byars [15] who briefly mentioned early stages of *Meloidogyne* parasitism. About thirty years later, four research papers were published [29, 54, 63, 77] specifically describing host-parasite interactions during the early stages of infection. Bird and colleagues at that time [8, 9, 10, 13] were interested in the changes that occurred in the nematode with the onset of parasitism and compared morphological and physiological features in pre-parasitic and parasitic second-stage juveniles (J2). Recently, using *Arabidopsis thaliana* as a host [76] and applying video enhanced light microscopy, Wyss *et al.* [87] were able to observe the parasitic behaviour of these infective juveniles during invasion, migration and the initiation of the giant cells. The above and additional research at the cellular level [35,

36] indicates that during these early stages in the development of root-knot nematodes important signalling events determine the compatibility of the host-parasite relationship.

The migratory behaviour of cyst-forming nematodes has been reviewed and described on *Heterodera schachtii* in great detail by Wyss and Zunke [86]. This publication will be the basis for a comparison of the migratory behaviour of cyst-forming with root-knot nematodes.

2. Invasion

After the first contact with the growing root, J2 of root-knot nematodes explore the root surface by pressing and rubbing their lips against the epidermal cells preferentially between the tip and the root hair zone. More mature areas are also explored, but the J2 eventually moves towards the elongation or meristematic area. The stylet is continuously moved forwards and backwards, rarely beyond the oral aperture, at a rate of one to two thrusts per second [51, 74, 87]. This gentle stylet probing is in contrast to the action during intercellular migration, when the stylet thrusts are rhythmical and the stylet is protruded for several seconds during the typical cycles of parasitic behaviour.

The preferred invasion site of the J2 is at the elongation zone of growing root tips for all hosts tested [16, 29, 31, 52, 64, 77, 87]. Invasion at regions of increasing age of cells was more difficult [87] and McClure and Robertson [54] provided evidence that J2 fail to penetrate mature root tissue. Juveniles also enter through damaged tissue at the base of lateral roots or through wounds, although to a lesser degree. Siddiqui [74] mentions heavy infection of root primordia. Attempts to enter root hair cells have been observed [51, 87] and whilst some juveniles invaded they became trapped [74]. Gravato Nobre *et al.* [35] state that the site of future root hair cells is preferably chosen for invasion. In general, the invasion of a root tip by one juvenile will attract others which then try to invade at the same point [7, 31, 35, 51, 64, 74, 77, 87]; this can cause extensive cell destruction and eventually lead to the maceration of the root tip.

Wyss *et al.* [87] noted that invasion of *M. incognita* into *Arabidopsis* roots is not successful if the juveniles enter at the apex of the tip, because too many meristematic cells are destroyed. However, juveniles of *M. javanica* and *M. arenaria* were able to enter *Arabidopsis* root tips at the apex and establish successfully. With several juveniles attacking a tip, both invasion sites (i.e. root apex and elongation zone) can be chosen at the same time (von Mende, unpubl.).

With the onset of the invasion process, J2 usually line up along the root facing the root tip (von Mende, unpubl.), become immobile and stay closely pressed against the root [52]. The rhythmical stylet thrusts and the stylet protrusions occur then for the first time followed by contractions of the metacorpal bulb [87]. Linford had described this earlier [49, 51], but believed that he was observing feeding. Siddiqui [74] only mentions that the nematode used rhythmical stylet thrusts to invade and migrate.

Juveniles penetrate the root epidermis both intra- or intercellularly and both ways have been observed by almost all investigators, although the first is the more common [42, 74, 85, 87]. During intracellular invasion, subepidermal cells were also destroyed [35, 42, 73, 85, 87] indicating that injury to the root can occur in this early stage yet, generally, there is little damage to the plant. The rupture of cells does not depend on the use of the stylet alone but also on the mechanical force exerted by the juveniles [87]. Change from initial intracellular invasion to intercellular migration is achieved by concentrated stylet and head movements at cell junctions; intercellular invasion is faster and seems easier for the juveniles than intracellular invasion [87].

Recorded entry rates vary due to different inoculation methods and times of sampling. Most juveniles invaded within 24-48 hours after inoculation [7, 25, 31, 33, 42, 56, 60, 85]. Godfrey and Oliveira [31] and Siddiqui [74] detected *Meloidogyne* juveniles in host roots within 6 hours after inoculation and Gourd *et al.* [33] found that invasion of soybeans by J2 of *M. incognita*, *M. arenaria* and *M. javanica* had begun within 3 hours and after 24 hours up to 12% had invaded; in comparison, *M. hapla* was slower, with no J2 having invaded within 3 hours and only 5% having penetrated the root 24 hours after inoculation. Such differences of invasion rate between nematode species were also noted by Ngundo and Taylor [60]. More *M. incognita* than *M. javanica* invaded bean roots, yet Arens *et al.* [2] found that more J2 of *M. javanica* than *M. incognita* invaded tobacco. These authors conclude that the ability to invade and induce galls appears to be related primarily to differences in the aggressiveness between nematode species. In addition, motility and infectivity is influenced by the age of the infective juveniles and temperature [5, 12, 60, 74, 80].

Like root-knot nematodes, cyst-forming nematodes invade preferentially in the zone of elongation of growing root tips or at sites where lateral roots emerge. The behaviour of exploration and selection of penetration sites has been described by Doncaster and Seymour [22] who also filmed these processes. The J2 of *Heterodera cruciferae* perforate the epidermal cell wall purely mechanically by repeated stylet thrusts along a line, producing a slit for the J2 to invade the cell. At sites where single J2 have successfully invaded, many more J2 are attracted to invade.

3. Migration

Only recently has the migratory behaviour of *Meloidogyne* juveniles been studied in detail. This was made possible by the use of *A. thaliana* as a host [76]. The thin, translucent roots of this plant allow easy observations, especially when applying video enhanced light microscopy [87]. In addition, the simple root anatomy with single layers of root tissues has led to complementary studies at the cellular level [35]. The migratory pathway in *A. thaliana* is as follows: after penetrating the epidermis juveniles orientate themselves towards the root tip, lining up parallel to the long axis of the root. They migrate intercellularly between cortical cells until they reach the meristematic tissue. They then turn and migrate within xylem tissue into the vascular cylinder where they stop at a suitable site in xylem parenchyma to establish their permanent feeding site [35,87].

The migration of cyst-forming nematodes differs in many aspects. Above all, the infective juveniles are physically much stronger and have a more robust stylet. This allows the nematode to apply strong mechanical forces and they invade and migrate intracellularly through the cortex straight into the vascular cylinder [86].

3.1. BEHAVIOUR

Two important observations made during *Meloidogyne* migration are that J2 have the ability to move intercellularly and orientate themselves during migration, firstly, immediately after invasion and, secondly, during the turning phase. This implies that complex recognition mechanisms operate throughout migration, possibly involving sensory organs e.g. the amphids, and the cuticle of the nematode, as the direct contact points between the nematode and plant cell surfaces. The fact that juveniles migrate intercellularly had been suggested much earlier by Némec in 1910 [59] and, since then, has been generally accepted.

Following observations on *Allium cepa*, Smith and Mai [77] suggested for the first time a distinct migratory pathway for *M. hapla*. The J2 were lying straight in the cortex heading distally, or inside the vascular cylinder heading proximally, yet, in the very root tip, J2 could be remarkably curved and randomly orientated. Orientation after invasion has also been observed in cotton by McClure and Robertson [54] and in *Portulaca grandiflora* by Linford [51]. Even though Godfrey and Oliveira [31] did not mention an orientation towards the root tip, one of their figures clearly illustrates invading juveniles turning distally with their tails still outside the root. The reason for this orientation behaviour after invasion is not clear. Wyss *et al.* [87] demonstrated that, occasionally, a J2 will get a wrong signal and move away from the tip into differentiated root tissue. The J2 will not succeed in making progress in this case and will leave the root or starve. From these observations, it can be inferred that J2 are able to recognize softer immature tissue. In a similar way, Siddiqui [74] reasoned that the juveniles take the path of least resistance after invasion. Other factors could play a role, such as a pH gradient, the direction of cell elongation or the flow of an electric current. It has been shown that *Meloidogyne* juveniles move preferentially within an electric field [81]. McClure and Robertson [54] suggested that directional migration within the root is promoted by the same factor(s) operating externally, i.e. one of the factors could be a CO_2 gradient, as CO_2 is found in the intercellular spaces. Wyss *et al.* [87] suggested that J2 are not able to cross the cortex and the developing endodermis. The Casparian strips may be the final insurmountable barrier, forcing the nematode to enter the vascular cylinder via the tip. However, Gravato Nobre *et al.* [35] clearly showed that the nematode is able to separate endodermal cells during the early stage of migration. These authors propose that the nematode might need to move toward the quiescent centre in the tip following a signal from the apical initials, which serve as a pattern for the development of all root tissues, allowing the nematode to eventually detect future xylem tissue and to enter the vascular cylinder. This would also explain the extensive curving of the nematode in the root tip during the turning phase, as if the nematode is searching for a signal or contact point. These observations suggest that recognition of host tissues may be important for a successful migration, as the nematode is always associated with the cortex on the way towards the tip and with the xylem when migrating onto the stele.

Turning in the root tip can occur at different levels. In *A. thaliana*, J2 mostly turn round in the meristematic region, about 50 µm above the apical initials at the quiescent centre. Usually they are then between the cortical and endodermal cell layers [35]. Occasionally, turning J2 are found in the root cap columella or even at some distance above the quiescent centre. Smith and Mai [77] never found *M. hapla* J2 in or below the apical initial region in onion root tips.

It is interesting that juveniles can also infect leaves, stems and inflorescences [47, 50, 66]. They penetrate young structures, preferably in groups and migrate intercellularly in a polar orientation until feeding sites are located [50]. Such orientation was found in a wide range of hosts and the J2 were in a straight posture when found establishing feeding sites. However, when found in undifferentiated tissue they were irregularly orientated in varied postures [50]. This again indicates that intercellular migration and orientation within young or meristematic tissue are essential features of the early stage of pathogenicity.

Migration is completed within about 24 hours in *A. thaliana* roots and the J2 become sedentary in the xylem parenchyma [87]. According to data from earlier work, migration probably takes longer in plants with bigger roots, e.g. McClure and Robertson [54] noted that 12 hr after invasion J2 are migrating distally in cotton and, at 32 hr, are just turning in the root tip, comparable to only 8-10 hours in *A. thaliana*. The duration of invasion plus migration within pineapple and cowpea was between 1 and 3 days [31] and in soybean and tomato about 2 days [16, 28]. Preliminary tests have shown that *M. javanica* juveniles move twice as fast in roots as juveniles of *M. incognita*, but are slower in sand. This might reflect different levels of aggressiveness which is not correlated to physical strength but rather to other factors such as the nature of the saliva, possibly due to different degrees of enzymatic activity [82].

Second-stage juveniles become established within the differentiating vascular cylinder [87], within the region of cell elongation [16, 77] or just above the original invasion point in *A. thaliana* (Gravato Nobre, unpubl.), generally with a proximal orientation. However, the final position and orientation of the nematodes can vary between hosts. Godfrey and Oliveira [31] found that in pineapple the heads of the J2 were embedded in the periphery of the stele with the bodies in the cortex compared to the more common case, already described by Bessey in 1911 [6], with the nematodes completely inside the central cylinder and appearing in surrounding tissue only when adult females break through the cortex. Occasionally one can find galls on infected root systems with females orientated distally, indicating that the migrating nematode was not able to turn but could still initiate giant cells (von Mende, unpubl.). *Meloidogyne incognita* juveniles which were inhibited from turning at the tip during a continuous distal migration in the mature cortex, either eventually emerged from the root or were able to establish a feeding site, orientated distally (von Mende, unpubl.); it is not clear how this was accomplished.

After locating the invasion site, the exploratory stylet probing of the J2 changes to a typical parasitic behaviour which will be maintained throughout its development in the root. Cycles of such parasitic behaviour which occur during invasion, migration, giant cell initiation, and feeding consist of repeated periods of stylet tip protrusions followed by a resting interval, in turn followed by metacorpal bulb pumping, and sequentially by stylet probing and by head movement [82, 87]. During invasion and migration, J2 do not take up any nutrients, but instead, glandular secretions are passed through the stylet and injected as saliva between plant cells [87]. Towards the end of the migration, the period of metacorpal bulb pumping increases from 10 to 20 sec within a cycle [87]. Feeding starts with the establishment of giant cells and the cycles of parasitic behaviour are then always correlated with the uptake of nutrients. The durations of stylet protrusion and metacorpal bulb pumping significantly increase with giant cell formation. Linford [49] has observed this 'feeding' behaviour in females which were removed from pea roots, in sedentary juveniles within pineapple roots and in migrating juveniles [51], in the latter case assuming that metacorpal bulb pumping is correlated with uptake of cell contents. The durations of stylet tip protrusion of 15-30 sec and metacorpal bulb pumping of 10-40 sec [51] are similar to those measured by Wyss *et al.* [87].

This parasitic behaviour does not always occur during migration. Second-stage juveniles which migrated distally in the intercellular spaces of differentiated cortex tissue progressed quickly without going through a single cycle of stylet protrusion and metacorpal bulb pumping (von Mende, unpubl.). This demonstrated that stylet activity and saliva production during migration is only initiated and necessary when cell walls have to be separated, again possibly indicating enzymatic activity of the saliva. Similarly Linford [49] noted that stylet thrusting by J2 occurred only when the lips were in contact with something firm.

Cyst-forming nematodes migrate exclusively intracellularly through root tissues to the site of syncytium induction in or near the vascular cylinder. The walls of cortical and endodermal cells are explored repeatedly by stylet probing, which becomes highly coordinated stylet thrusting at suitable sites [22, 86]. The stylet can be under great physical stress which can cause its tip to bend. Again, slits are cut into the cell wall, with less effort than during invasion, to allow the nematode to move into the next cell. Breaking through a cell wall can take up to 15 min [86]. The speed of progress depends on factors such as rigidity of the cell wall, conditions for applying sufficient lip pressure to the wall and the vigour of the invading juvenile [86].

During invasion and migration, cyst-forming nematodes do not exhibit the repeated cycles of stylet probing and pumping of the metacorpal bulb as observed in root-knot nematodes. It appears that they do not produce any saliva during migration although the three oesophageal glands are active, especially the subventral glands: their ampullae are filled with secretory granules which decrease in concentration only when the nematode initiates the first syncytial cell, at the same time the concentration of dorsal gland granules increases.

3.2. SECRETIONS AND EXCRETIONS

Secretions and excretions (ES) of plant parasitic nematodes are produced in amphids, oesophageal glands, excretory glands, phasmids, rectal glands [67] and by the cuticle and a range of nematode antigens have been identified with potential roles during host-parasite interactions [19, 39]. The ES products from at least three origins, the excretory gland, the amphids [29] and the oesophageal glands [87] appear to be involved in the processes of intercellular migration of root-knot nematodes. Preliminary immunocytological studies indicated that a glycoprotein, which covers the surface of the nematode, is probably produced in the excretory glands (Gravato Nobre and McClure unpubl.). During migration, this surface coat is continuously sloughed off and left behind adhering to the plant cell surfaces along the migratory path [36]. The electron microscopy work of Endo and Wergin [29] illustrates an electron dense layer around the nematode and along the middle lamellae of cells near the lip region. This material may represent the surface coat of the nematode as described above, because similar images were obtained by labelling cross sections of migrating nematodes with antibodies specific to the surface coat [35, 36]. Paulson and Webster [63] found that intercellular spaces between cells away from the nematode were filled with electron-dense material within 24 hours of exposure, and assumed that it was probably cellular debris and/or nematode ES products. The ultrastructural investigations by Endo and Wergin [29] also identified a homogeneous, electron-dense material in and around the amphids. This material, which does not appear to be of plant origin, extends outward from the amphid to the outer layer of the primary wall of the host cell. The third ES product, the saliva, is produced in the oesophageal glands, secreted through the stylet and injected between cells during intercellular migration [87]. Bird [11] suggested that the oesophageal granules may produce cellulase-like enzymes. Such enzymes and the physical movement of the nematode will break down the middle lamella to separate plant cells.

Cell-wall degrading enzymes such as cellulases and pectinases have been detected in various plant parasitic nematodes, including *Meloidogyne* sp. [14, 23, 32, 57, 58, 61]. As these enzymes are not known to exist in free-living nematodes, it seems likely that for plant parasitic nematodes they may play an important role in establishing and maintaining a suitable host-parasite relationship [14]. Immunocytological studies, using an antibody specific to polygalacturonic acid, indicate that migrating juveniles produce enzymes which appear to affect the middle lamella [35]. The origin for cellulase in *Meloidogyne* is not yet known. Cytochemical tests for this enzyme were negative for the subventral and dorsal glands in juveniles and females [79]. Other enzymes produced by J2 are amylase, invertase and proteolytic enzymes [88], but their function in pathogenicity is not known.

The nature and role of secretions originating from the oesophageal glands have been investigated extensively [39]. Both types of glands, one dorsal and two subventral, are believed to be involved directly in the pathogenicity of root-knot nematodes. The evidence relies on the morphological and physiological changes observed in these glands and their contents throughout the development of the nematode in the plant [9, 10, 41]. During invasion and migration, the subventral glands are clearly active and, therefore, may play a role in the penetration of plant tissue [11], whereas the dorsal gland becomes active towards the end of migration and during feeding site initiation and is therefore associated with the development and maintenance of the giant cells [11, 41].

Before invasion, the subventral glands are at their maximum size and presumably at their peak of activity [10]. The ducts and ampullae are filled with secretory granules, which are 0.6-0.8 µm in diameter, are irregular in shape, have a distinct outer membrane [9] and contain an electron transparent core with minute sphericle vesicles [41, 53]. Ultrastructural cytochemical analyses of these granules for a range of enzymes and nucleic acid showed that acid phosphate was present, but peroxidase, cellulase, DNase and RNase, and nucleic acid were absent [79]. During migration, the density of the granules greatly decreases by the end of migration when the feeding site is initiated. The granules then appear smaller, the outer membrane becomes less distinct and the core is gone [9, 41]. Histochemical tests indicated that the protein component detected in the ducts changed with parasitism, staining more strongly with the periodic-acid Schiff (PAS) reagent, an indication for carbohydrates [13]. Generally the subventral glands and granules appear to deteriorate at this stage [9, 41]. In contrast, the density of the granules in the ducts and ampulla of the dorsal gland was low initially [29, 41] but increased during migration until the gland reached its peak of activity after moulting [10].

It is not yet known whether subventral gland secretions are important during invasion and migration of root-knot nematodes but several observations indicate a function during the early stages of the host-parasite relationship. Firstly, the high activity of the glands during invasion and early migration, secondly, linking the discharge of granules with the intermittent pumping of the metacorpus [87], and lastly, that secretions from the subventral glands can be detected in the stylet using immunofluorescence microscopy [21]. However, certain features are at variance with this hypothesis. The proximal position of the subventral gland valves at the base of the metacorpus pump chamber supports the idea that the granules are forced posteriorly toward the intestine during the periods of metacorpal bulb pumping, which has clearly been shown for the cyst-forming nematode *Heterodera schachtii* [86], yet subventral proteins were found to be secreted *in vitro* through the stylet of *H. glycines* [34]. In addition, the presence of some granules in the ampulla and gland extension of the dorsal gland before invasion and at the beginning of migration might suggest that the dorsal gland is involved in the intercellular migration.

3.3. CHANGES IN THE NEMATODE

In addition to the morphological and physiological changes of the oesophageal granules, the glands, the cuticle, the muscles and the surface of the J2 of root-knot nematodes are dynamic and change in appearance.

A clear positive PAS reaction in preparasitic J2 is found in the rectal glands, yet in parasitic (2-3 days) J2 they do not stain [13]. Reported morphological changes in the oesophageal glands in the early stages of parasitism (2-3 days) [8], were demonstrated by Bird as a significant increase in the length of these glands [11]. He argues that the increase may be due to the changes in muscle tone during this time resulting in loss of mobility. It is likely that there would be a change in the hydrostatic pressure within the nematode as the somatic muscle starts to atrophy and this may have some effect on the shape of these glands.

The early parasitic stage of the nematode is closely associated with a loss of mobility and the ability to reinfect [11]. The mobility of 1-2 day old parasitic juveniles dropped by 88%, and the infectivity dropped five-fold within 24 hrs. These changes in physiology are highly significant and take place within a day of penetration into the host.

The cuticle changes morphologically with the onset of parasitism. The striated basal and median zones disappear and the cuticle becomes homogeneous in appearance. Collagens also appear to be more evenly distributed in the cuticle of parasitic juveniles [40]. It is believed that the surface of the nematode cuticle plays an important role in the ability of the nematode to invade its hosts [20]. Ibrahim [43] noted that Con A and WGA binding was minimal with freshly hatched *M. javanica* J2, but strong with J2 which had emerged from the host root. This indicates that, after contact with the plant, the concentration of mannose/glucose and N-acetyl-D-glucosamine sugars increased on the surface. Spiegel and McClure [78] also observed differences in Con A binding between infective and sedentary J2 of *M. incognita* (2 wk old). In cyst-forming nematodes stage-specific labelling with Con A was observed in *Heterodera schachtii* [3, 4] and in *Globodera rostochiensis* [30]. Enzymatic tests by Dalmasso and Tournay [18] demonstrated that protease strongly reduces invasion and galactosidase incubation inhibits invasion by 50%.

3.4. PLANT RESPONSES

Intercellular migration of root-knot nematodes causes very little damage to the plant. No obvious plant responses occur during the early stages of invasion and migration [16, 29, 35]. As the J2 progresses, adjoining cells become distended and compressed and conform to the contour of the nematode [29, 77] or sometimes the entire cell shrinks [51]. Changes occur in the protoplast of meristematic and pericyclic cells which are in contact with the head of the nematode [29]. The more intense staining of these cells on electron micrographs might be due to the significantly increased density of free ribosomes and the appearance of a fine granular cytoplasmic matrix [29]. This was also noticed along the pathway of migrating *M. incognita* and *M. javanica* J2 in roots of *Impatiens balsamina* by Jones and Payne [44], who suggested that it might be due to the break of symplastic continuity, and by Paulson and Webster [63] in resistant tomato plants, who believed that they are the earliest indications (12 hours after inoculation) of a hypersensitive reaction. During the turning phase several meristematic cells are destroyed [16, 87] and a necrotic cell occasionally appeared next to a J2 in onion roots [77].

Morphological changes of the nuclei were observed by Endo and Wergin [29] and Linford [51]. The nucleoplasm of deformed nuclei is more electron dense and has ribosome like particles of about 45nm instead of 15nm. Occasionally, the nuclear envelope appears dilated. Endo and Wergin [29] also mention alterations in membranes such as dilated cisternae of the endoplasmic reticulum and a highly convoluted or interfolded plasmalemma. Histochemical studies by Endo and Veech [28] revealed that cells next to the nematode respond to infection, possibly to the pressure exerted by the nematode or stylet activity, by having higher levels of oxidoreductase activity in addition to the increase in numbers of ribosomes.

Growth retardation or inhibition of the root tip depends on the level of infestation and to some degree on the host plant [16, 31, 87]. Even so, it is surprising how many juveniles can enter a plant root and migrate within the root tip without causing any damage, e.g. up to 120 J2 in an oat root tip without impairment of root growth [73].

Hypertrophy of cortical cells shortly after invasion, in general within 24 hours, has been reported for many host plants [7, 16, 24, 29, 31, 51, 74, 85]. Loewenberg *et al.* [52] noted such galling effects in tomato roots even when the J2 left the root again, so this type of galling cannot be due to giant cell formation especially so early after invasion. The degree of galling depends on the host plant. Heavy galling was reported in *Asparagus officinalis* [45] whereas very little galling has been noticed with *I. balsamina* (von Mende, unpubl.). In *A. thaliana*, premature galling appeared in about half of the infected root tips (von Mende, unpubl.). Premature galls in *A. thaliana* were due to hyperplasia followed by hypertrophy of the epidermal, cortical and endodermal cells. Such galls were always associated with roots where endodermal cells had been separated by the nematode [35].

3.5. BEHAVIOUR IN RESISTANT PLANTS

Resistance is not necessarily of the same nature in all plants [17], each case reflecting a recognition event important to the compatible host-parasite interaction. A hypersensitive reaction which causes necrosis and results in the death of the nematode can occur early during invasion [27, 45, 48, 64] or towards the end of migration or at the initiation of the giant cells [69].

Compared to a susceptible host cultivar, the penetration rate into a resistant one can be less [26], equal [26, 68, 69, 75] or higher [38, 72] depending on the *Meloidogyne* species and the host. At high penetration rates, juvenile invasion peaked at the fourth or fifth day and then declined by which time juveniles had usually emerged or died [38, 71]. Re-infection with J2 which emerged from resistant plants was less successful than with J2 which had emerged from susceptible varieties but the reason is not known. At moderate penetration rates, J2 were found in resistant hosts in a migratory state for periods of up to 21 days [1, 28, 55, 62].

3.6. MOLECULAR ASPECTS

Very little information exists on migration at the molecular level. Williamson and colleagues [46, 84] have identified genes that are expressed early during root-knot nematode infection of resistant tomato plants. Genes are induced in the first 12 hours of nematode infestation and their levels of expression increase with the duration of infestation. These genes are also induced to various degrees in susceptible tomato plants. One gene which is specific to the resistant tomato shows homology to miraculin and another to extensin.

Hansen *et al.* [37] studied the expression of the reporter gene ß-glucuronidase directed by the wound-inducible promoter *wun1* in potato plants infested with sedentary plant parasitic nematodes. A strong reaction was found with *Globodera*, but *Meloidogyne* did not trigger a wound response during migration.

4. Conclusions

All of the research approaches that have been taken to study the migratory behaviour of root-knot nematodes after invasion and the responses of plants to invasion have been summarised. This topic has not received as much attention as research on the initiation and maintenance of the nematode feeding site, which has proved more fascinating to scientists because of the complex host-parasite interactions involved. It is a challenge to understand the apparent ability of the nematode to manipulate plant gene expression during feeding site formation and to apply this knowledge in the development of novel control strategies. By contrast, the migratory stage is short and apparently less intriguing. However, the latest observations suggest the occurrence of important signalling events which allow the nematode to find a suitable site for giant cell formation. It would be of great interest to understand these events, which could lead to the development of resistant plants in which the resistance mechanism acts before the nematode has reached the sedentary stage.

Two abilities of the nematode become apparent during intercellular migration: the ability to orientate within the root and the ability to separate plant cells. This probably applies to all *Meloidogyne* species, as the observations mentioned in this review were made on *M. incognita, M. incognita acrita, M. javanica, M. arenaria, M. hapla* and *M. naasi* and also seems to be independent of the host as it occurs in monocotyledons as well as in dicotyledons. Directed migration may even be necessary when *Meloidogyne* J2 are applied to carrot discs, although it is not clear how the J2 'burrow' into the plant tissue to settle and induce giant cells [70]. Clearly, future research should be directed toward the understanding of such recognition events; for instance, very little is known about chemoreception during migration [65] and the role of the amphids as chemoreceptors.

Phytoparasitic bacteria and fungi are known to produce cell wall degrading enzymes during their attack and invasion of plants [83]. These microorganisms can have more than just one enzyme to cleave a particular polymer and these are considered to be virulence factors. In addition, they can be elicitors for the induction of plant defense responses. Comparative future studies with plant nematodes on the identification and characterisation of pectinases and cellulases and their induction would further help to understand host-parasite interactions at this early stage of pathogenicity.

ACKNOWLEDGEMENTS
I thank Maria Joao Gravato Nobre, Kenneth Evans and Hugh D. Loxdale for critical review of the manuscript. IACR receives grant-aided support from the Biotechnology and Biological Sciences Research Council of the United Kingdom.

References

1. Anwar, S.A., Trudgill, D.L. and Phillips, M.S. (1994) The contribution of variation in invasion and development rates of *Meloidogyne incognita* to host status differences, *Nematologica* **40**, 579-586.
2. Arens, M.L., Rich, J.R. and Dickson, D.W. (1981) Comparative studies on root invasion, root galling, and fecundity of three *Meloidogyne* spp. on a susceptible tobacco cultivar, *J. Nematol.* **13**, 201-205.
3. Aumann, J. and Wyss, U. (1987) Lectin binding sites on mobile stages of *Heterodera schachtii* Schmidt (Nematoda: Heteroderidae), *Nematologica* **33**, 410-418.
4. Aumann, J., Robertson, W.M. and Wyss, U. (1991) Lectin binding to cuticle exudates of sedentary *Heterodera schachtii* (Nematoda: Heteroderidae) second stage juveniles. *Rev. Nématol.* **14**, 113-118.
5. Balmer, E. and Cairns, E.J. (1963) Relation of physiological age of root-knot larvae to infectivity, (Abstr.) *Phytopathology* **53**, 621.
6. Bessey, E.A. (1911) *Root-knot and its control. U.S. Dept. Agr. Bur. Plant Ind. Bul.* 217.
7. Bird, A.F. (1961) The ultrastructure and histochemistry of a nematode-induced giant cell, *J. Biophys. Biochem. Cytol.* **11**, 701-715.
8. Bird, A.F. (1967) Changes associated with parasitism in nematodes. I. Morphology and physiology of preparasitic and parsitic larvae of *Meloidogyne javanica*, *J. Parasit.* **53**, 768-776.

9. Bird, A.F. (1968) Changes associated with parasitism in nematodes. III. Ultrastructure of the egg shell, larval cuticle, and contents of the subventral esophageal glands in *Meloidogyne javanica*, with some observations on hatching, *J. Parasitol.* **54**, 475-489.
10. Bird, A.F. (1968) Changes associated with parasitism in nematodes. IV Cytochemical studies on the ampulla of the dorsal esophageal gland of *Meloidogyne javanica* and on exudations from the buccal stylet, *J. Parasitol.* **54**, 879-890.
11. Bird, A.F. (1983) Changes in the dimensions of the oesophageal glands in root-knot nematodes during the onset of parasitism, *Int. J. Parasitol.* **13**, 343-348.
12. Bird, A.F. and Wallace, H.R. (1965) The influence of temperature on *Meloidogyne hapla* and *M. javanica*, *Nematologica* **11**, 581-589.
13. Bird, A.F. and Saurer, W. (1967) Changes associated with parasitism in nematodes. II. Histochemical and microspectrophotometric analyses of preparasitic and parasitic larvae of *Meloidogyne javanica*, *J. Parasitol.* **53**, 1262-1269.
14. Bird, A.F., Downton W.J.S. and Hawker J.S. (1975) Cellulase secretion by second stage larvae of the root-knot nematode (*Meloidogyne javanica*), *Marcellia.* **38**, 165-169.
15. Byars, L.P. (1914) Preliminary notes on the cultivation of the plant parasitc nematode, *Heterodera radicicola*, *Phytopathology* **4**, 323-326.
16. Christie, J.R. (1936) The development of root-knot nematode galls, *Phytopathology* **26**, 1-22.
17. Christie, J.R. (1949) Host-parasite relationships of the root-knot nematodes, *Meloidogyne* spp. III. The nature of resistance in plants to root-knot, *Proc. Helm. Soc. Wash.* **16**, 104-108.
18. Dalmasso, A. and Tournay, O. (1990) Effects of enzyme incubation on penetration of *Meloidogyne arenaria* (*Nematoda*) in *Lycopersicon esculentum.*, *Nematologica* **36**, 321-323.
19. Davies, K.G. (1994) A nematode case study focusing on the application of serology. In: *The identification and characterization of pest organisms*. Ed. D.L. Hawksworth, 395-413.
20. Davis, E.L., Kaplan, D.T., Dickson, D.W. and Mitchell, D.J. (1989) Root tissue response of two related soybean cultivars to infection by lectin-treated *Meloidogyne* spp., *J. Nematol.* **21**, 219-228.
21. Davis, E.L., Allen, R. and Hussey, R.S. (1994) Developmental expression of esophageal gland antigens and their detection in stylet secretions of *Meloidogyne incognita*, *Fundam. appl. Nematol.* **17**, 255-262.
22. Doncaster, C.C. and Seymour, M.K. (1973) Exploration and selection of penetration site by Tylenchida, *Nematologica* **19**, 137-145.
23. Dropkin V.H. (1963) Cellulase in phytoparasitic nematodes, *Nematologica* **9**, 444-454.
24. Dropkin ,V.H. and Boone, W.R. (1966) Analysis of host-parasite relationships of root-knot nematodes by single-larva inoculations of excised tomato roots, *Nematologica* **12**, 225-236.
25. Dropkin, V.H. and Nelson P.E. (1960) The histopathology of root-knot nematode infections in soybeans, *Phytopathology* **50**, 442-447.
26. Dropkin, V.H. and Webb R.E. (1967) Resistance of axenic tomato seedlings to *Meloidogyne incognita acrita* and to *M. hapla*, *Phytopathology* **57**, 584-587.
27. Dropkin, V.H., Helgeson, J.P. and Upper, C.D. (1969) The hypersensitivity reaction of tomatoes resistant to *Meloidogyne incognita*: reversal by cytokinins, *J. Nematol.* **1**, 55-61.
28. Endo, B.Y. and Veech, J.A. (1969) The histochemical localization of oxidoreductive enzymes of soybeans infected with the root-knot nematode *Meloidogyne incognita acrita*, *Phytopathology* **59**, 418-425.
29. Endo, B.Y. and Wergin, W.P. (1973) Ultrastructural investigation of clover roots during early stages of infection by the root-knot nematode, *Meloidogyne incognita*, *Protoplasma* **78**, 365-379.
30. Forrest ,J.M.S., Robertson, W.M. and Milne, E.W. (1988) Changes in the structure of amphidial exudate and the nature of lectin labelling on freshly hatched, invaded and emigrant second stage juveniles of *Globodera rostochiensis*, *Nematologica* **34**, 422-431.
31. Godfrey, G.H. and Oliveira, J. (1932) The development of the root-knot nematode in relation to root tissues of pineapple and cowpea, *Phytopathology* **22**, 325-348.
32. Goffart, H. and Heiling, A. (1962) Beobachtungen über die enzymatische Wirkung von Speicheldrüsensekreten pflanzenparasitärer Nematoden, *Nematologica* **7**, 173-176.
33. Gourd, T.R., Schmitt, D.P. and Barker, K.R. (1993) Penetration rates by second-stage juveniles of *Meloidogyne* spp. and *Heterodera glycines* into soybean roots, *J. Nematol.* **25**, 38-41.
34. Goverse, A., Davis, E.L. and Hussey, R.S. (1994) Monoclonal antibodies to the esophageal glands and stylet secretions of *Heterodera glycines*, *J. Nematol.* **26**, 251-259.
35. Gravato Nobre M.J., von Mende N., Dolan L., Schmidt K.P., Evans K. and Mulligan B. (1995)

Immunolabelling of cell surfaces of *Arabidopsis thaliana* roots following infection by *Meloidogyne incognita* (Nematoda), *J. Exp. Botany* **46**, 1711-1720.
36. Gravato Nobre M.J., Dolan L., Calder G., McClure M.A., Davies K.G., von Mende N., Mulligan B. and K. Evans. (1996) Giant cells in *Arabidopsis thaliana* roots expressing *Meloidogyne* surface associated molecules, Abstract: *Third International Nematology Congress, Guadeloupe* 84-85.
37. Hansen, E., Harper, G., McPherson, M.J. and Atkinson, H.J. (1996) Differential expression patterns of the wound-inducible transgene *wun1-uidA* in potato roots following infection with either cyst or root-knot nematodes, *Physiol. Mol. Plant Pathol.* **48**, 161-170.
38. Herman, M., Hussey, R.S. and Boerma, H.R. (1991) Penetration and development of *Meloidogyne incognita* on roots of resistant soybean genotypes, *J. Nematol.* **23**, 155-161.
39. Hussey, R.S. (1989) Disease-incuding secretions of plant-parasitic nematodes. *Ann. Rev. Phytopathol.* **27**, 123-141
40. Hussey, R.S. and Jansma, P. (1988) Immunogold localization of collagen in the cuticle of different life stages of *Meloidogyne incognita, J. Nematol.* **20**, 641-642.
41. Hussey, R.S. and Mims, C.W. (1990) Ultrastructure of esophageal glands and their secretory granules in the root-knot nematode *Meloidogyne incognita, Protoplasma* **156**, 9-18.
42. Ibrahim, I.K.A. and Massoud, S.I. (1974) Development and pathogenesis of a root-knot nematode, *Meloidogyne javanica, Proc. Helm. Soc. Wash.* **41**, 68-72.
43. Ibrahim, S.K. (1991) Distribution of carbohydrates on the cuticle of several developmental stages of *Meloidogyne javanica, Nematologica* **37**, 275-284.
44. Jones, M.G.K. and Payne, H.L. (1978) Early stages of nematode-induced giant-cell formation in roots of *Impatiens balsamina, J. Nematol.* **10**, 70-84
45. Koen, H. (1966) Observations on plant-parasitic relationship between the root-knot nematode *Meloidogyne javanica* and some resistant and susceptible plants, *S. Afr. J. agric. Sci.* **9**, 981-992.
46. Lambert, K.N. and Williamson, V.M. (1994) Isolation and characterization of nematode induced cDNAs in tomato, Abstr. 4th International Congress of Plant Molecular Biology, Amsterdam., 1742.
47. Lehman, P.S. and MacGowan, J.B. (1986) Inflorescence and leaf galls on *Palisota barteri* caused by *Meloidogyne javanica, J. Nematol.* **18**, 583-586.
48. Liao, S.C. and Dunlap, A.A. (1950) Arrested invasion of *Lycopersicon peruvianum* roots by the root-knot nematode, *Phytopathology* **40**, 216-218.
49. Linford, M.B. (1937) The feeding of the root-knot nematode in root tissue and nutrient solution, *Phytopathology* **27**, 824-835.
50. Linford, M.B. (1941) Parasitism of the root-knot nematode in leaves and stems. *Phytopathology* **31**, 634- 648.
51. Linford, M.B. (1942), The transient feeding of root-knot nematode larvae, *Phytopathology* **32** 580-589.
52. Loewenberg, J.R., Sullivan, T. and Schuster, M.L. (1960), Gall induction by *Meloidogyne incognita incognita* by surface feeding and factors affecting the behavior pattern of the second-stage larvae, *Phytopathology* **50** 322-323.
53. McClure, M.A. and von Mende, N. (1987) Induced salivation in plant-parasitic nematodes, *Phytopathology* **77**, 1463-1469.
54. McClure M.A. and Robertson J. (1973) Infection of cotton seedlings by *Meloidogyne incognita* and a method of producing uniformly infected root segments, *Nematologica* **19**, 428-434.
55. McClure, M.A., Ellis, K.C. and Nigh, E.L. (1974) Post-infection development and histopathology of *Meloidogyne incognita* in resistant cotton, *J. Nematology* **6**, 21-26.
56. Mildenberger, G. and Wartenberg, H. (1958) Histologische Untersuchungen der Nematodengallen in den Wurzeln der Kartoffelpflanze, (*Heterodera rostochiensis* Woll. und *Meloidogyne* spec. in *Solanum tuberosum* L.), *Z. Pflanzenkr. (Pflanzenpathol.) Pflanzensch.* **65**, 449-464.
57. Myers, R.F. (1963) Materials discharged by plant-parasitic nematodes, Abstr. *Phytopathology* **53**, 884.
58. Myers, R.F. (1965) Amylase, cellulase, invertase and pectinase in several free-living, mycophagous, and plant-parasitic nematodes, *Nematologica* **11**, 441-448.
59. Némec, B. (1910) Das Problem der Befruchtungsvorgänge und andere zytologische Fragen. VI.Vielkernige Riesenzellen in *Heterodera* Gallen., 151-173., Gebrüder Borntrăger.
60. Ngundo, B.W. and Taylor, D.P. (1975) Some factors affecting penetration of bean roots by larvae of *Meloidogyne incognita* and *M. javanica, Phytopathology* **65**, 175-178.
61. Okopnyi, N.S. and Spasskii, A.A. (1977) The role of enzymes in the host-parasite system in *Meloidogyne* infestation. *Izvestiya Akademii Nauk Moldavskoi SSR (Buletinul Akademiei de Shtiintsee a*

RSS Moldovenesht, Biologicheskie Nauki No. 4., 59-65. {Ru}. (Ref. *Helm. Abstr.* **849** (1980), 118.)
62. Oteifa, B.A. (1953) Development of the root-knot nematode, *Meloidogyne incognita* as affected by potassium nutrition in the host, *Phytopathology* **43**, 717-174.
63. Paulson, R.E. and Webster, J.M. (1972) Ultrastructure of the hypersensitive reaction in roots of tomato, *Lycopersicon esculentum* L., to infection by the root-knot nematode, *Meloidogyne incognita, Physiol. Plant Pathol.* **2**, 227-234.
64. Peacock, F.C. (1959) The development of a technique for studying the host parasite relationship of the root-knot nematode *Meloidogyne incognita* under controlled conditions, *Nematologica* **4**, 43-55.
65. Perry, R.N. (1996), Chemoreception in plant parasitic nematodes. *Annu. Rev. Phytopathol.* **34** 181-199.
66. Powell, N.T. and Moore, E.L. (1961) A technique for inoculating leaves with root-knot nematodes, *Phytopathology* **51**, 201-202.
67. Premachandran, D., von Mende, N., Hussey, R.S. and McClure, M.A. (1988) A method for staining nematode secretions and structures, *J. Nematol.* **20**, 70-78.
68. Reynolds, H.W., Carter, W.W. and O'Bannon, J.H. (1970) Symptomless resistance of alfalfa to *Meloidogyne incognita acrita, J. Nematol.* **2**, 131-134.
69. Riggs, R.D. and Winstead, N.N. (1959) Studies on resistance in tomato to root-knot nematodes and on the occurrence of pathogenic biotypes, *Phytopathology* **49**, 716-724.
70. Sandstedt, R. and Schuster, M.L. (1963) Nematode-induced callus on carrot discs grown *in vitro*, *Phytopathology* **53**, 1309-1312.
71. Sarr, E. and Prot, J.C. (1985) Penetration and development of juveniles of a strain of *Meloidogyne javanica* and of a B race of *M. incognita* in the roots of fonio (*Digitaria exilis* Stapf), *Rev. Nématol.* **8**, 59-65.
72. Schneider, S.M. (1991) Penetration of susceptible and resistant tobacco cultivars by *Meloidogyne* juveniles, *J. Nematol.* **23**, 225-228.
73. Siddiqui, I.A. (1971) Histopathogenesis of galls induced by *Meloidogyne naasi* in oat roots, *Nematologica* **17**, 237-242.
74. Siddiqui, I.A. (1971) Comparative penetration and development of *Meloidogyne naasi* in wheat and oat roots, *Nematologica* **17**, 566-574.
75. Siddiqui, I.A. and Taylor D.P. (1970) Histopathogenesis of galls induced by *Meloidogyne naasi* in wheat roots, *J. Nematol.* **2**, 239-247.
76. Sijmons, P., Grundler, F.M.W., von Mende, N., Burrows, P.R, and Wyss, U. (1991) *Arabidopsis thaliana* as a new model host for plant parasitic nematodes, *The Plant Journal* **1**, 245-254.
77. Smith, J.J. and Mai, W.F. (1965) Host-parasite relationships of *Allium cepa* and *Meloidogyne hapla*, *Phytopathology* **55**, 693-697.
78. Spiegel, Y. and McClure, M.A. (1991) Stage-specific differences in lectin binding to the surface of *Anguina tritici* and *Meloidogyne incognita, J. Nematol.* **23**, 259-263.
79. Sundermann, C.A. and Hussey, R.S. (1988) Ultrastructural cytochemistry of secretory granules of esophageal glands of *Meloidogyne incognita, J. Nematol.* **20**, 141-149.
80. Thomason, I.J., Van Gundy, S.D., and Kirkpatrick, J.D. (1964) Motility and infectivity of *Meloidogyne javanica* as affected by storage time and temperature in water, *Phytopathology* **54**, 192-195.
81. Viglierchio, D.R. and Yu, P.K. (1983) On nematode behavior in an electric field. *Rev. Nématol.* **6**, 171-178.
82. Von Mende, N. (1996) Comparison of the migratory behaviour in *Arabidopsis* roots of three *Meloidogyne* species, Abstract: *Third International Nematology Congress, Guadeloupe* 146.
83. Walton, J.D. (1994) Deconstructing the cell wall, *Plant Physiol.* **104**, 1113-1118.
84. Williamson, V.M., Lambert, K.N., and Kaloshian, I. (1994) Molecular biology of nematode resistance in tomato. In: Advances in molecular plant nematology. Ed. F. Lamberti, C. De Giorgi and D. Mck. Bird., NATO ASI Series., 211-219.
85. Wergin, W.P. and Orion, D. (1981) Scanning electron microscope study of the root-knot nematode (*Meloidogyne incognita*) on tomato root, *J. Nematol.* **13**, 358-367.
86. Wyss, U. and Zunke, U. (1986) Observations on the behaviour of second stage juveniles of *Heterodera schachtii* inside host roots, *Rev. Nématol.* **9**, 153-165.
87. Wyss, U., Grundler, F.M.W., and Münch, A. (1992) The parasitic behaviour of second-stage juveniles of *Meloidogyne incognita* in roots of *Arabidopsis thaliana, Nematologica* **38**, 98-111.
88. Zinoviev V.G. (1957) Enzymatic activity of the nematodes parasitising plants, *Zool. Zhurnal.* **36**, 617-620. (Ref. *Helm. Abstr.* **26** (1957), 109-110).

THE BIOLOGY OF GIANT CELLS

Teresa BLEVE-ZACHEO and Maria T. MELILLO
Istituto di Nematologia Agraria, CNR, Via Amendola 165/A, 70126 Bari, Italy

Abstract

When endoparasitic root-knot nematodes infect plants, they induce complex feeding sites (giant cells) within the root tissues of their host. Induction and maintenance of giant cells involve physiological changes in the root including enlargement of cells, nuclear division, cell wall ingrowth and subcellular membrane and organelle proliferation. Histological observations strongly indicate that these nematodes in some way regulate specific host genes. Molecular techniques and antibody availability now offer unprecedented opportunities to detect at the cytological level either specific nematode proteins or defense-related host metabolic products. In particular, the overexpression of HMGR in giant cells is reported and the possible function is suggested.

1. Introduction

Although recent advances in molecular biological techniques have expanded our knowledge and understanding of the genic background of host metabolism triggered by nematode infection, the exact relationship between gene expression and biochemical and/or cytological phenomena, remains obscure. Since any single, visible and/or detectable response results from a series of complex metabolic processes, it is almost impossible to envisage that only a single gene is responsible for the expression of any host response to nematode infection. The activation of multiple genes may be necessary to produce the visible and/or detectable phenomena. Therefore it is necessary to define the processes of activation of multiple genes to interpret any single phenomenon at the molecular level.

Cytological analysis has increased our understanding of the interactions of *Meloidogyne* species with their hosts. Structural studies have provided important information about the spatial relationship between host and pathogen, the timing of events such as nematode penetration and infection, and the nature of host resistance response. We are now entering an era in which cytology promises to add a new perspective to molecular studies of pathogenesis by nematode species. Modern techniques such as immunocytochemistry and *in situ* hybridization make it possible to determine the cellular location of specific macromolecules. This means that researchers should be able to determine when and where important genes, such as host resistance genes or nematode pathogenicity genes, are functioning and where gene products are accumulating.

This review focuses on the latest results in the area of host cell response which relate to the contribution that some enzymes are thought to involve important mechanisms for membrane targeting and for cell cycle regulation.

2. Morphology and structure of giant cells

The most obvious morphological response to infection by *Meloidogyne* is the characteristic galling of the host plant roots. The amount of galling in roots varies depending on the species of both plant and nematode involved in the host-parasite relationship. Galling starts relatively rapidly, several hours after invasion and progresses during a few days after the infection process. Detailed *in vitro* observations of *M. incognita* infection on *Arabidopsis* roots [83] and on tomato root explants [38] resulted in the most practical way to obtain step-by-step information on development of galling tissues. Changes in morphology of the root tip are accompanied by severe physiological changes of cells become feeding sites of second stage juveniles. The juvenile reaches the root meristem passing between cells and minimizing destruction, and stops its migration at the level of the developing vascular cylinder, where a parenchymatic xylem cell is selected as the initial feeding cell. Then the nematode becomes sedentary, and the initial feeding site, that constitutes the permanent nutrient source for the parasite, enlarges in a system of multinucleate transfer cells.

Early stages of multinucleate transfer cells, as examined by electron microscopy, indicate that binucleate cells are found to be associated with juveniles 24 h after nematode inoculation of a very good host (tomato) *in vitro*. Within 48 h, fed cells show 4 to 8 nuclei. No signs of cell plate indicate that there is cytokynesis without cytodieris and the multinucleate stage is apparently due to synchronous mitoses [46]. Nuclear and nucleolar volumes increase, normal shape of the nucleus becoming highly amoeboid. Vacuolization of nucleoli in giant cells seems to be related to ribosomal RNA transcription and then to the intense metabolism of the giant cells. Soluble assimilates are indeed continuously removed by the developing nematode and replenished by the host. Increase of ground cytoplasm, cellular organelles, and small vacuoles in 3-day old giant cells, all suggest that there is an active protein biosynthetic machinery. Vascular elements are differentiated and formed giant cells tend to increase their surface through development of wall ingrowths, followed by the plasma membrane, on the site of the xylem vessels. The increase in the surface of plasma membrane is related to the intense activity of Golgi stacks with a large presence of transitional vesicles. Some of these vesicles are coated vesicles and/or spiny vesicles [14], their outer surface carrying an investment of protein chains.

Six days after induction, lateral expansion of giant cells is considerable. Numerous nuclei are present within the cytoplasm, the central vacuole is further reduced, and differentiation of abnormal vascular elements around the giant cells is advanced. Transfer cell-wall ingrowths become quite extensive next to the to xylem elements, and it is remarkable that cytoplasmic vacuoles are absent from such regions. Numerous mitochondria are present in that region because of the energy required for the selective flow across the plasma membrane. The characteristic "wall-membrane apparatus" provides evidence in such cells of a versatile mechanism that facilitate transmembrane transport of solutes. Transfer cells occur in normal plant systems, and can be both absorptive and secretory. Thus, possession of wall ingrowth implies that enhanced short-distance solute transport between apoplast (cell wall compartment) and symplast (cytoplasmic compartment) is occurring in multinucleate transfer cells [45].

Giant cell enlargement continues for 2 weeks, occupying the most part of the central cylinder, where vascular elements are crushed and deformed, with severe restriction of longitudinal water movement. Final giant cell dimensions may reach 600 µm long by 200 µm in diameter [45]. Cytoplasmic contents increase greatly, and wall ingrowth development continues with the formation of an extensive wall labyrinth. Nuclei with dispersed chromatin, reflecting intensive transcription of genes, as is to be expected in an actively growing cell, are highly amoeboid. The distribution of endoplasmic reticulum is relatively sparse, the ground cytoplasm containing more ribosomes and polysomes. This implies that the most part of protein synthesis will be utilized *in loco* rather than exported. The demands made on the modified cells are considerable, since the mature female transformed nematode becomes larger than the giant cells from which it feeds. The cytoplasmic appearance of the giant cells reflects this modification for solute transport, with stimulated synthesis of cytoplasmic organelles such as mitochondria and plastids. Wichever is the pathway of solute entry into modified giant cells and the mode of production of a large volume of cytoplasm, the end result is similar. The multinucleate transfer cells are specialized for rapid solute uptake and these solutes are destined for nematode [31]. These highly specialized cells are induced, maintained, and completely dependent on a continuous stimulus from the nematode [9]. After completion of its life cycle, the nematode dies and giant cells and gall tissue degenerate [73].

Induction and physiology of the giant cells will be discussed in more detail below in this chapter.

3. Feeding Site Selection

Root-knot nematodes have a very broad host range, encompassing over 2,000 plant species, and most cultivated crops are attacked by at least one species of *Meloidogyne* [68].

Root-knot nematode parasitism of plants is a complex and dynamic interaction and involves hatching stimuli, attraction to the host, penetration of host tissue, recognition of tissues suitable for feeding site formation, modification of host tissues, and an active response from the host. Most research on nematode-host interaction has focused on tissue modification at the feeding site. In addition a few studies have examined gland secretions of nematodes and determined how these secretions modify host tissues [40].

Infective second-stage juveniles (J2) of root-knot nematodes migrate in the soil and are attracted to root tips were they penetrate directly behind the root cap. Juveniles migrate intercellularly in the root to the region of cell differentiation and locate cells in the vascular cylinder where permanent feeding sites are established [32, 83].

Evidence supporting the concept that amphidial exudates may play a role in chemotaxis and host recognition has been reviewed [85]. Lectin binding to the head of several *Meloidogyne* species has shown the presence of various carbohydrate residues in the amphidial secretions and qualitative differences in lectin binding ability [11, 53]. Recently, indirect immunofluorescence studies using polyclonal antisera have confirmed the presence of a glycoprotein, termed gp 32, in the region of amphids of J2 of *M. incognita* [76]. Furthermore, the immunoreactivity found in five other species of *Meloidogyne* appears to be genus-specific, suggesting a very specialized function for this protein. Gp 32 is expressed in all stages of the *Meloidogyne* life cycle, including males, but not in the sedentary adult females where the amphids appear to be non-functional [62]. This glycoprotein appears to be very important as it seems to be involved directly or indirectly in the primary transduction of chemical stimuli. It is known that infective J2 of root-knot nematodes are attracted to host roots [63], and that J2 inside the roots are highly selective in their choice of a host cell with which to establish a feeding site [30].

Differences in the composition of amphidial secretions between J2 and females of *M. incognita* have also been demonstrated by using monoclonal antibodies which reacted with amphids of adult females but not with those of J2 [24]. Assuming that the amphids serve a chemosensory role in *Meloidogyne*, the specific localization of gp32 in juveniles could be involved in location of host roots and in selecting the feeding site.

Once a juvenile reaches its feeding site, electron microscopy observations have revealed the presence of a viscous material, probably secreted from the amphidial pouches, filling the intercellular space between the nematode and host cells (Fig. 1, 1). The presence of this secretory material could be related to feeding site recognition during the moulting phases of juveniles or with the production of protective and/or lubricating glycoproteins.

4. Induction of Giant Cells

In response to repeated stimulation from the parasite, cells in the stele of the root undergo repeated nuclear division without cytokinesis, and also assume a new differentiated state termed the giant cell. Nuclei in the giant cells are enlarged and lobate, and may contain 14-16 times more DNA than do normal root tip nuclei [45, 80].

Despite the fact that it has never been proven that giant cells are actively induced by the infecting nematode, it is widely accepted that the induction of giant cell formation is due to glandular secretions injected via the nematode stylet into several root cells surrounding the head [39]. In fact, it is difficult to comprehend how the early events could occur in the absence of nematode-derived products. Later events might be an active plant response to a passively feeding nematode. Bird [9] showed that the nematode alone was necessary and sufficient for giant cell production, that an as yet unidentified plant-growth-promoting substance, absent in uninfected root tissue, is present in galls, and continuous presence of the female is required for maintenance of the giant cell. There is no direct evidence of continuous stimulation in terms of nematode-derived factors that modify the plant cell to the nematode's benefit. Indeed, during the first days of giant cell induction, giant cells can bear a striking resemblance to developing xylem. Histological evidence suggests that the normal route for giant cell formation involves large parts of the developmental pathway that is similar to differentiating metaxylem elements [13]. If this is the case, it seems likely that the nematode-encoded factors used to initiate nurse cell formation might closely resemble normal plant effectors, and might work in concert with endogenous host signals [12].

Induction and maintenance of giant cells can be considered separate phenomena, the induction mediated by J2 involving differentiation of a root cell into a giant cell in which the number of nuclei is increased by synchronous and non-synchronous mitosis or amitosis [14, see also Gheysen *et al.*, this volume]. The maintenance phase, mediated by the female, involves an increase in growth and the number of giant cells with accompanying nuclear division and extensive invagination and thickening of cell walls, similar to cell walls observed in transfer cells that are not usually present in normal root tissues [45,46].

Giant cells have been demonstrated to have a proton-coupled transport system located between the plasma membrane at wall ingrowths and xylem vessels ([29]; Grundler and Böckenhoff, this volume). Amino acid and sugar transport into the higher plant cell is generally thought to be mediated by a proton motive force [67] and there is evidence in support of a chemiosmotic model of proton-amino acid symport [18, 28]. It is now recognized that plant plasma membrane H^+-ATPases are encoded by a multigene family that are subject to differential, temporal, and spatial gene regulation and encode isoproteins with different kinetic and regulatory properties [61]. The large increase of the plasma membrane in giant cells suggests an increase in the proton pumping system and molecule carriers involved in the translocation of nutrients for the nematode into the giant cells. As a consequence, it seems likely that the genes encoding certain constitutive enzymes and structural proteins may be up-regulated or, at least, altered in their expression in order to support the increased cellular metabolic activity that is related to nematode feeding. More likely, the substantial alteration of cells of the root vascular system means that some plant genes have their normal expression patterns either quantitatively or qualitatively altered to meet the demands of the nematode [57, 73]. Although not yet proven, it is an intriguing possibility that the nematode induces these changes directly either by modifying preexisting plant transcription factors or by introducing proteins that can function as specific plant transcription factors (see Fenoll *et al.*, this volume).

Giant cell induction is associated with a specific tube-like structure, the feeding tube, that is formed rapidly after nematode stylet penetration of the selected cell wall [82, Grundler and Böckenhoff, this volume]. *Meloidogyne* and *Heterodera* spp. both induce feeding tubes, though with some differences [31, 43] the former being strictly interconnected with the endomembrane system. While the morphology of feeding tubes and their direct relation with nematode secretions has been described [65, 66], no conclusive evidence is available on the nature of nematode secretions themselves. Interaction of the secretion with host cytoplasm or with differences in pH has been suggested as a way in which polymerization of secretions forms feeding tubes [43]. The structure of feeding tubes induced by *Meloidogyne* suggests that they are mostly constituted of host membranes. Overexpression and modulation of these membranes might be more linked to a signal transduction of plasma membrane receptors at the level secretory antigens during parasitism by *Meloidogyne* species indicate a changing role for the oesophageal glands and their secretions at different stages of the nematode's life cycle [7, 8, 10, 39, 41, 83] (see also Stiekema *et al.*, this volume). Development of monoclonal antibodies [4] to secretory granules present in the oesophageal glands provides evidence that there are similarities and differences in the antigens sequestered in the secretory granules formed in subventral and dorsal glands, and in stylet secretions [24]. Furthermore, important evidence has been obtained on the temporal synthesis of secretory components by the two types of glands in *Meloidogyne* during the process of parasitism. An antibody binding both subventral and dorsal glands, and also the stylet secretions of preparasitic and parasitic juveniles (J2), did not bind subventral glands of the developing nematode and strongly labelled only the dorsal gland of the female [25]. This could indicate the presence of an antigen that may be important either in establishing the feeding site of the J2 or in maintaining the induced giant cells [43].

A novel method of immunogen preparation to produce monoclonal antibodies specific to oesophageal gland antigens of *Globodera rostochiensis* is of particular interest [26]. These MAbs on Western blots of second stage juveniles identified specific proteins of the subventral glands. In addition, ultrastructural immunolocalization of these polypeptides showed an intense binding of the antibody to secretory granules of subventral glands, and with *in vitro* experiments the same proteins were retrieved in stylet exudates of preparasitic J2 [26]. These recent results indicate that secretory components from subventral oesophageal glands of J2 may be responsible for the induction of the feeding site.

A MAb binding to the subventral oesophageal glands of adult females of *M. incognita* used to isolate a gene from a cDNA expression library has also been found to bind somatic muscles of J2 *M. incognita*. Because of the homology of the isolated gene with myosin heavy chains, it has been suggested that this antigen may be involved in the movement of nematode secretions [64]. Energy dependent transport is reported to provide the force to move intracellular components, to deliver and concentrate these components at specific locations, and to transport these components faster and over long distances [70].

Immunocytochemical localization of 3-hydroxy-3-methylglutaryl CoA reductase (HMGR), a key enzyme for sterol synthesis, in oesophageal glands of *M. incognita* prompted the conclusion that the protein co-localizes with microtubules and kinesin (motor protein) in the dorsal oesophageal gland of the female (Fig. 1, 3) [Bleve-Zacheo *et al.*, unpublished]. Kinesin and perhaps other motor proteins [1, 36] may play an important role in secretion movement during the parasitism.

Figure 1. Amphidial secretion (->) in the intercellular space of giant cells. x1700.
Figure 2. Feeding tube in a syncytial cell of BFA treated pea root. x30000.
Figure 3. HMGRase activity on the oesophageal dorsal gland of *M. incognita* female. a) Double immunogold labeling for HMGRase and tubulin, and b) HMGRase and kinesin. x95000.
Figure 4. Catalase in peroxisomes. x46000.
Figure 5. Amitotic (->) nucleus of a giant cell. x3500.
Figure 6. Immunolocalization of HMGRase in a giant cell, a) Co-localization of HMGRase and tubulin, and b) HMGRase and kinesin x80000.

6. Metabolism in Multinucleate Giant Cell

Recent work indicates that *Meloidogyne* spp. are able to activate or inactivate those plant promoters that are beneficial or detrimental to the development and maintenance of a suitable feeding cell structure (see Fenoll *et al.*, this volume).

Several genes are systematically induced in *Meloidogyne*-host compatible interactions. One of these genes *Cat2St* encodes catalase [56]. Catalase inactivates hydrogen peroxide by dismutation into water and oxygen, hydrogen peroxide being a signal transducer triggering the plant defense mechanism [50]. The induction of catalase suggests that the nematode has evolved mechanisms to increase catalase levels that can remove hydrogen peroxide thereby favouring the physiological development of the giant cell. This finding is confirmed by ultrastructural observations of susceptible tomato roots infected *in vitro* with *M. incognita* and resistant pepper invaded by an avirulent *Meloidogyne* line. Parenchyma cells adjacent to giant cells show large amounts of peroxisomes with crystalline catalase (Fig. 1, 4), which is very unusual in normal root tissue [17, 38]. In addition, biochemical analysis of resistant tomato heat-treated and infected with *M. incognita* demonstrated that induced plant susceptibility was accompanied by the activation of catalase and superoxide dismutase, both enzymes known to be involved in scavenging superoxides (see Zacheo *et al.*, this volume). All these data provide evidence that catalase is perhaps an important pathogen-induced protein.

Structural protein genes related to extensin have been identified in tomato roots infected with root-knot nematode [77, 78]. Extensin is normally induced in actively dividing cells, and quickly appears in the incompatible host-nematode interaction (see Zacheo *et al.*, this volume). This could explain the reported [77, 78] temporal shift of extensin gene expression in the giant cells at a later stage of infection.

A tobacco gene (*TobRB7*) encoding a membrane channel protein, possibly implicated in water or small solute uptake, has recently been found to be induced in giant cells by different species of *Meloidogyne* [57]. Expression of *TobRB7* has also been reported to be limited to the root meristematic cells and immature vascular tissue [84]. The ability of root-knot nematodes to activate this gene promoter could be indicative of a mechanism triggered for giant cell specialization *versus* meristematic cell metabolism. Another example of up-regulated genes is given by Niebel *et al*. [55] who used root-knot and cyst nematodes to infect *Arabidopsis* roots that were transformed with the GUS gene fused to the promoter of two essential cell cycle genes, *cdc2aAt* and *cyc1At* (see Gheysen *et al.*, this volume). They observed a strong correlation between the initiation of feeding cells and the synthesis of DNA that increased around 3 days after nematode infection. The induction of the two cell cycle genes seems to be linked to the process of endo-reduplication, induced by root-knot nematodes at the early stage of giant cell formation. Cytokinesis has also been shown to occur in giant cells without a previous nuclear division (Fig. 1, 5) [14]. The nuclei increase in size and volume and are highly chromophilic, except in the region corresponding to the plane of constriction. The largely amoeboid nucleus, containing one or several nucleoli, divides by a distinct cleavage developing from a constriction perpendicular or oblique to its longitudinal axis. The division usually results in an asymmetric formation of two daughter nuclei.

Amitosis has been reported to occur in animals and plants and is often related to critical phases in development (e.g. changes in growth phase and gametogenesis) [35, 71]. Moreover, amitosis has been related to a high auxin-cytokinin ratio [21] and to the level of endogenous hormones and other growth regulators [2]. Cells undergoing amitosis appear to resume the mitotic cycle, thus maintaining their totipotency. Increase in the number of nuclei is a good indicator of the progression of the cell cycle in mitotically dividing cells [72]. Their increase in volume, coinciding with ^{3}H-timidine incorporation, and flow cytometry after DAPI staining also precedes the amitotic events, after which follows the S phase [2, 72]. Root-knot nematodes seem able to utilize the synchronous cell cycle seen in plant organs in nature [75].

Another promoter that has been found to be strongly expressed in developing giant cells very early after *M. incognita* or *M. hapla* infection is the promoter of *hmg2* gene, one of the four genes that code for HMGR in tomato [23]. This enzyme catalyses the synthesis of mevalonate, the committed precursor of a great variety of isoprenoid compounds and derivatives synthesized in higher plants [20]. This biosynthetic strategy allows the synthesis of an astonishing variety of plant secondary metabolites that have many different roles [20, 54]. The tomato *hmg2* gene is a defense related gene that is induced by both fungal and bacterial pathogens [23]. Moreover, high levels of GUS activity in galling tissue of tomato seedlings containing the *hmg2*-GUS gene and inoculated with *M. incognita* or *M. hapla* suggests that *hmg2* may be a nematode-responsive gene mediating terpenoid phytoalexin production toxic to nematodes [79]. In contrast, tomato *hmg1* gene expression is not induced by defense elicitors but is detected in tissue undergoing division and associated with sterol biosynthesis [23]. However, phytoalexin production seems to be in contrast with giant cell induction, because susceptibility would be based on failure of the feeding cells to defend themselves after nematode recognition. For example, the timing and rate of HMGR accumulation and *de novo* enzyme synthesis leads to a burst of phytoalexin production sufficient to halt the spread of *Verticilium* on resistant cotton, while the delayed response permits the fungus to stay a step ahead of susceptible cotton defense [47]. This lends support to the suggestion that HMGR gene induction in giant cells may be related primarily to the high rate of sterol biosynthesis required to sustain the active demand from the nematode [74] which is thought to depend on plant sterols [22].

Recently, two genes encoding HMGR in *Arabidopsis*, *hmg1* and *hmg2*, the former corresponding to a housekeeping form of the enzyme, the latter having a more specific role in the synthesis of specific isoprenoid have been reported [33]. In transgenic tobacco plants containing both *hmg1* and *hmg2*-GUS fusion, *hmg2* expression has been found to be restricted to meristematic and floral tissues [34], and the high level of GUS expression related to an abundant supply of sterols obviously required for the synthesis of membranes in actively dividing cells and inflorescences. *Arabidopsis hmg2* gene promoter is not induced in wounded leaves of transgenic tobacco suggesting that this gene is not involved in the synthesis of specific isoprenoids in response to wounding [34]. Root-knot nematodes have been shown to activate both promoters in transgenic tobacco roots, the strong induction of *hmg1* being restricted to the giant cells [3, Fenoll *et al.*, this volume]. Besides promoter activation in tobacco, in *Arabidopsis* roots infected with *M.incognita*, high level of HMGR protein have been detected in the giant cells.

Specific antibodies directed against *Arabidopsis* HMGR provide evidence that the protein accumulates in the cytoplasm of giant cells and resides in endoplasmic reticulum-derived vesicles [15]. It must be noted that there exists a considerable controversy as to the exact intracellular location of HMGR in plants [6], although recently the sub-compartmentalization of plant HMGR in the endomembrane system has been suggested [19]. Ultrastructural immonolocalization also indicates the presence of ER vesicles near to the plasma membrane, suggesting that they are *en route* to the cell surface. HMGR has been found to co-localize with tubulin and to a lesser extent with kinesin (Fig. 1, 6). This association of HMGR with cytoskeleton elements and motor protein may be an indication of ER vesicle translocation to specific cellular locations such as the plasma membrane, and suggests that exocytic processes may be implicated not only in cell surface growth but also in nematode feeding [16]. A complex arrangement of immunoreactive systems homogenously distributed through the cytoplasm of the nematode female dorsal gland, following incubation with anti-reductase and anti tubulin IgG, shows the great activity of HMGR at this stage of pathogenesis. The broad expression of HMGR in the female dorsal oesophageal gland could suggest that the enzyme, actively transported through the gland, regulates phytosterol de-alkylation into sterols, coinciding with the extensive feeding requirement of the developing female.

7. Feeding and Plant HMGR-Related Response

It is widely accepted that the plant HMGR catalyzing the NADPH-dependent double-reduction of HMG-CoA to mevalonic acid, plays a crucial role in the part of the isoprenoid pathway leading to phytosterols [5]. A regulatory role of HMGR in phytosterol accumulation derives from the observation that in triazole-resistant tobacco mutant cells overproducing sterols [52], HMGR activity had increased 3 to 4 fold [37].

Moreover, overexpression of HMGR in yeast and hamster cells has been reported to induce an overproduction of membranes. These observations indicate the existence of cellular mechanisms that monitor the level of membrane proteins and reduce synthesis of particular types of membranes [59, 81].

Structures similar to paracrystalline membranes arranged as typical stacks and packed together in regular hexagonal arrays have also been observed in resistant tomato giant cells induced by a virulent line of *M. incognita* [14] as well as in *Arabidopsis* giant cells overexpressing HMGR [Bleve-Zacheo *et al*, unpublished]. The apparently identical pattern of membrane overproduction in yeast, animal, and *Arabidopsis* cells could indicate that HMGR levels may control membrane production. This overproduction seems to fit well with the tremendous amount of secondary wall apposition in "transfer giant cells" and with the sterol requirement of the feeding nematode.

8. Conclusions

The striking modifications occurring in *Arabidopsis* during the establishment and development of *Meloidogyne* spp. make this relationship an interesting model for

37. Gondet, L., Weber, T., Maillot-Vernier, P., Benveniste, P., and Bach, T.J. (1992) Regulatory role of microsomal 3-hydroxy-3-methylglutaryl Coenzyme A reductase a tobacco mutant that overproduces sterols, *Bioch. Biophys. Res. Commun.* **186**, 888-893.
38. Guida, G., Bleve-Zacheo, T., and Melillo, M.T. (1991) Galls induced by nematodes on tomato roots *in vivo* and *in vitro*, *Giorn. Bot. Ital.* **125**, 968-969.
39. Hussey, R.S. (1989) Disease-inducing secretions of plant parasitic nematodes, *Annu. Rev. Phytopathol.* **27**, 123-141.
40. Hussey, R.S. (1992) Secretions of esophageal glands in root-nematodes, in F.J. Gommers and P.W.T. Maas (eds), *Nematology from Molecule to Ecosystem*, Eur. Soc. Nematologists Inc., Dundee, pp. 41-50.
41. Hussey, R.S., and Mims, C.W. (1990) Ultrastructure of esophageal glands and their secretory granules in the root-knot nematode *Meloidogyne incognita*, *Protoplasma* **165**, 9-18.
42. Hussey, R.S., Mims, C.W., and Westcott (1992) Immunocytochemical localization of callose in root cortical cells parasitized by the nematode *Criconemella xenoplax*, *Protoplasma* **171**, 1-6.
43. Hussey, R.S., Davis, E.L., and Ray, C. (1994) *Meloidogyne* stylet secretions, in F. Lamberti, D.J. Chitwood and C. De Giorgi (eds), *Advances in Molecular Plant Nematology*, Plenum Press, New York, pp. 233-276.
44. Jasmer, D.P. (1993) *Trichinella spiralis* infected skeletal muscle cells arrest in G2/M and cease muscle gene expression, *J. Cell Biol.* **121**, 785-793.
45. Jones J.M.K. (1981) Host cell responses to endoparasitic nematode attack: Structure and function of giant cells and syncytia, *Ann. Appl. Biol.* **97**, 353-372.
46. Jones, M.G.K., and Payne, H.L. (1978) Early stages of nematode-induced giant-cell formation in roots of *Impatiens balsamina*, *J. Nematol.* **10**, 70-76.
47. Joost, O., Bianchini, G., Bell, A.A., Benedict, C.R., and Magill, C.W. (1995) Differential induction of 3-hydroxy-3-methylglutaryl CoA reductase in two cotton species following inoculation with *Verticillium*, *Mol. Plant-Microbe Interac.* **6**, 880-885.
48. Klausner, R.D., Donaldson, J.G., and Lippincot-Schwartz, J. (1992) Brefeldin A: insights into the control of membrane traffic and organelle structure, *J. Cell Biol.* **116**, 1071-1080.
49. Ko, R.C., Fan, L, and Lee D.L. (1992) Experimental reorganization of host muscle cells by excretory/secretory products of infective *Trichinella spiralis* larvae, *Trans. R. Soc. Trop. Med. Hyg.* **86**, 77-78.
50. Lamb, C.J. (1994) Plant disease resistance genes in signal perception and transduction, *Cell* **76**, 419-422.
51. Lee, L., Ko, R.C., Yi, X.Y., and Yueng, M.H.F. (1991) *Trichinella spiralis* antigenic epitopes from the stichocytes detected in the hypertrophic nuclei and cytoplasm of the parasitized muscle fiber (nurse cell) of the host, *Parasitology* **102**, 117-123.
52. Maillot-Vernier, P., Gondet, L., Schaller, H., Benveniste, P., and Belliard, G. (1991) Genetic study and further biochemical characterization of a tobacco mutant that overproduces sterols, *Mol. Gen. Genet.* **231**, 33-40.
53. McClure, M.A., and Stynes, B. A. (1988) Lectin binding sites on the amphidial exudates of *Meloidogyne*, *J. Nematol.* **20**, 321-326.
54. McGarvey, D.J., and Croteau, R. (1995) Terpenoid metabolism, *Plant Cell* **7**, 1015-1026.
55. Niebel, A., Barthels N., de Almeida-Engler, J., Karimi, M., Vercauteren, I., Van Montagu, M., and Gheysen, G. (1994) *Arabidopsis thaliana* as a model host plant to study molecular interactions with root-knot and cyst nematodes, in F. Lamberti, D.J. Chitwood and C. De Giorgi (eds), *Advances in Molecular Plant Nematology*, Plenum Press, New York, pp. 161-169.
56. Niebel, A., Heungens, K., Barthels N., Inzè, D., Van Montagu, M., and Gheysen, G. (1995) Characterization of a pathogen induced potato catalase and its systemic expression upon nematode and bacterial infection, *Mol. Plant-Microbe Interac.* **8**, 371-378.
57. Opperman, C.H., Taylor, C.G., and Conkling, M.A. (1994) Root-knot nematode-directed expression of a plant root-specific gene, *Science* **263**, 221-223.
58. Opperman, C.H., Acedo, G., Saravitz, D., Skantar, A., Song, W., Taylor, C., and Conkling, M. (1994) Bioenginering resistance to sedentary endoparasitic nematodes, in F. Lamberti, D.J. Chitwood and C. De Giorgi (eds), *Advances in Molecular Plant Nematology*, Plenum Press, New York, pp. 221-230
59. Orci. L., Brown, M., Goldstein, J., Garcia-Segura, L., and Anderson, R. (1984) Increase in membrane cholesterol: A possible trigger for degradation of HMG-CoA reductase and crystalloid endoplasmic reticulum in Ut-1 cells, *Cell* **36**, 835-845.

60. Orci, L., Perrelet., A., Ravazzola, M., Wieland, F., Schekman, R., and Rothman, J. (1994) "BFA bodies": A subcompartment of the endoplasmic reticulum, *Proc. Natl. Acad. Sci. USA* **90**, 11089-11093.
61. Palmgren, M.G., and Christensen, G. (1994) Functional comparison between plant plasma membrane H+-ATPase isoforms expressed in yeast, *J. Biol. Chem.* **269**, 3027-3033.
62. Perry, R.N. (1994) Studies on nematode sensory perception as a basis for novel control strategies, *Fundam. Appl. Nematol.* **17**, 199-202.
63. Prot, J.C. (1980) Migration of plant-parasitic nematodes towards plant roots. *Rev. Nematol.* **3**, 305-318.
64. Ray, C., Abbott, A:G., and Hussey, R.S. (1994) *Trans*-splicing of a *Meloidogyne incognita* mRNA encoding a putative esophageal gland protein, *Mol. Biochem. Parasitol.***68**, 77-85.
65. Razak, A.R., and Evans, A.A.F. (1976) An intracellular tube associated with feeding by *Rotylenchus reniformis* on cowpea root, *Nematologica* **22**, 182-191.
66. Rebois, R.V. (1980) Ultrastructure of a feeding peg and tube associated with *Rotylenchus reniformis* in cotton, *Nematologica* **26**, 396-406.
67. Rheinhold, L., and Kaplan, A. (1984) Membrane transport into the vacuole of oat roots, *Annu. Rev. Plant Physiol.* **35**, 45-83.
68. Sasser, J.N. (1980) Root-knot nematodes: a global menace to crop production, *Plant Dis.* **64**, 36-41.
69. Satiat-Jeunemaitre, B., Cole,L., Bouret, T., Howard, R., and Hawes, C. (1996) Brefeldin A effects in plant and fungal cells: something new about vesicle traffincking?, *J. Microsc.***181**, 162-177.
70. Schroer, T.A., and Sheetz, M.P. (1991) Functions of microtubule-based motors, *Annu. Rev. Plant Physiol.* **53**, 629-652.
71. Serafini-Fracassini, D., Bagni, N., and Torrigiani, P. (1980) *Nicotiana glauca x Nicotiana langsdorffii* tumor hybrid: growth, morphology, polyamines and nucleic acid *in vitro*, *Can. J. Bot.* **58**, 2285-2293.
72. Serafini-Fracassini, D., Del Duca, S., and Bregoli, A.M. (1995) Comparison between amitotic and mitotic cell cycles, *J. Exp. Clinic. Res.* **23**, 146-148.
73. Sijmons, P.C. (1993) Plant-nematode interaction, *Plant Mol. Biol.* **23**, 917-931.
74. Sijmons, P.C., Atkinson, H.J., and Wyss, U. (1994) Parasitic strategies of root nematodes and associated host cell responses, *Annu. Rev. Phytopathol.* **32**, 235-259.
75. Slocum, R.D., and Flores, H.E. (1991) *Biochemistry and Physiology of Polyamines in Plants*, CRC Press, London.
76. Stewart, G.R., Perry, R.N., and Wright, D.J. (1993) Studies on the amphid specific glycoprotein gp32 in different life.cycle stages of *Meloidogyne* species, *Parasitology* **107**, 573-578.
77. Van der Eycken, W., de Almeida-Engler, J., Gheysen, G., and Van Montagu, M. (1995) Molecular study of the compatible interaction : *Meloidogyne incognita*-tomato. Proc. 22nd Int. Symp. E.S.N., Ghent, Belgium, 7-12 August 1994, *Nematologica* **41**, 350.
78. Van der Eycken, W., de Almeida-Engler, J., Inzè, D.,Van Montagu, M., and Gheysen, G. (1996) A molecular study of root-knot nematode-induced feeding sites, *Plant J.* **9**, 45-54.
79. Weissenborn, D.L., Zhang, X., Eissenback, J.D- Radin, D.N., and Cramer, C.L. (1994) Induction of the tomato *hmg2* gene in response to endoparasitic nematodes, Proc. 4th Int. Cong. of Plant Mol. Biol., Amsterdam, The Netherlands, June 19-24, p. 1744.
80. Wiggers,R.J., Starr, J.C., and Price, H.J. (1990) DNA content and variation in chromosome number in plant cells affected by the parasitic nematodes *Meloidogyne incognita* and *M. arenaria*, *Phytopathology* **80**, 1391-1395.
81. Wright, R., Basson, M., D'Ari, L., and Rine, J. (1988) Increased amounts of HMG-CoA reductase induce "karmellae": a proliferation of stacked membrane pairs surrounding the yeast nucleus, *J. Cell Biol.* **107**, 101-114.
82. Wyss, U., and Zunke, U. (1986) Observation on the behaviour of second stage juveniles of *Heterodera schachtii* inside host roots, *Rev. Nematol.* **9**, 153-165.
83. Wyss, U:, Grundler, F.M.W., and Much, A. (1992) The parasitic behaviour of 2nd-stage juveniles of *Meloidogyne incognita* in roots of *Arabidopsis thaliana*, *Nematologica* **38**, 98-11.
84. Yamamoto, Y.T., Taylor, C.G., Acedo, G:N., Cheng, C.I., and Conkling, M.A. (1991) Characterization of *cis*-acting sequences regulating root-specific gene expression in tobacco, *The Plant Cell* **3**, 371-382.
85. Zuckermann, B.M., and Jansson, H.B. (1984) Nematode chemotaxis and mechanism of host/prey recognition, *Annu. Rev. Phytophatol.* **22**, 95-113.

THE STRUCTURE OF SYNCYTIA

Wladyslaw GOLINOWSKI, Miroslaw SOBCZAK, Wojciech KUREK and Grazyna GRYMASZEWSKA
Department of Botany, Warsaw Agricultural University, Rakowiecka 26/30, 02528 Warszawa, POLAND

Abstract

Infective second stage juveniles of the cyst forming genera of plant parasitic nematodes (*Heterodera* and *Globodera*) induce the formation of feeding sites in roots of resistant and susceptible plants. Under special experimental conditions in *Arabidopsis thaliana*, future males select the initial syncytial cell (ISC) in the pericycle and future females select their ISC in the procambium. After a preparation phase the nematode feeding site is still unicellular. Cytoplasmic streaming and cytoplasm density, amount of endoplasmic reticulum (ER) and volume of nucleus are increased while the volume of central vacuole is decreased and new small cytoplasmic vacuoles are formed. The cell walls are thickened and covered by a layer of callose-like material. The developing syncytia expand along the vascular cylinder. The expansion of the syncytium triggers the proliferation of cambial and peridermal tissues, in a manner similar to secondary growth. Compared with syncytia associated with females, syncytia of males are less hypertrophied and are composed of more cells. Distinctive cell wall openings are mostly found between the few strongly hypertrophied pericyclic syncytial elements. The ultrastructure of both types of syncytia is very similar but shows conspicuous differences in the structure and localisation of cell wall ingrowths.

Cultivars resistant to nematodes share some common reactions upon nematode infection. At the stage of migration and ISC selection there are no essential differences between susceptible and resistant plants. However, at the stage of syncytium induction some resistant plants show a very quick defence based on the hypersensitive response (HR). Many other resistant plants do not exhibit the HR. In these cases resistance develops more gradually during syncytium formation. The resistance can be correlated with changes in symplast and apoplast. In the symplast they usually deal with the ER. Another characteristic change is gradual reduction of cytoplasm density indicating loss of its physiological activity. Changes in apoplast can be responsible for the obstruction of syncytial development. The syncytial cell walls are abnormally thickened and restructured. The cell wall openings are smaller and less numerous. The syncytia are often surrounded by a conspicuous layer of necrotic cells. Syncytia developing in roots of resistant plants necrotise prematurely.

1. Introduction

The formation of the nematode feeding site is one of the most interesting events among plant-pathogen interactions. The ability of some nematode species to induce and to maintain specific feeding sites in roots of different plant host species suggests interference of nematodes in some basic stages of root morphogenesis. Our review is focused on the anatomy and cytology of feeding sites (so called syncytia) induced by cyst forming nematodes. These sites are composed of many plant cells, usually enlarged and connected each other via cell wall openings. This structure, a kind of plant organ, serves as a source of nutrients to the pathogen. To the cyst forming nematodes belong some genera (e.g. *Heterodera* and *Globodera*) from the family Heteroderidae [31].

The cyst forming nematodes undergo four juvenile stages before reaching the adult stage. The first-stage juveniles (J1) moult inside the egg shell and emerge as infective migratory second-stage juveniles (J2). The J2 look for host roots. Further development of the juveniles is possible only after induction of the specific feeding sites inside host roots. After several days of feeding the juveniles moult to the third-stage (J3) and continue to feed. After J3 stage the future males undergo inside the cuticle two subsequent moults and adults re-acquire vermiform shape and mobility. With the beginning of the moult to the fourth-stage juvenile (J4) the male juveniles cease the nutrient uptake. The female J4 juveniles continue to feed and acquire citron-like shape. They moult to adults and after mating start to produce eggs. During both these stages female J4 and adults actively withdraw nutrients from the feeding sites [16, 39].

The mechanisms of sex inheritance among cyst forming nematodes are not finally resolved. Many evidences indicate that J2 are not sexually differentiated and the sex is epigenetically regulated among these obligatory amphimictic pathogens [18, 36, 37, 38]. The first anatomical and morphological evidences for sexual differentiation appear at the end of J2 stage [16]. The amount and quality of nutrients seem to be crucial factors for nematode sex differentiation. One can suppose that the differences in the efficiency of the male's and female's feeding sites are reflected in their anatomy and ultrastructure.

2. Induction of the Initial Syncytial Cell

Infective second-stage juveniles of cyst forming nematodes penetrate roots preferentially in the root elongation and root-hair zone [43]. They also invade roots sometimes in the area of lateral root formation, where the roots are already in the state of secondary growth. A very detailed description of the penetration process in roots of *Brassica silvestris* var. *campestris* has been published by Wyss and Zunke [52]. This process is also described by von Mende in this volume. The penetration appears to be purely mechanical and occurs close to cell wall junctions. With the aid of its robust stylet the juveniles make multiple cell wall perforations very close to each other. The protoplast of the plant cell collapses after the first perforation of the cell wall. The juvenile then mechanically breaks through the cell wall and pushes its head into the cell lumen. After entering the first epidermal cell, the juveniles similarly perforate the walls of underlying cortical and endodermal cells. The juveniles usually migrate for some distance along the

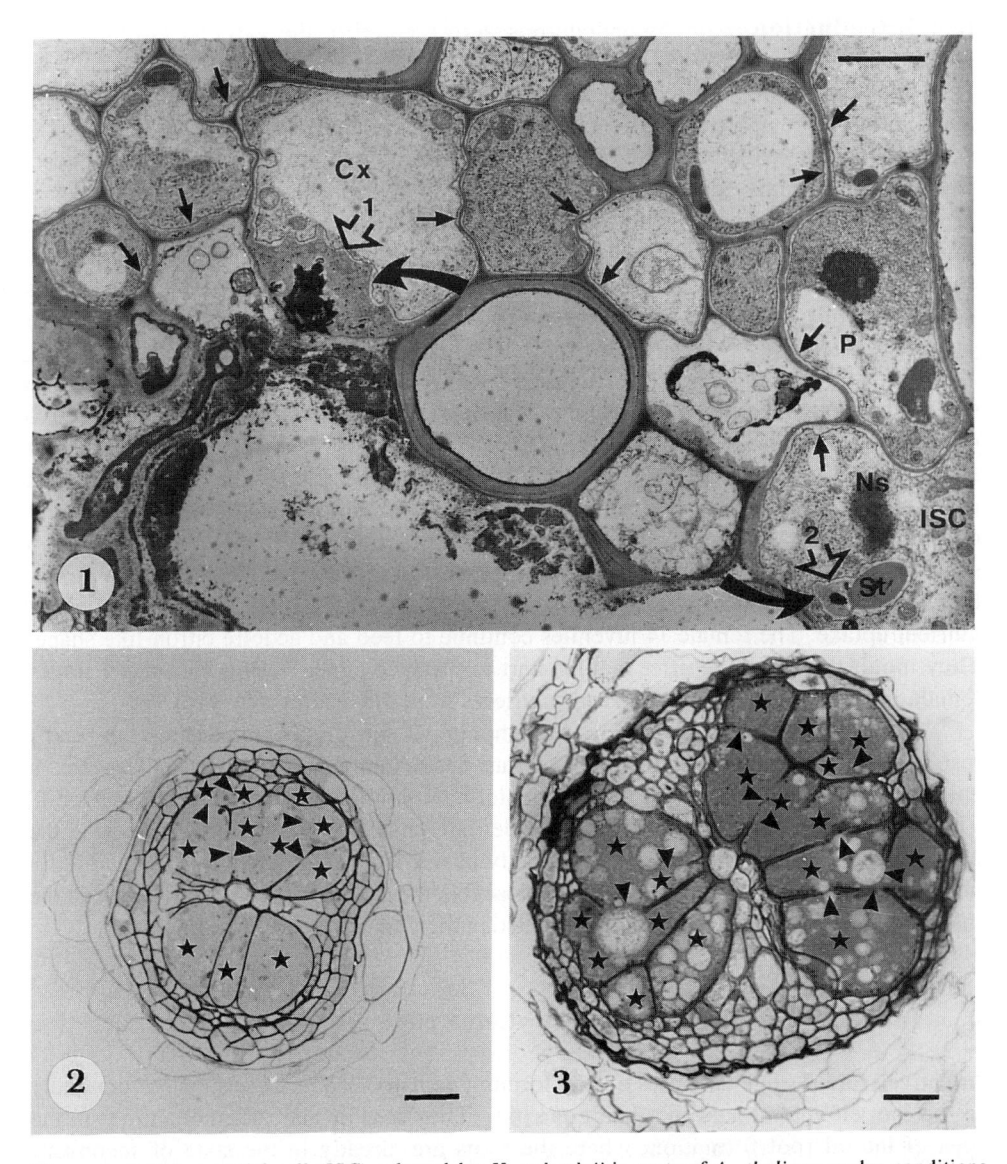

Figure 1. Initial syncytial cell (ISC) selected by *H. schachtii* in roots of *A. thaliana* under conditions favouring development of males (6h after ISC induction). The juvenile's attempts to induce an ISC and to inject its secretions first into the central metaxylem precursor cell (Cx) and then into a pericyclic cell (P) are marked with open arrows in the order of attempts as indicated by the numbers. All vascular cylinder cells deposit a new electron translucent cell wall layer (arrows), which is especially pronounced around places of stylet insertions (curved arrows). After a preparation phase the nematode's stylet (St) is withdrawn and then reinserted and nematode secretions (Ns) are injected into the ISC for the first time. Bar 2 μm.
Figure 2. Syncytium (asterisks) of female J3 juvenile of *H. schachtii* in roots of *A. thaliana*. Cell wall openings are marked with arrowheads. Bar 20 μm.
Figure 3. Syncytium (asterisks) of female J3 juvenile of *H. schachtii* in roots of *R. sativus* cv. "Siletina". Cell wall openings are marked with arrowheads. Bar 20 μm.

root within the cortex or the endodermis before entering the vascular cylinder. The destructive behaviour expressed during migration is changed to a more subtle and exploratory one when the juveniles enter the vascular cylinder. They carefully insert their stylet into cells searching for the appropriate cell to initiate a feeding site. The process of initial syncytial cell (ISC) selection sometimes lasts for several hours [50, 52]. When the cytoplasm of this cell collapses, the juvenile usually enters this cell and then carefully turns to the next cell. In some cases juveniles probe adjacent cells without moving forward [45]. At the end of migration they insert their stylet into a selected cell that is turned into the ISC (Fig. 1). The cytoplasm of this cell does not collapse after stylet insertion [52]. When the ISC is found the juvenile remains motionless with the stylet inserted into the ISC for the next 6-8 hours [50]. Within this period changes in nematode physiology occur that adapt it to the sedentary mode of live. When this preparation period is finished, the juvenile retracts its stylet and reinserts it again into the same site (Fig. 1). Then secretory material originating from the oesophageal glands [23, 24] is injected through the hollow stylet into the cytoplasm of the ISC and forms a structure similar to a feeding tube [45].

Using *Arabidopsis thaliana* as a host and changing culture conditions it was possible to create conditions leading to the development of an abundance of either males or females [45]. Which type of cell is selected as the ISC is still debated. Under conditions favouring the development of female juveniles procambial cells abutting primary tracheary elements are selected as ISCs [14]. These cells occupy the same position as so called "xylem parenchyma cells" that are suggested to be crucial for the compatible response of susceptible cultivars of white mustard and rape to cyst nematodes [15, 32]. Under conditions favouring the development of male juveniles usually pericyclic cells are selected for ISCs [45]. If procambial cells are selected as ISCs by juveniles under conditions favouring development of males they react to nematode stimulation with deposition of callose-like material around inserted nematode's stylet (Fig. 1) or with collapse of the entire cytoplasm. Only in few cases were the juveniles able to enter the preparation phase and induce the syncytia in procambial cells. Thus two situations were observed: (i) the ISC degenerates just after completion of the preparation phase when the juvenile injects oesophageal glands secretions; (ii) the ISC develops into multicellular syncytium and degenerates within 48 hours [45]. The ISCs selected among procambial and pericyclic cells exhibit a similar sequence of cytological changes during preparation phase (Fig. 1). Callose-like material is deposited on the cell wall. The area of the central vacuole (not surrounded by the tonoplast) decreases. The ER and ribosomes proliferate and small vacuoles are formed. Myelin bodies are frequently formed in the cytoplasm and sometimes enclosed in the material covering the cell wall or nematode stylet [45].

In general, also in other cruciferous plants infected with *H. schachtii*, a procambial cell abutting protoxylem vessels [14, Grymaszewska and Golinowski unpubl.] or a pericycle cell [15, 32] is selected as an ISC. Also in the wheat/*H. avenae* interaction syncytia are induced in the pericycle [21]. According to Endo [11] in soybean infected with *H. glycines* cortical, endodermal, pericycle or phloem parenchyma cells may be selected for the ISC.

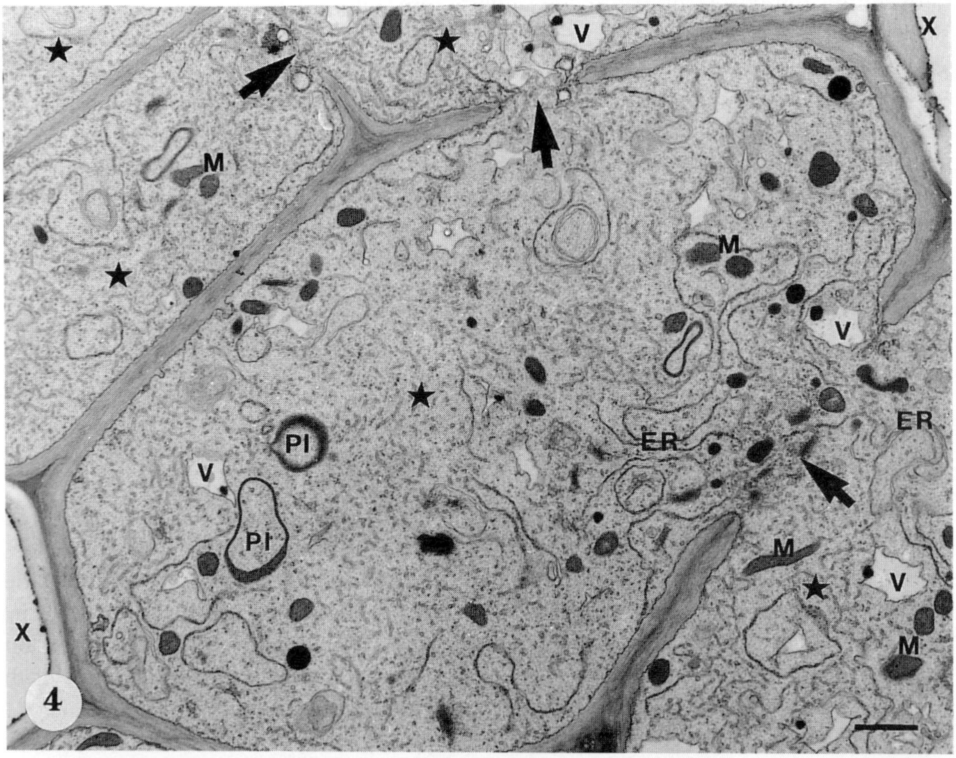
4
★
V
X
M
ER
Pl

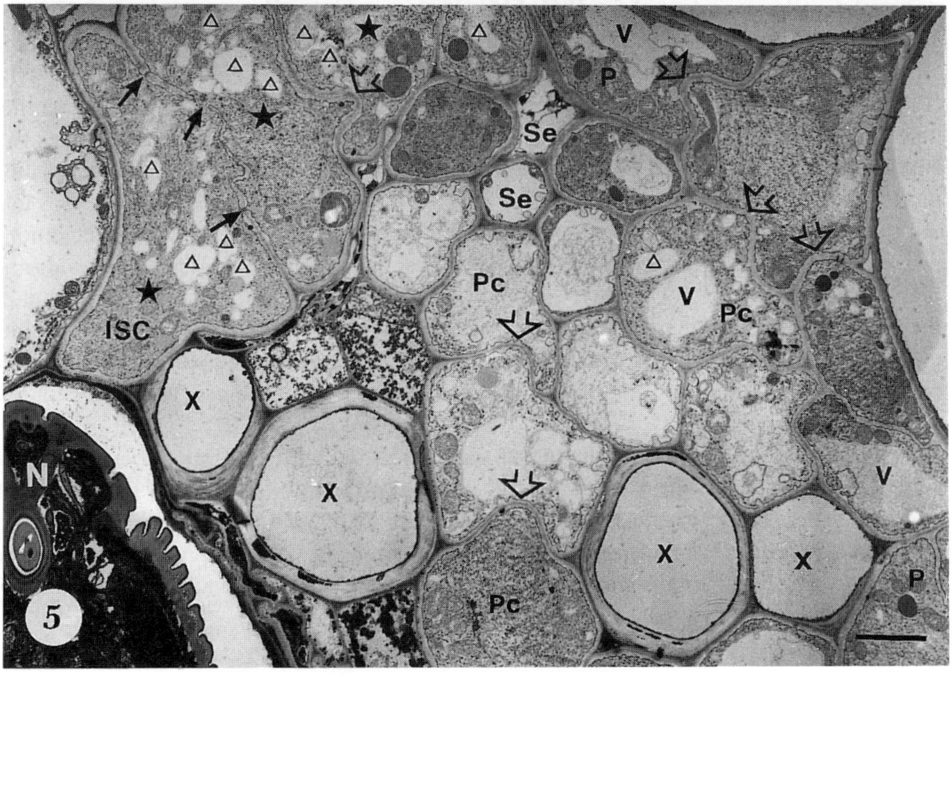
5
V
P
Se
Pc
ISC
X
N

The extent of destruction caused during the juvenile's migration in roots of susceptible plants is variable. When juveniles migrate through the cortical cell layers, only cells directly punctured by the nematode's stylet necrotise. More destruction is caused at the end of migration when juveniles enter the vascular cylinder and search for a cell to form the ISC. In these regions usually several pericyclic, procambial, and eventually xylem and phloem cells are destroyed [14, 45, 46]. However, this destruction is generally limited.

3. Development of Syncytia Induced by Female Juveniles

More distinct physiological and ultrastructural changes occur in ISC a few hours after commencement of food uptake by the juvenile. These changes include intensification of cytoplasmic streaming and increased density of the cytoplasm and nuclear volume [50]. Ultrastructural investigations on soybean infected with *H. glycines* reveal thickening of syncytial cell walls within 24h after inoculation [11]. An electron dense structure called "feeding plug" is deposited around the stylet inserted into the ISC. The feeding plug seals the site of cell wall perforation [10, 11, 51] and is always covered by a thin layer of electron translucent cell wall material. In *Brassica silvestris* var. *campestris* this structure is formed 12-15h after beginning of feeding by the nematode [50].

Eighteen hours after inoculation the ISC is only slightly hypertrophied and distinct cell wall openings are formed enabling protoplasts of ISC and neighbouring cells to fuse. By means of progressive cell wall lysis and coalescence of protoplasts a syncytium spreads within the stele towards vascular tissues [26, 27, 28, 29]. In soybean/*H. glycines* interactions numerous small vacuoles arise in the ISC [11]. Nematode secretions that form feeding tubes accumulate close to the stylet. Accumulation of smooth ER occurs near the feeding tube. Rough ER and a number of other organelles are more prevalent in other areas of syncytium. Similarly in *A. thaliana* infected with *H. schachtii* reorganisation of vacuolar system in cells first incorporated into the syncytium occurs [14, 17, 19]. Fibrillar material and granular inclusions are present in the numerous vacuoles. Within the cytoplasm proliferation of organelles occurs: free ribosomes, mitochondria, plastids, dictyosomes and ER. Openings already existing in cell walls are enlarged. Numerous paramural bodies are formed in their vicinity. The outer syncytial wall is also thickened. Scarce plasmodesmata present in thickened syncytial walls are sealed off from the syncytial side [19, 27, 45].

Stimulation of cell divisions within the stele occurs quite early in the vicinity of

---

Figure 4. Ultrastructure of a syncytium (asterisks) of a female J3 of *H. schachtii* induced in roots of *A. thaliana*. Arrows point to cell wall openings. ER endoplasmic reticulum; Pl plastids; M mitochondria; X xylem; V cytoplasmic vacuoles. Bar 1 µm.

Figure 5. Syncytium (asterisks) induced by juvenile of *H. schachtii* (N) in roots of *A. thaliana* under conditions favouring development of males (18h after ISC induction). At this stage of development the syncytium is composed of several pericyclic cells (P) only and surrounds underlying procambial cells (Pc), and xylem (X), and sieve elements (Se). Cells destined to be syncytial elements are separated by bent cell wall (open arrows) at which cell wall openings (arrows) are formed. The central vacuoles (V) disappear and are substituted in syncytial elements by small cytoplasmic vacuoles (triangles). Bar 3 µm.

the developing syncytium [14, 20, 21]. Procambial, cambial and pericyclic derivatives are incorporated into the syncytium or differentiate into parenchyma encircling the nematode feeding site (Fig. 2). In wheat/*H. avenae* interaction dividing pericyclic cells give rise to numerous lateral roots [20, 21]. Swelling of the root with abundantly emerging lateral roots is a characteristic feature of infection of wheat with *H. avenae* [49]. In the interaction of white mustard/*H. schachtii* the syncytium spreads into lateral roots and develops in contact with the vascular tissues. Syncytia formed in *A. thaliana* and susceptible *Raphanus sativus* cv. "Siletina" do not invade the emerging lateral roots. In these plants an osmiophilic substance is deposited on the inner surface of tracheary elements ([14] and Grymaszewska and Golinowski, unpubl.). At the time of moulting to J3 stage, syncytia of juveniles of *H. schachtii* are twice as long as the nematode's body [50].

Syncytia of the J3 are characterised by further condensation of cytoplasm and by proliferation of organelles (Fig. 4) [4, 14, 27, 29]. The density of ribosomes and polysomes increases [47] and the ER system expands [13, 14, 51]. Abundant mitochondria, dictyosomes, plastids and lipid bodies are present [4, 14, 21, 27, 51]. Mitochondria and plastids are spherical and distributed along syncytial walls [14]. Numerous syncytial nuclei are strongly enlarged, acquire ameboid shape [4, 8, 13, 29, 47], and do not divide [6, 7, 27, 29]. The syncytium continuously enlarges by means of hypertrophy of syncytial elements and by incorporation of new cells [14]. It spreads along the root, mainly basipetally [14, 32]. The cells abutting tracheary elements are basic components of the syncytium (Figs. 2 and 3). They are incorporated into the syncytium throughout its length and create a major part of the volume of the feeding site [15, 32]. Changes occurring in these cells during their differentiation into syncytial elements are of fundamental importance for proper functioning of the plant/nematode system. In oil-seed rape infected with *H. schachtii* increase in cytoplasm density of cells destined for the syncytium occurs yet before openings in cell walls are created [14, 32]. In these cells the number of free ribosomes increases, the nucleus enlarges and acquires amoeboid shape, and ER accumulates, features which indicate that the intensive protein synthesis is occurring. Amount of tubular structures of ER probably involved in intracellular transport as well as amount of cisternal ER, mitochondria, plastids and vacuoles increases.

At the beginning of the J3 stage cell wall ingrowths develop on syncytial walls abutting tracheary elements [14, 19]. The cell wall ingrowths are composed of wall matrix and cellulose fibrils [29] and are never lignified. According to Jones and Northcote [29] during their formation multivesicular and paramural bodies are involved. Their localisation indicates that they are specialised in quick solute uptake from the vascular tissues. Since the cell wall ingrowths are delineated with the continuous plasmalemma the surface of the plasmalemma is increased by many times in these regions. The expanded plasmalemma implies high surface/volume ratio for the syncytium that intensifies short-distance transport of solutes from apoplast to the syncytial cytoplasm.

In *A. thaliana* a syncytium of the J4 enlarges mainly through cell hypertrophy. Its ultrastructure is similar as observed in syncytium formed by J3. Proliferation of smooth ER is characteristic for it [14]. Accumulation of smooth ER occurs in advanced stages also in syncytia formed in other susceptible plants invaded by different *Heterodera* species

[4, 13, 29, 47, 51] and Grymaszewska and Golinowski, unpubl.). Lipid bodies are still present but they are less abundant than in the syncytia formed by J3. Numerous small vacuoles sometimes containing osmiophilic substance emerge in different parts of protoplast of the active syncytium [14]. Movement of the vacuoles within the syncytium can be observed by means of video enhanced microscopy [50].

After a female matures to an adult and completes eggs production the syncytium collapses gradually. Degenerative changes within syncytia are expressed as decrease in cytoplasm density and increase of vacuolar volume [14, 19]. The sequence of changes during syncytium formation in a susceptible model system *A. thaliana*/*H. schachtii* is summarised in Table 1.

TABLE 1. Sequence of changes during syncytium formation in an *A. thaliana*/*H. schachtii* susceptible model system

Nematode stage	Events inside the root

J2	* J2 proceeds through the cortex towards the stele
	* Procambial cells are selected as ISC
	  ** breakdown of the tonoplast
	  ** first openings in cell walls of ISC
	** Hypertrophy of first syncytial elements

J3	* Syncytium increases in volume via incorporation of abutting procambial cells and synchronised hypertrophy
	* Procambial cells surrounding the syncytium divide intensively in a synchronised manner
	* Divisions within pericycle lead to formation of a protective layer encircling the syncytium
	* Restructuring of root apoplast
	  ** lysis of parts of inner cell walls
	  ** formation of cell wall ingrowths
	  ** thickening of some parts of outer syncytial cell walls
	* Restructuring of root symplast
	  ** fusion of protoplasts of cells incorporated into syncytium
	  ** proliferation of smooth ER, ribosomes, mitochondria and plastids
	  ** syncytial nuclei enlarge and become amoeboid
	  ** formation of numerous lipid bodies

J4/A	* Syncytium stabilises in volume * In some parts of the syncytium numerous vacuoles with osmiophilic inclusions appear

4. Development and Structure of Syncytia Induced by Male Juveniles

Very little is known about the structure and ultrastructure of the syncytia supporting development of male juveniles. The older reports concerning this problem are restricted only to three short notes [6, 8, 9]. Recently using *A. thaliana* and *in vitro* culture techniques it was possible to give continuous and detailed description of the structure of syncytia of male J3 [45, 46].

The J2 developmental stage of *H. schachtii* on roots of *A. thaliana* lasts for about 5 days. During this period the length of syncytia induced among pericyclic cells readily exceeds 1 mm. Within the first 18 h after syncytium induction only pericyclic cells are incorporated and the syncytium forms a half-ring surrounding underlying procambial and vascular cells (Fig. 5) [45]. During the next two days of syncytium development the procambial cells are also incorporated. The cells bound to be syncytial elements undergo the same sequence of changes as described above for development of syncytia of female juveniles. The most obvious changes occur in the organisation of the vacuolar system and the structure of syncytial cell walls [45]. The vacuoles are de-differentiated and infiltrated with the cytoplasm. The electron translucent material is deposited on the cell walls that become slightly bent. The first cell wall openings are then formed at these regions. Except for differences in cellular composition at the ultrastructural level there are no obvious structural differences in comparison to the syncytia of female juveniles developing on *A. thaliana* [14].

After the nematode moults to the J3 developmental stage, the syncytium continues to grow. At the end of J3 (about 10 days after ISC induction) the syncytium length readily exceeds 2 mm [45]. The distant acro- and basipetal syncytial elements are formed by precursor cells of tracheary elements (Fig. 6). Only in cases where these cells have been already differentiated the terminal syncytial elements are formed by procambial cells abutting tracheary elements. Close to the nematode head the larger part of syncytium is formed by pericyclic cells (Fig. 7). These cells have been incorporated into the syncytium very early and are the most hypertrophied elements in the entire syncytium. In general, the syncytial elements are less enlarged in syncytia of male juveniles than in female ones (Figs. 2 and 3 versus 6 and 7). The number and extent of cell wall openings are much lower in syncytia of male juveniles. Ultrastructurally the syncytia of male juveniles are very similar to the syncytia of females (Fig. 8). The syncytial cytoplasm is strongly condensed and its cyclosis is increased. The number of mitochondria, plastids and ribosomes is increased. The cisternae of ER are arranged parallelly or into concentric swirls. The most obvious changes appear in the amount of cytoplasmic vacuoles. They are more abundant in syncytia of male juveniles. They have usually round outline and do not contain osmiophilic depositions. The mechanism of cell wall ingrowths formation seems to be disturbed. Cell wall ingrowths are formed at the beginning of the J3 stage. However, they are very weakly developed and tend to fuse and to form irregular masses of wall material instead of the typical cell wall ingrowths observed in the syncytia induced by female juveniles. They are formed not only at the cell walls abutting tracheary elements but are also frequently formed at the outer syncytial walls and walls between neighbouring syncytial elements [17, 44, 45, 46].

At the beginning of J4 syncytial cytoplasm retains still functional shape. No abrupt necrosis of syncytial cytoplasm occurs. The degradative changes appear slowly. The amount of vacuoles increases and they occupy gradually larger part of syncytia. Later on, the cytoplasm becomes electron translucent and retracted from the cell walls. The tubules and cisternae of ER dilate strongly. The nucleus, mitochondria and plastids lose their internal structure and become also electron translucent [45, 46].

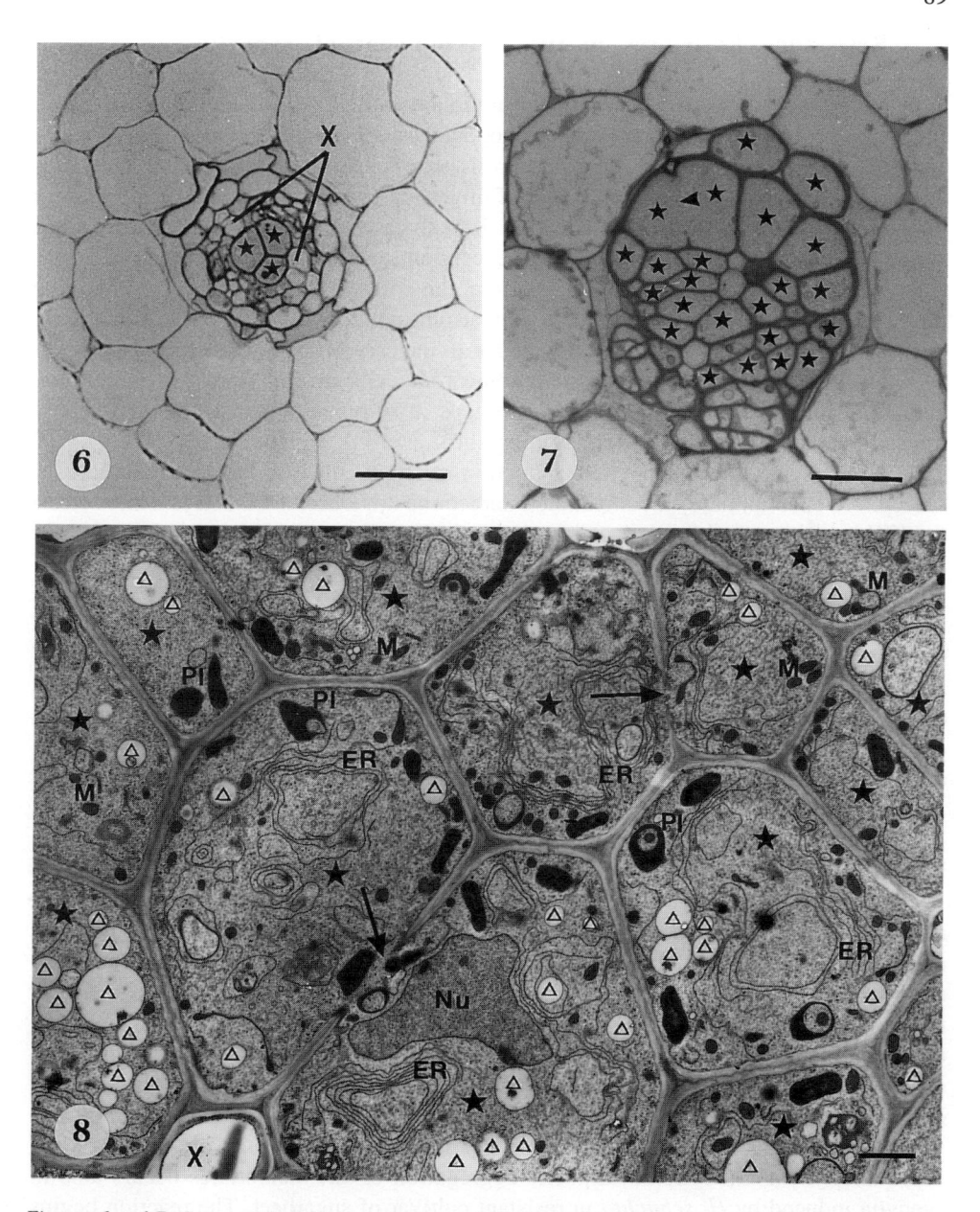

Figures 6 and 7. Cross sections of syncytia (asterisks) of male J3 of *H. schachtii* induced in roots of *A. thaliana*. Bars 20 μm. *Fig. 6.* Section through terminal part of the syncytium. X xylem. *Fig. 7.* Section taken close to the nematode's head. The arrowhead indicates cell wall opening. *Figure 8.* Ultrastructure of a syncytium (asterisks) of a male J3 of *H. schachtii* induced in roots of *A. thaliana*. Arrows point to cell wall openings and triangles indicate cytoplasmic vacuoles. ER endoplasmic reticulum; M mitochondria; Nu nucleus; Pl plastids; X xylem. Bar 2 μm.

5. Resistance Responses in the Development of Syncytia

Plant cultivars resistant to cyst nematodes are found within diverse families: Solanaceae, Brassicaceae, Fabaceae, Poaceae and Chenopodiaceae. In spite of taxonomic divergence and different genetic sources of resistance, resistant plants share some common responses to the invading nematodes. At the stage of migration there are no essential differences between resistant and susceptible cultivars. In both cases J2 of cyst-forming nematodes penetrate the root intracellularly. The only exception is mentioned by Bleve-Zacheo et al. [3] who report that mechanical damage along the pathway of *Globodera* J2 migrating through the cortex includes more cells in resistant potato cultivars than in susceptible ones. The other interesting observation is that in the case of local resistance response in the *A. thaliana* / *H. schachtii* system some root cells distant from the penetrating nematode start to necrotise very early, perhaps in response to some unknown signal transmitted along the root (Golinowski, unpubl.).

At the stage of the syncytium induction resistant cultivars react still similarly to susceptible ones. In many incompatible interactions it is observed that a juvenile selects the ISC in a tissue and in the location similar as in a susceptible plant cultivar. At this stage some resistant cultivars trigger a very quick defence system based on a hypersensitive reaction (HR). The reaction includes necrosis of the induced ISC or first syncytial elements. A HR occurs in white mustard cv. "Maxi" infected with *H. schachtii* (Figures 9 and 10) (Golinowski, unpubl.) and prohibits syncytium development. HR may be also exhibited by cells encircling the site of syncytium induction. The cells degenerate within 1 to 6 days after inoculation [12, 22, 30, 40]. HR also may occur in cells located in the vicinity of the invading nematode although this phenomenon is not correlated with resistance [40]. Importance of HR is stressed by an investigation of a non-host interaction *A. thaliana*/*H. glycines* (F.M.W. Grundler, M. Sobczak and S. Lange, unpubl.) where features apparently characteristic for HR: extensive necroses, browning, autofluorescence and depositions of cell wall, were observed.

Many other resistant plants do not exhibit the HR. In these cases resistance develops more gradually during syncytium formation. Here, the first differences between resistant and susceptible cultivars are visible 18 hours-2 days after inoculation at the earliest. The changes in symplast and apoplast can be correlated with the resistant reaction. In the symplast they usually deal with the ER. In resistant cultivars of soybean accumulation of wide cisternae of rough ER was seen as early as 18h or 42h (depending on the cultivar) after inoculation [11]. Similar phenomenon was observed in resistant *Raphanus sativus* cv. "Pegletta" 3 days after inoculation [51]. Predominance of rough ER was also reported in syncytia of a partially resistant potato cultivar 12 days after inoculation with *Globodera* [33]. The aggregation of ER membranes occurs also in syncytia induced by *H. schachtii* in resistant cultivar of sugarbeet. The reaction begins about 4 days after syncytium induction and leads to the syncytium degradation (B. Holtmann, unpubl.). Another characteristic change in the symplast is gradual reduction of cytoplasm density indicating loss of its physiological activity. It was observed in resistant wheat 15 days after inoculation [48] and in pea 10 days after inoculation [34]. Even earlier (7 days after inoculation) disintegration of syncytial cytoplasm begins in potatoes carrying polygene resistance from *Solanum vernei* [41]. Signs of cytoplasm

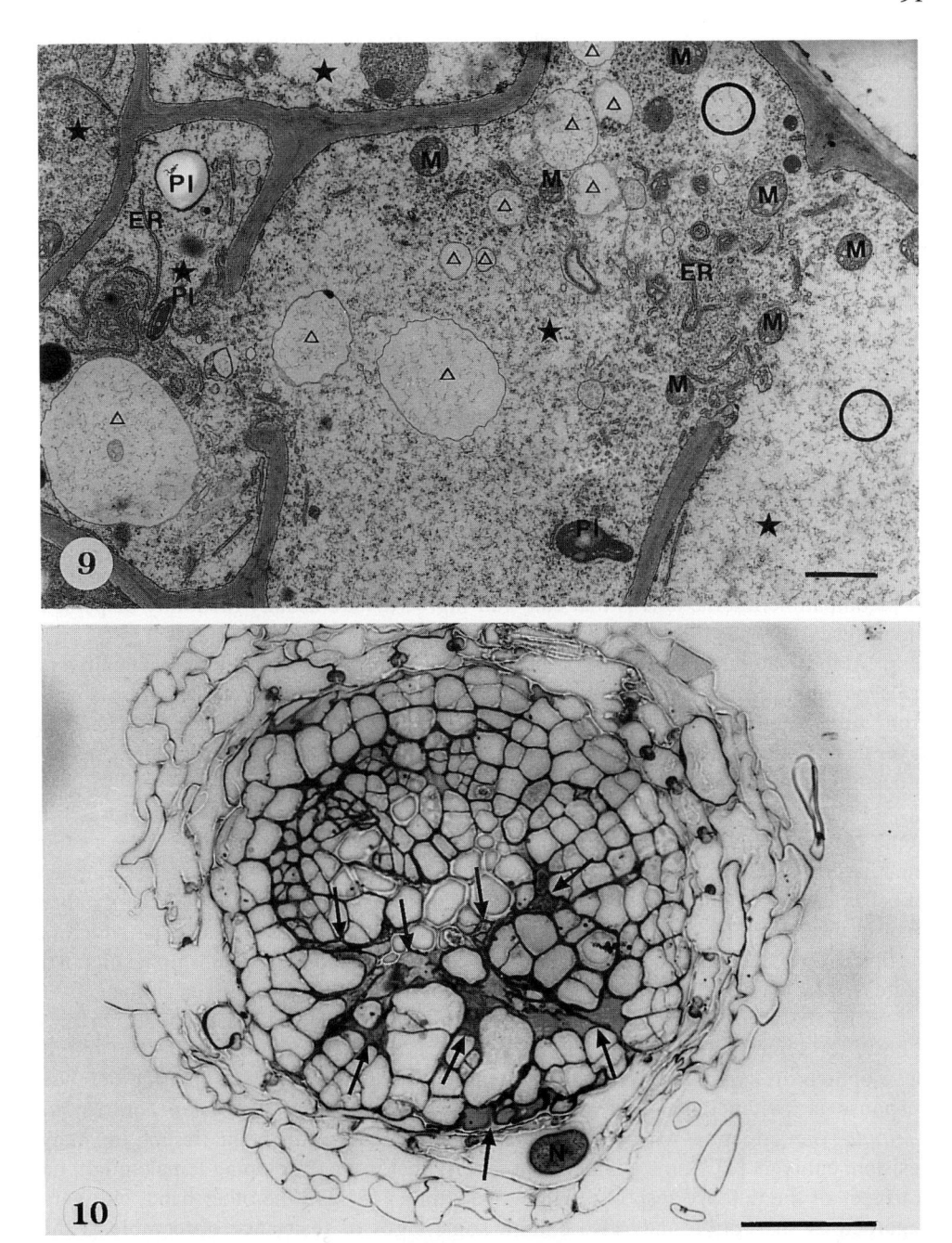

Figures 9 and 10. Cross sections of roots of *Sinapis alba* cv. "Maxi" infected with *H. schachtii*. *Fig. 9.* Ultrastructure of 4 days-old syncytium (asterisks). The syncytial cytoplasm is strongly electron translucent. The endoplasmic reticulum (ER) is weakly developed. Many cytoplasmic vacuoles (triangles) are present and large parts of syncytium contain only strongly dispersed fibrillar material (rings). M mitochondria; Pl plastids. Bar 1 µm. *Fig. 10.* Section of root showing extensive necrosis (arrows) in the vascular cylinder induced by a juvenile (N). Bar 20 µm.

disintegration are reported even as early as 5 days after inoculation in *Raphanus sativus* cv. "Pegletta" [51]. In some cases the degradation of cytoplasm results in formation of osmiophilic globules [3, 51]. These changes in cytoplasm are slightly preceded by the expansion of the vacuolar system. In resistant cultivars already at early stages of syncytium development syncytial vacuoles progressively enlarge [3, 33, 34, 41, 48]. Sometimes this trend is visible already 3 days after inoculation [51]. At later stages these vacuoles may contain granular material and myelin bodies [3]. As a result of vacuolar expansion degenerating cytoplasm is in many places reduced to a thin layer compressed against the cell wall [48, 51]. Other degradative features in syncytia include the degeneration of mitochondria and nuclei and the deposition of large starch grains in plastids [2, 3, 33, 40].

Also some changes in the apoplast in resistant cultivars can be responsible for the obstruction of syncytia development. The cell wall openings connecting syncytial cells are smaller and less numerous than in susceptible plants. The syncytial cell walls are abnormally thickened and restructured. Osmiophlic depositions are present therein [2]. Microtubules [42] and paramural bodies [3, 40, 42] are observed in vicinity of the thickened walls indicating intensive rebuilding of apoplast. Extensive thickening of outer syncytial walls observed in some interactions may be very important for resistance [2] although sometimes it does not occur [3, 42, 51]. Another important feature of apoplast in incompatible interactions is that cell wall ingrowths are usually randomly dispersed along syncytial walls [Grymaszewska and Golinowski unpubl.] or even absent [40].

At the anatomical level some important differences also occur. In resistant cultivars the syncytium is often surrounded by a conspicuous layer of necrotic cells that isolates it from neighbouring healthy tissues [2, 3, 5, 11, 12, 40, 42]. Even in the case when the layer of necrotic cells is absent a syncytium may be isolated from tracheary elements by procambial cells that have resisted incorporation into it [32].

Finally, syncytia developing in roots of resistant cultivars necrotise prematurely. In some instances they are already non-functional 10 days [3, 53] or even 7 days after inoculation [20, 42]. Such short-living syncytia are often composed of smaller number of cells than in susceptible plants [12, 21] and sometimes are located in the cortex rather than in the stele [33]. Cortical syncytium creates a poor source of nutrients, which is sufficient only to support development of males. Such a feeding relationship results in a decreased female/male ratio that is an indication of the resistance of a given cultivar.

The above discussion of differences between susceptible and resistant plants creates a basis for some more general statements. The similarity in the induction and initial development of syncytia in resistant and susceptible plants indicates that the defence response is switched on only after ISC selection [41, 51]. This is in contrast to *Meloidogyne* infection where a quick HR is induced within plant tissues in many resistant cultivars [1]. Some authors even specify that just a developing female might be a trigger of the defence response within syncytium [48]. On the other hand, cultivars regarded as susceptible may also exhibit some signs of resistance observable at the microscopical level as it is in a case of some potato cultivars [33]. It is also worth noting that the degree of resistance/susceptibility varies with pathotypes of nematodes associated with a given plant cultivar.

The formation of syncytia may be regarded as an example of modifications of the morphogenetic program of root development. In Table 2 three possible pathways of root morphogenesis are shown. In the first column normal root development is outlined. Procambial and pericyclic cells are here involved in the formation of secondary roots. These two meristematic tissues play key roles in the syncytial development in a pathological interaction. In susceptible cultivars (second column) procambial cells create the basic structure of the syncytium while pericycle forms protective tissue for it. On the contrary, in incompatible interactions (third column) involvement of procambium in syncytium formation is blocked or limited. Potentials of pericycle as meristematic tissue serve in this case as a rescue for nematode development. Peripheral syncytium formed on the basis of pericycle is much less effective. These changes in the roles of procambium and pericycle in syncytium formation seem to be the structural basis of plant resistance to nematode attack.

TABLE 2. Modifications of the morphogenetic programme of root development induced by the nematode

Normal root development, formation of the secondary root structure	Nematode invasion	
	Compatible interaction	Incompatible interaction
* Procambial cells: divisions and differentiation into xylem and phloem elements * Cytodifferentiation leading to formation of tracheary or sieve elements	* Procambial cell selected as ISC * Incorporation of neigh-bouring procambial cells and their derivatives into the syncytium * Changes in normal program of cell differen-tiation leading to the development of structures and functions characteristic for syncytial elements	* Procambial cell selected as ISC * Abortion or severe restriction of syncytium development (in some cases numerous necroses) * In some cases regene-rative divisions within the stele
* Pericycle: divisions and differentiation into the secondary cover tissue-periderm	* Divisions in pericycle leading to formation of a protective tissue encircling the syncytium	* Hypertrophy and/or divisions of pericyclic cells; formation of a small syncytium mainly of pericyclic origin at the periphery of the stele

The anatomical and ultrastructural characteristics outlined in Table 2 may be related to mechanisms underlying resistance against the nematode attack. Within symplastic reactions an increase in the amount of rough ER may result in synthesis of lytic enzymes. The enzymes released to the lumen of the rough ER may be further transported to the enlarging vacuoles. Successive breakdown of their tonoplasts might lead to the observed rapid lysis of syncytial cytoplasm. The lower nutrient status in resistant cultivars may arise from fewer contacts with tracheary elements or may be the result of developing syncytium being pushed away from the xylem by procambial cells. In susceptible plants solute uptake is facilitated by cell wall ingrowths located predominantly on syncytial walls abutting tracheary elements. In resistant cultivars these ingrowths are rather randomly dispersed at syncytial walls adjacent to different types of surrounding cells. In some cases the ingrowths are even absent. Other barrier against import of nutrients not only from xylem but also from other neighbouring tissues is

20. Grymaszewska, G. and Golinowski, W. (1987) Changes in the structure of wheat (*Triticum aestivum* L.) roots in varieties susceptible and resistant to infection by *Heterodera avenae* Woll., *Acta Soc. Bot. Pol.* **56**, 381-383.
21. Grymaszewska, G. and Golinowski, W. (1991) Structure of syncytia induced by *Heterodera avenae* Woll. in roots of susceptible and resistant wheat (*Triticum aestivum* L.), *J. Phytopathol.* **133**, 307-319.
22. Hoopes, R.W., Anderson, R.E. and Mai, W.F. (1978) Internal response of resistant and susceptible potato clones to invasion by potato cyst nematode *Heterodera rostochiensis*, *Nematropica* **8**, 13-20.
23. Hussey, R.S. (1989) Disease-inducing secretions of plant parasitic nematodes. *Annu. Rev. Phytopathol*, **27**, 123-141.
24. Hussey, R.S., Davis, E.L. and Ray, C. (1994) *Meloidogyne* stylet secretions, in F. Lamberti, C. De Giorgi, and D.McK. Bird (eds.), *Advances in Molecular Plant Nematology,* NATO ASI Series A: no. 268, Plenum Press, N.Y., pp. *233-248.*
25. Jones, A.M. and Dangl, J.L. (1996) Longjam at the Styx: programmed cell death in plants, *Trends Plant Sci.* **1**, 114-119.
26. Jones, M.G.K. (1981a) Host cell responses to endoparasitic nematode attack: Structure and function of giant cells and syncytia, *Ann. Appl. Biol.* **97**, 353-372.
27. Jones, M.G.K. (1981b) The development and function of plant cells modified by endoparasitic nematodes, in B.M. Zukerman and R.A. Rohde (eds.) *Plant Parasitic Nematodes*, vol. III, Academic Press, N.Y., pp. 255-279.
28. Jones, M.G.K. and Dropkin, V.H. (1975) Scanning electron microscopy of syncytial transfer cells induced in roots by cyst nematodes, *Physiol. Plant Pathol.* **7**, 259-263.
29. Jones, M.G.K. and Northcote, D.H. (1972) Nematode-induced syncytium-a multinucleate transfer cell, *J. Cell Sci.* **10**, 789- 809.
30. Kim, Y.H., Riggs, R.D. and Kim, K.S. (1987) Structural changes associated with resistance of soybean to *Heterodera glycines*, *J. Nematol.* **19**, 177-187.
31. Luc, M., Maggenti, A.R. and Fortuner, R. (1988) A reappraisal of Tylenchina (Nemata). 9. The family Heteroderidae Filip'ev & Schuurmans Stekhoven, 1941, *Rev. N(matol.* **11**, 159-176.
32. Magnusson, C. and Golinowski, W. (1991) Ultrastructural relationships of the developing syncytium induced by *Heterodera schachtii* (Nematoda) in root tissue of rape, *Can. J. Bot.* **69**, 44-52.
33. Melillo, M.T., Bleve-Zacheo, T. and Zacheo, G. (1990a) Ultrastructural response of potato roots susceptible to cyst nematode *Globodera pallida* pathotype Pa3, *Rev. N(matol.* **13**, 17-28.
34. Melillo, M.T., Bleve-Zacheo, T., Zacheo, G. and Perrino, P. (1990b) Morphology and enzyme histochemistry in germplasm pea roots attacked by *Heterodera goettingiana*, *Nematol. mediter.* **18**, 83-91.
35. Mittler, R. and Lam, E. (1996) Sacrifice in the face of foes: pathogen-induced programmed cell death in plants, *Trends Microbiol.* **4**, 10-15.
36. Mugniery, D. and Bossis, M. (1985) Influence de l'hote, de la compétition et de l'état physiologique des juvéniles sur la pénétration, le développement et la sexe d'*Heterodera carotae* Jones. *Nematologica* **31**, 335-343.
37. Mugniery, D. and Fayet, G. (1981) Determination du sexe chez *Globodera pallida* Stone, *Rev. N(matol.* **4**, 41-45.
38. Mugniery, D. and Fayet, G. (1984) Détermination du sexe de *Globodera rostochiensis* Woll. et influence des niveaux d'infestacion sur la pénétration le développement et le sexe de ce nématode, *Rev. N(matol.* **7**, 233-238.
39. Raski, D.J. (1950) The life history and morphology of the sugar beet nematode, *Heterodera schachtii* Schmidt, *Phytopathology* **40**, 135-150.
40. Rice, S.L., Leadbeater, B.S.C. and Stone, A.R. (1985) Changes in cell structure in roots of resistant potatoes parasitized by potato cyst nematodes. 1. Potatoes with resistance gene H_1 derived from *Solanum tuberosum* ssp. *andigena*, *Physiol. Plant Pathol.* **27**, 219-234.
41. Rice, S.L., Stone, A.R. and Leadbeater, B.S.C. (1987) Changes in cell structure in roots of resistant potatoes parasitized by potatoes cyst nematodes. 2. Potatoes with resistance derived from *Solanum vernei*, *Physiol. Mol. Plant Pathol.* **31**, 1-14.
42. Riggs, R.D., Kim, K.S. and Gipson, I. (1973) Ultrastructural changes in Peking soybeans infected with *Heterodera glycines*, *Phytopathology* **63**, 76-84.
43. Sijmons, P.C., Grundler, F.M.W., von Mende, N., Burrows, P.R. and Wyss, U. (1991) *Arabidopsis thaliana* as a new model host for plant-parasitic nematodes, *Plant J.* **1**, 245-254.

44. Sijmons, P.C., Atkinson, H.J. and Wyss, U. (1994) Parasitic strategies of root nematodes and associated host cell responses, *Annu. Rev. Phytopathol.* **32**, 235-259.
45. Sobczak, M. (1996) Investigations on the structure of syncytia in roots of *Arabidopsis thaliana* induced by the beet cyst nematode *Heterodera schachtii* and its relevance to the sex of the nematode. PhD Thesis, University of Kiel.
46. Sobczak, M., Golinowski, W. and Grundler, F.M.W. (1996) Changes in the structure of *Arabidopsis thaliana* roots induced during development of males of the plant parasitic nematode *Heterodera schachtii, Europ. J. Plant Pathol.* in press.
47. Stender, C., Lehmann, H. and Wyss, U. (1982) Feinstrukturelle Untersuchungen zur Entwicklung von Wurzelriesenzellen (Syncytien) induziert durch Rübenzystennematoden *Heterodera schachtii, Flora* **172**, 223-233.
48. Williams, K.J. and Fisher, J.M. (1993) Development of *Heterodera avenae* Woll. and host cellular responses in susceptible and resistant wheat, *Fundam. Appl. Nematol.* **16**, 417-423.
49. Wilski, A. (1977) M1twik zbożowy *Heterodera avenae* Woll. (1924), PAN, Komitet Ochrony Roœlin.
50. Wyss, U. (1992) Observations on the feeding behaviour of *Heterodera schachtii* throughout development, including events during moulting, *Fundam. Appl. Nematol.* **15**, 75-89.
51. Wyss, U., Stender, C. and Lehmann, H. (1984) Ultrastructure of feeding sites of the cyst nematode *Heterodera schachtii* Schmidt in roots of susceptible and resistant *Raphanus sativus* L. var. *oleiformis* Pers. cultivars, *Physiol. Plant Pathol.* **25**, 21-37.
52. Wyss, U. and Zunke, U. (1986) Observations on the behaviour of second stage juveniles of *Heterodera schachtii* inside host roots, *Rev. Nématol.* **9**, 153-165.
53. Yu, M.H. and Steele, A.E. (1981) Host-parasite interaction of resistant sugarbeet and *Heterodera schachtii, J. Nematol.* **13**, 206-212.

NEMATODE SECRETIONS

John T. JONES and Walter M. ROBERTSON
Nematology Department, Scottish Crop Research Institute, Invergowrie, Dundee, DD2 5DA

Abstract

Secretions of plant parasitic nematodes are thought to have important roles in various aspects of the host-parasite relationship. In contrast to studies using much larger animal parasitic nematodes, few secreted molecules from plant parasitic nematodes have previously been characterised in detail. However, the use of antibodies raised against nematode secretory components has now allowed the identification and biochemical characterisation of some of the secretions of plant parasitic nematodes. This article reviews progress in this area.

1. Introduction

Nematode secretions originate from a range of body structures including the oesophageal gland cells, the amphids, the excretory/secretory system and the cuticle and are of interest for a variety of reasons. Molecules secreted from parasitic nematodes are in more intimate contact with the host than any other parasite molecules and thus are likely to have important roles in the host-parasite relationship. In plant parasitic nematodes various functions have been suggested for almost all secretions. It has been suggested that those present on the cuticle surface may serve to mask the nematode from its host or to anchor the nematode at its feeding site [19] and that the amphidial secretions form a feeding plug [16]. The secretions of the subventral and dorsal oesophageal gland cells have been implicated in a range of processes associated with invasion and feeding including penetration of plant tissues [8], inducing the formation of the feeding site [28] and production of the feeding tube [42]. However, no functional evidence has yet been found to support these claims and indeed no nematode secretions have been localised to the feeding site in the plant. However, there is strong observational evidence to suggest that the secretions of the oesophageal gland cells are involved in the induction of feeding sites (see below).

Early work on the nature of nematode secretions, and indeed much present work of this kind, has been hampered by the small size of plant parasitic nematodes and the consequent difficulties in obtaining sufficient quantities of biological material for biochemical analysis. Developments in molecular biological techniques and an increased awareness of the benefits of such techniques [32] mean that progress is now being made by a number of groups towards identifying components of nematode secretions at a molecular level. This article reviews these developments.

2. Secretions of Free Living and Animal Parasitic Nematodes

Little is known about the secretions of free-living nematodes despite the enormous amount of research on *Caenorhabditis elegans*. Acetylcholine esterase has been identified in the sense organ secretions of *C. elegans* [38] and other nematodes [4, 36]. In other nematodes this enzyme has been identified in the secretions of the subventral gland cells [37]. Some analysis of *C. elegans srf* mutants, which have abnormal reactivities to surface-specific antibodies and/or lectins has been carried out [9] although the nature of the molecules in which the mutations reside has yet to be determined.

There have been studies carried out on the secretions of animal parasitic nematodes but a comprehensive review of this area is beyond the scope of this article. Molecules which help the parasite avoid the host immune response are produced in several parasites [e.g. 33]. Others which induce changes in host tissues are produced by *Trichinella spiralis* [13, 34] and anticoagulants are produced by the sense organs of many blood-feeding nematodes [e.g. 36]. Thus, there is precedence in animal parasitic nematodes for many of the phenomena attributed to secretions of plant parasitic nematodes including avoidance of detection by the host, modification of host gene expression and facilitation of the nematode feeding process.

3. Gland Cell Secretions

Most work on the secretions of plant parasitic nematodes has been aimed at the secretions produced by the dorsal and subventral oesophageal gland cells. This interest derives from the assumption that the secretions of these cells are involved in the establishment of the nematode feeding site. Gland cell secretions have never been localised to the feeding site and there is no functional evidence that these secretions induce changes in the host plant. However, observations made using video-enhanced light microscopy clearly show that secretions from the gland cells are pumped into the plant prior to the onset of the changes in the plant [56, 57]. Furthermore, ectoparasites such as *Xiphimema* and *Longidorus* remain outside the root and pierce a cell deep within the root using a long stylet. This chosen cell and those around it which are not in direct contact with the nematode surface respond by becoming enlarged, multinucleate and show increased rates of DNA and RNA synthesis [21, 22]. However, in endoparasites the possibility remains that secretions present on the nematode surface, as well as those from the oesophageal gland cells, are involved in feeding site formation.

Some debate has also arisen as to whether the secretions of the dorsal or subventral gland cells are more likely to be involved in feeding site establishment. It has been argued on anatomical grounds [14] that the secretions of the subventral gland cells are unlikely to be moved forward and secreted via the stylet and hence that the dorsal gland cell is likely to produce the secretions most important in parasitism. It has now been demonstrated using antibodies that secretions of the subventral gland cell can be exuded from the stylet of *Meloidogyne incognita* [11] and that a role in parasitism for these secretions is therefore feasible. Observations on the morphology of the subventral and dorsal gland cells at different stages in the life cycle of *M. incognita* suggest the secretions of the subventral gland cells may be important during induction and establishment of the feeding site while the secretions of the dorsal gland cell may have a role in maintenance of the feeding site, in the actual feeding process or in production of the feeding tube (below). The subventral gland cells are large and full of granules immediately prior to invasion, but decrease in size and contain fewer granules in older, feeding nematodes [25, 56]. By contrast, the dorsal gland cell which is small and contains few granules prior to invasion, begins to fill with granules soon after juveniles enter the root and predominates in the adult female [5, 25].

In addition to containing substances responsible for initiating the feeding site, the oesophageal gland cells are also thought to be the site of origin of the feeding tube. Although no direct evidence that oesophageal gland cell secretions form the feeding tube is available, various observations taken together provide compelling evidence that feeding tubes form directly from gland cell secretions [24]. This evidence includes the fact that feeding tube-like structures are not observed in healthy plants or plants infected with any other pathogens, that feeding tubes form in parasitised cells immediately after the insertion of the tip of the nematode stylet and that the feeding tubes formed by one species of nematode are dissimilar to those formed by other species whereas the feeding tube produced by any one species of nematode always has the same fine structure regardless of the host plant. The biochemical nature of the feeding tube remains to be discovered and there is debate as to the precise role of this structure and whether it is essential for nematode feeding to occur [see 24].

Since invasive stage juveniles of plant parasitic nematodes are too small to allow the gland cells to be separated from the rest of the body contents for analysis, attempts to characterise secretions of the oesophageal gland cells have centred mainly on the use of antibodies raised against secretory components [1, 12, 20, 23, 29]. Some studies have attempted direct analysis of nematode gland cell products. Reddigari, Sundermann & Hussey [43] used Percoll gradients to purify secretory granules from *M. incognita* and showed the granules contained at least 15 major proteins. Although acid phosphatase activity was found in the collected protein and was also localised to the subventral gland cells of this nematode, it was not possible to identify which of the other protein components originated from the gland cells. Using a more direct approach, Veech, Starr & Nordgren [50] were able to collect stylet secretions from adult females of *M. incognita* and demonstrate that at least 9 proteins were present. Further analysis of the secretions collected was not possible as insufficient protein was generated by the nematodes to allow for N-terminal sequencing. The discovery that a range of exogenously applied compounds can induce the secretion of oesophageal gland cell products [15, 20, 35, 39] may in the future enable sufficient secretions to be collected for direct sequence analysis. The observation that some of the elicitors of secretion may be plant hormones [15] is an intriguing one and suggests a mechanism linking commencement of secretion to the onset of a parasitic mode of existence.

Despite several studies describing the raising of antibodies against nematode secretions, little progress has been made in characterising the proteins which they recognise. An antibody reactive against the dorsal and subventral oesophageal gland cell secretions of *M. incognita* was shown to recognise a glycoprotein with a relative molecular mass (Mr) of greater than 212 kDa but no function has been suggested for this protein [26]. De Boer *et al* [12], used a different approach to produce antibodies which recognised the subventral gland cells of *Globodera rostochiensis*. Proteins of second stage juveniles were first fractionated and the fractions corresponding to Mr range 38 to 40.5 kDa were pooled and used to produce antibodies. These antibodies bound to the subventral gland cells and recognised four major protein bands on western blots ranging in size from 30 to 49 kDa including one protein with an Mr of 39 kDa corresponding with that of the proteins used for immunisation. The results suggest that several distinct but related proteins are produced by the subventral gland cells. It was suggested that the proteins may be variants of a single gene product produced by post-translational modifications or by alternative splicing pathways.

Antibodies have now been used to isolate genes by screening of cDNA expression libraries although given the enormous number of antibodies raised against nematode secretions surprisingly few genes have been sequenced. A monoclonal antibody which recognised subventral glands of *M. incognita* was used to isolate a gene from an expression library [41]. Unfortunately, this antibody turned out to recognise a myosin-like molecule thought to be involved in secretory granule transport rather than a secreted molecule *per se*. This molecule is clearly extremely antigenic as other groups have obtained antibodies against subventral gland cells which recognise the same epitope [e.g. 55].

Recently an antibody which recognises the subventral gland cells and surface coat of *G. rostochiensis* and *G. pallida* has been used to screen an expression library [29]. The isolated gene has extensive homology to secreted molecules from *Onchocerca volvulus* and *C. elegans* and appears to be expressed differentially at different stages of the PCN life cycle (J.T. Jones, unpublished observations). This molecule is thought to function as a lipid-binding protein in *O. volvulus* and functional experiments to determine the range of ligands the *Globodera* homologue is capable of binding *in vitro* are currently being undertaken.

In the future the approaches described above are likely to reveal much more about the range of molecules produced in the gland cells of plant parasitic nematodes as several groups are currently isolating further genes from a range of nematodes. Hopefully in the near future such studies will uncover the mechanisms by which nematodes induce the range of changes in their host plants which accompany feeding site establishment.

4. Other Secretions

4.1. EXCRETORY-SECRETORY SYSTEM

The word excretory in the description given to this organ is perhaps a misnomer as no evidence for an excretory function has been shown [6]. An osmoregulatory function has been suggested for the Excretory/Secretory (E/S) system in animal parasitic nematodes [10] and it is possible that the E/S system of plant parasites fulfils a similar role. In *Meloidogyne* and *Anguina* the exudates of the E/S system have been shown to contain proteins [7, 39] some of which may be glycosylated. This led to the suggestion that the E/S system is the source of the surface coat, covering the cuticle of plant parasitic nematodes [7]. However, no secreted molecules from the E/S system have been characterised at a molecular level and no labelling studies have been carried out to determine whether the E/S system is the source of the surface coat. The role of the E/S system and the nature of the molecules it secretes therefore remain unknown.

4.2. SENSE ORGANS

The sense organs of plant parasitic nematodes have gland cells surrounding their nerve processes and consequently produce secretions. The secretions of the largest sense organs (the amphids) have been shown to have a role in host-parasite relationships in some blood-feeding nematodes, producing anticoagulants to aid the feeding process [e.g. 36]. Endo [16] suggested the amphids of *Heterodera schactii* may also have a role in feeding, producing the feeding plug which surrounds the point at which the nematode pierces the syncitium with its stylet. However, work with *G. rostochiensis* suggested that the feeding plug did not originate in the amphids but was more likely to be secreted via the cuticle [31]. The role, if any, of the amphids of plant parasitic nematodes in the feeding process therefore remains uncertain. Given the observation that all the sense organs, not simply the amphids, have associated gland cells and secretions it seems more likely that the secretions of these gland cells are involved in chemoreception. Precedents for this exist in many other groups of organisms, including insects. In the secretions of insect sense organs two main groups of proteins exist which have important roles in chemoreception [53]. Olfactory binding proteins (OBPs) are small and abundant proteins which are thought to solubilise hydrophobic odorant molecules and allow them to cross the receptor cavity and reach the nerve processes. Different OBPs bind different odorant molecules [51, 54] and may act as selective signal filters peripheral to the actual receptor proteins, although odorant discrimination is thought to be a property of receptor proteins themselves. The secretions of insect sense organs also contain a range of odorant degrading enzymes (ODEs) which destroy odorant molecules released from the receptors at the nerve processes and thus prevent repeated restimulation of the nerve by a single odorant molecule [45, 49, 52]. Although the sense organs of most nematodes are much too small to allow for collection of secretions and biochemical analysis for degrading enzymes it seems plausible that similar molecules, or molecules which fulfil a similar role are present.

No secretions of plant parasitic nematode sense organs have been characterised in detail although a glycoprotein with an Mr of 32 kDa has been identified in the amphids of *Meloidogyne* species [47, 48]. This protein was found in a number of different species and was present in a range of life cycle stages. Antibodies against this protein blocked chemoattraction, suggesting a role in chemoreception for this molecule. Other studies have focused on the possible role of carbohydrates in the chemoreception process following the suggestion that the amphidial secretions are the source of the carbohydrates located on the anterior end of many plant parasitic nematodes [58]. The binding of lectins to the amphidial secretions and cuticle surface of plant parasitic nematodes has been used as evidence to suggest carbohydrate residues are present in amphidial secretions in the form of glycoproteins [3, 18]. However, contrasting results on the effects of lectins on the ability of nematodes to respond to chemoattractants have been obtained [e.g. 2, 27] leaving doubts as to the involvement of carbohydrate residues in the chemoreception process. A full understanding of the role of the secretions of the sense organs of plant parasitic nematodes in the infection process or in chemoreception awaits further characterisation of these secretions at the molecular level.

4.3. CUTICLE

In this section it is not the chemical nature of the cuticle itself that is considered but that of the surface coat. The surface coat is observed as a fuzzy coating in some ultrathin sections of nematodes from a range of habitats [6]. Its chemical nature in plant parasites remains undetermined although it has been shown to contain various carbohydrates, probably in the form of glycoproteins [e.g. 44, 46]. Interestingly, the cuticle surface and surface coats some animal parasites have been shown to change, both physically and chemically, upon invasion of their hosts [40] and similar changes have been observed in different life cycle stages of *C. elegans* [40, 59]. It is likely therefore the surface secretions of plant parasitic nematodes varies between stages and that this variation may have a role in pathogenesis. Changes in the cuticle occur once second stage juveniles of *G. rostochiensis* enter the root and establish feeding sites [19, 30]. Whilst some of these changes are likely to be structural and associated with the onset of moulting, other changes in the cuticle surface were also observed which may be associated with parasitism. In *Heterodera schachtii*, fibrillar material, which was shown to originate in the hypodermis, has been found on the cuticle surface shortly after establishment of a feeding site [17] but no function has been suggested for this material and no information about its chemical nature has been presented. The cuticle of *G. rostochiensis* was also found to change during hatching as a result of exposure to host root diffusate [30]. Such host induced changes may be associated with the onset of a parasitic mode of existence in this nematode.

Although little is known about the chemical nature of secretions present on the cuticle surface of plant parasitic nematodes it seems likely given the range of molecules identified from the surfaces of animal parasites that they may have an important role in pathogenesis.

5. Conclusions

Despite their obvious importance in the host-parasite relationship little progress has been made towards characterising nematode secretions at a molecular level. As molecular biological techniques are applied by a greater range of groups it seems likely that this will not remain the case for much longer and that the next few years should see our understanding of this area increase dramatically. Hopefully this information will reveal how plant parasitic nematodes are able to maintain such intimate relationships with their hosts.

References

1. Atkinson, H.J., Harris, P.D., Halk, E.J., Novitsk, C. Leighton-Sands, J., Nolan, P. & Fox, P.C. (1988) Monoclonal antibodies to the soya bean cyst nematode, *Heterodera glycines. Annals of Applied Biology*, **112**, 459-469.
2. Aumann, J., Clemens, C.D. & Wyss, U. (1990) Influence of lectins on female sex pheromone reception by *Heterodera schactii* (Nematoda: Heteroderidae) males, *Journal of Chemical Ecology*, **16**, 2371-2380.
3. Aumann, J. & Wyss, U. (1989) Histochemical studies on exudates of *Heterodera schactii* (Nematoda: Heteroderidae) males. *Revue de Nematologie*, **12**, 309-315.
4. Bird, A.F. (1966) Esterases in the genus *Meloidogyne*, *Nematologica*, **12**, 359-361.
5. Bird, A.F. (1969) Changes associated with parasitism in nematodes V. Ultrastructure of the stylet exudation and dorsal oesophageal gland contents of female *Meloidogyne javanica*, *Journal of Parasitology*, **55**, 337.
6. Bird, A.F. & Bird, J. (1991) The structure of nematodes, 2nd Edn. Academic Press, London.
7. Bird, A.F., Bonig, I. & Bacic, A. (1988) A role for the 'Excretory' system in Secernentean nematodes. *Journal of Nematology*, **20**, 493-496.
8. Bird, A.F., Downton, W.J.S. & Hawker, J.S. (1975) Cellulase secretion by second stage larvae of the root knot nematode *Meloidogyne javanica. Marcellia*, **38**, 165-169.
9. Blaxter, M.L. (1993) Cuticle surface proteins of wild type and mutant *Caenorhabditis elegans*. *Journal of Biological Chemistry*, **268**, 6600-6609.
10. Croll, N.A., Slater, L. & Smith, J.M (1972) *Experimental Parasitology*, **31**, 356.
11. Davis, E.L., Allen, R. & Hussey, R.S. (1994) Developmental expression of oesophageal gland antigens and their detection in stylet secretions of *Meloidogyne incognita*, *Fundamental and Applied Nematology* **17**, 255-262.
12. De Boer, J.M., Smant, G., Goverse, A., Davis, E.L., Overmars, H.A., Pomp, H. R., Gent-Pelzer, M. van, Zilverentant, J.F., Stokkermans, J.P.W.G., Hussey, R.S., Gommers, F.J., Bakker, J & Schots, A. (1996) Secretory granule proteins from the subventral oesophageal glands of the potato cyst nematode identified by monoclonal antibodies to a protein fraction from second stage juveniles, *Molecular Plant Microbe Interactions*, **9**, 39-46.
13. Despommier, D.D., Gold, A.M., Buck, S.W., Capo, V. & Silberstein, D. (1990) *Trichinella spiralis*: secreted antigen of the infective L1 larva localises to the cytoplasm and nucleoplasm of infected host cells. *Experimental Parasitology*, **71**, 27.
14. Doncaster, C.C. (1971) Feeding in plant parasitic nematodes: mechanisms and behaviour. *In*: Plant Parasitic Nematodes, B.M. Zuckerman, Mai, W.F. & Rhodes, R.A., eds. Academic Press; New York.
15. Duncan, L.H., Robertson, W.M. & Kusel, J.R. (1995) Induction of secretions in *Globodera pallida*, *Abstracts of the BSP Spring Meeting,* Edinburgh 1995. p13.
16. Endo, B.Y. (1978) Feeding plug formation in soybean roots infected with the soybean cyst nematode, *Phytopathology,* **68**, 1022-1031.
17. Endo, B.Y. & Wyss, U. (1992) Ultrastructure of cuticular exudations in parasitic juvenile *Heterodera schachtii*, as related to cuticle structure, *Protoplasma,* **166**, 67-77.
18. Forrest, J.M.S. & Robertson, W.M. (1986) Characterisation and localisation of saccharides on the head region of four populations of the potato cyst nematode *Globodera rostochiensis* and *G. pallida, Journal of Nematology*, **18**, 23-26.
19. Forrest J.M.S., Robertson, W.M. & Milne, E.W. (1989) Observations on the cuticle surface of second stage juveniles of *Globodera rostochiensis* and *Meloidogyne incognita. Revue de Nematologie*, **12**, 337-341.
20. Goverse, A., Davis, E.L. & Hussey, R.S. (1994) Monoclonal antibodies to the oesophageal glands and stylet secretions of *Heterodera glycines, Journal of Nematology*, **26**, 251-259.
21. Griffiths, B. S. & Robertson, W.M. (1984a) Morphological and histochemical changes occurring during the lifespan of root tip galls on *Lolium perenne* induced by *Longidorus elongatus*, *Journal of*

Nematology, **16**, 223-229.
22. Griffiths, B. S. & Robertson, W.M. (1984b) Nuclear changes induced by the nematode *Xiphinema diversicaudatum* in root-tips of strawberry, *Histochemical Journal*, **16**, 265-173.
23. Hussey, R.S. (1989) Monoclonal antibodies to secretory granules in oesophageal glands of *Meloidogyne* species, *Journal of Nematology*, **21**, 392-398.
24. Hussey, R.S., Davis, E.L. & Ray, C. (1994) *Meloidogyne* stylet secretions. *In:* Advances in Molecular Plant Nematology, Lamberti, F, De Giorgi, C. & Bird, D. McK Eds., pp233-249, Plenum Press, New York.
25. Hussey, R.S. & Mimms C.W. (1990) Ultrastructure of oesophageal glands and their secretory granules in the root-knot nematode *Meloidogyne incognita*, *Protoplasma*, **165**, 9.
26. Hussey, R.S., Paguio, O.R. & Seabury, F. (1990) Localisation and purification of a secretory protein from the oesophageal glands of *Meloidogyne incognita* with a monoclonal antibody. *Phytopathology*, **80**, 709-714.
27. Jansson, H.B. & Nordbring-Hertz, B. (1984) Involvement of sialic acid in nematode chemotaxis and infection by an endoparasitic nematophagous fungus, *Journal of General Microbiology*, **130**, 39-43.
28. Jones, M.G.K. (1981) Host cell responses to endoparasitic nematode attack: structure and function of giant cells and syncitia, *Annals of Applied Biology*, **97**, 353-372.
29. Jones, J.T., Burrows, P.R., Wightman, P.J., Forrest, J.M.S. & Robertson, W.M. (1996) Isolation and characterisation of a cDNA clone encoding a secreted molecule from the subventral gland cells of *Globodera pallida*, *Abstracts of the BSP Spring Meeting*, Bangor 1996. p71.
30. Jones, J.T., Perry, R.N. & Johnston, M.R.L. (1993) Changes in the ultrastructure of the cuticle of the potato cyst nematode *Globodera rostochiensis* during dvelopment and infection, *Fundamental and Applied Nematology*, **16**, 433-455.
31. Jones, J.T., Perry, R.N. & Johnston, M.R.L. (1994) Changes in the ultrastructure of the amphids of the potato cyst nematode *Globodera rostochiensis* during dvelopment and infection, *Fundamental and Applied Nematology*, **17**, 369-382.
32. Jones, J.T., Phillips, M.S. & Armstrong, M.R. (1996) Molecular approaches to plant nematology, *Fundamental and Applied Nematology, in press.*
33. Kennedy, M.W. (Ed) (1991) Parasitic nematodes - antigens, membranes and genes, Taylor & Francis, London.
34. Ko, R.C., Fan, L. & Lee, D.L. (1992) Experimental reorganisation of host muscle cells by excretory/secretory products of infective *Trichinella spiralis* larvae, *Transactions of the Royal Society for Tropical Medicine and Hygeine*, **86**,77.
35. McClure, M.A. & Von Mende, N. (1987) Induced salivation in plant parasitic nematodes, *Phytopathology*, **77**, 1463-1469.
36. McLaren, D.J., Burt, J.S. & Ogilvie, B.M. (1974) The anterior glands of adult *Necator americanus* (Nematoda: Strogyloidea): II. Cytochemical and functional studies, *International Journal of Parasitology*, **4**, 39-46.
37. Nakazawa. M., Yamada, M, Uchikawa, R, Arizono, N. (1995) Immunocytochemical localisation of secretory acetylcholinesterase of the parasitic nematode *Nippostrogylus brasiliensis*, *Cell and Tissue Research*, **280**, 59-64.
38. Pertel, R., Paran, N. & Mattern, C.F.T. (1976) *Caenorhabditis elegans*: localisation of cholinesterase associated with anterior nematode structures, *Experimental Parasitology*, **39**, 401-414.
39. Premachandran, D., Von Mende, N., Hussey, R.S. & McClure, M.A. (1988) A method for staining nematode secretions and structures, *Journal of Nematology*, **20**, 70-78.
40. Proudfoot, L., Kusel, J.R., Smith, H.V. & Kennedy, M.W. (1991) Biophysical properties of the nematode surface. *In*: Parasitic nematodes - antigens, membranes and genes, M.W. Kennedy Ed. Taylor & Francis, London.
41. Ray, C., Abbot, A.G. & Hussey, R.S. (1994) *Trans*-splicing of a *Meloidogyne incognita* mRNA encoding a putative oesophageal gland protein, *Molecular and Biochemical Parasitology*, **68**, 93-101.
42. Rumpenhorst, H.J. (1984) Intracellular feeding tubes associated with sedentary plant parasitic nematodes. *Nematologica*, **30**, 77-85.
43. Reddigari, S.R., Sundermann, C.A. & Hussey, R.S. (1985) Isolation of subcellular granules from second stage juveniles of *Meloidogyne incognita*, *Journal of Nematology*, **17**, 482-488.
44. Robertson, W.M., Spiegel, Y., Jansson, H-B. , Marban-Mendoza, M & Zuckerman, B.M. (1989) Surface carbohydrates of plant parasitic nematodes, *Nematologica*, **35**, 180-187.

45. Rybczynski, R., Reagan, J. & Lerner, M.R. (1989) A pheromone degrading aldehyde oxidase in the antennae of the moth *Manduca sexta*, *Journal of Neuroscience*, **9**, 1341-1353.
46. Spiegel, Y., Robertson, W.M., Himmelhoch, S. & Zuckerman, B.M. (1983) Electron microscope characterisation of carbohydrate residues on the body wall of *Xiphinema index*, *Journal of Nematology*, **15**, 528-535.
47. Stewart, G.R., Perry, R.N., Alexander, J. & Wright, D.J. (1993) A glycoprotein specific to the amphids of *Meloidogyne* species, *Parasitology*, **106**, 405-412.
48. Stewart, G.R., Perry, R.N. & Wright, D.J. (1993) Studies on the amphid specific glycoprotein gp32 in different life cycle stages of *Meloidogyne* species, *Parasitology*, **107**, 573-578.
49. Tascayo, M.L., & Prestwich, G.D. (1990) Aldehyde oxidases and dehydrogenases in antennae of five moth species, *Insect Biochemistry*, **20**, 691-700.
50. Veech, J.A., Starr, J.L. & Nordgren, R.M. (1987) Production and partial characterisation of stylet exudate from adult females of *Meloidogyne incognita*, *Journal of Nematology*, **19**, 463-468.
51. Vogt, R.G., Prestwich, G.D., & Lerner, M.R. (1991) Odorant binding protein subfamilies associate with distinct classes of olfactory receptor neurones in insects, *Journal of Neurobiology*, **22**, 74-84.
52. Vogt, R.G. & Riddiford, L.M. (1981) Pheromone binding and inactivation by moth antennae, *Nature*, **293**, 161-163.
53. Vogt, R.G., Riddiford, L.M. & Prestwich, G.D. (1985) Kinetic properties of a sex pheromone degrading enzyme, the sensillar esterase of *Antherea polyphemus*, *Proceedings of the National Acaademy of Science of the USA*. **82**, 8827-8831.
54. Vogt, Rybczynski & Lerner, 1991; The biochemistry of odorant reception and transduction. In: Chemosensory Information Processing, Schild D. (Ed) p33-76. Springer-Verlag, Berlin.
55. Willats, W.G.T., Atkinson, H.J., & Perry, R.N. (1995) The immunoflourescent localisation of subventral pharyngeal gland epitopes of preparasitic juveniles of *Heterodera glycines* using laser scanning confocal microscopy, *Journal of Nematology*, **27**, 135-142.
56. Wyss, U., Grundler, F.M.W. & Munch, A. (1992) The parasitic behaviour of second stage juveniles of *Meloidogyne incognita* in roots of *Arabidopsis thaliana*, *Nematologica*, **38**, 98.
57. Wyss, U. & Zunke, U. (1986) Observations on the behaviour of second stage juveniles of *Heterodera schachtii* inside host roots, *Revue de Nematologie*, **9**, 153.
58. Zuckerman, B.M. & Jansson, H.B. (1984) Nematode chemotaxis and possible mechanisms of host/prey recognition, *Annual Review of Phytopathology*, **22**, 95-113.
59. Zuckerman, B.M. & Kahane, I. (1983) *Caenorhabditis elegans*: stage specific differences in cuticle surface carbohydrates, *Journal of Nematology*, **15**, 535-539.

PHYSIOLOGY OF NEMATODE FEEDING AND FEEDING SITES

Florian M.W. GRUNDLER and Annette BÖCKENHOFF
Institut für Phytopathologie, Christian-Albrechts-Universität Kiel, 24098 Kiel, Germany

Abstract

Physiological studies of the interaction between plants and sedentary nematodes mostly focus on transfer processes of water and solutes into the specific feeding structures and further to the feeding nematodes. The description given here includes the anatomical and behavioural basis of nematode feeding as well as the most important physiological features of the feeding sites. Finally, specific adaptations of the nematode to restrictions in the nutrient supply, namely the concept of environmentally controlled sex determination is presented.

1. Introduction

Sedentary nematodes damage their host plants by causing profound physiological changes that concern not only the infection site but the entire system of water, mineral and assimilate transport. In the field these changes become obvious by the typical symptoms caused by nematode infection such as reduced growth and enhanced proneness to wilting.

The basis of all the processes involved is the formation of the specific feeding structures in the plant root and the withdrawal of nutrients from these structures by the associated nematodes.

This chapter gives an overview of the anatomical and behavioural basis of nematode feeding and reviews the current knowledge of the transfer of solutes from the plant vascular tissue into the feeding structure and from the feeding structure into the developing nematodes. Finally, the effects of reduced nutrient supply on nematode development, either induced by restricted plant nutrition or caused by resistance responses, is discussed.

2. Feeding Apparatus

Tylenchid nematodes possess a feeding apparatus that is relatively uniform in structure and function throughout the different members of the order (see Wyss, this volume). The anterior part of the feeding apparatus consists of a hollow stylet with a subterminal orifice. The stylet lumen is connected to an oesophageal duct that merges into a muscular bulb, often designated as median or metacorpal bulb. A dorsal and two subventral gland cells form the posterior part of the feeding apparatus. The dorsal gland has a long extension that terminates in an ampulla in which the secretory granules are collected. The ampulla has an orifice into the oesophageal duct close to the nematode stylet. The ampullae of the subventral glands are placed at the basis of the median bulb where a duct leads into the oesophageal lumen.

3. Feeding Behaviour

Most of the informations available on the feeding behaviour of sedentary nematodes is based on *in vivo* observations. Early work in this field was performed by Linford [23], who observed the feeding behaviour of root-knot nematodes in unfixed sections of galls and dissected nematode specimens. He first stated that for feeding the nematodes have to protrude their stylet into a modified cell, then to release secretions (see Jones and Robertson, this volume) and subsequently to take up nutrients from this cell by pulsation of the median bulb. This basic pattern has been confirmed in more recent studies of the feeding behaviour of root-knot and cyst nematodes.

More detailed analyses became available by *in vivo* observations that were performed with improved optics and video contrast enhancement. In this way the feeding behaviour of *Meloidogyne incognita* and *Heterodera schachtii* could be followed inside the root tissue of cruciferous hosts, such as *Arabidopsis thaliana* and *Brassica rapa*.

The parasitic development of *H. schachtii* starts with the perforation of the cell wall of an initial syncytial cell with the stylet (see Golinowski *et al.*, this volume). The stylet remains protruded for several hours whereas the juveniles do not exhibit feeding activity [48]. This "preparatory phase" was also observed by Steinbach [43] with *Globodera rostochiensis* parasitising on tomato roots but could not be seen with freshly invading *M. incognita* juveniles [46].

In *H. schachtii*, Wyss and Zunke [48] observed three distinct phases of feeding repeated in cycles. The cycles were only interrupted during moulting, when no feeding occurred. During phase I nutrients are drawn from the syncytium by continous pumping of the median bulb. In phase II the stylet is retracted from the syncytium and re-inserted, while during phase III secretory granules were observed to move forward in the extension of the dorsal gland, suggesting the release of secretions into the syncytium. The cyclic repetition of these three phases is maintained during the entire development of the female juvenile and the feeding stages (J2 and J3) of the male juveniles. The duration of the phases II and III remains constant during nematode development, whereas phase I, during which nutrients are ingested, increases. Interestingly, the total time of feeding activity of male J3 exceeds that of female J3 [45]. The site of stylet insertion usually is the same during nematode development. However, sometimes the syncytial cell wall responds with depositions of material that may obstruct the re-insertion. In these cases a prolongation of phase II or a complete failure of nematode development can be observed.

After establishing *A. thaliana* as a model nematode host plant it became easier to study the feeding behaviour of root-knot nematodes *in vivo* [46]. Nevertheless, a detailed study of the behaviour during induction and early feeding still could not be performed. After ceasing migration within the differentiating vascular cylinder, the juveniles of *M. incognita* induce the formation of several separate giant cells. The sedentary juveniles move their head from one cell to the other, piercing the cell walls and protruding the stylet tip through it. The periods of stylet tip protrusions last from 45 sec in the early J2 up to 100 sec in advanced J2. This phase is comparable to phase III in cyst nematodes. The periods of nutrient uptake as indicated by the activity of the median bulb last from 60 to 200 sec reflecting a gradual increase in nutrient withdrawal. After feeding the stylet is redrawn and another giant cell is selected by specific movements of the head region of the nematode.

4. Feeding Tubes

4.1. NATURE AND FORMATION

The current knowledge on the nature and formation of feeding tubes is very vague (see also Jones and Robertson, this volume). Since their first detailed description [35], feeding tubes were found in several host-parasite interactions and a number of ultrastructural and *in vivo* studies were performed [12, 18, 34, 47]. These studies clearly showed that the formation of feeding tubes is based on the release of secretions and that they are functionally involved in the feeding process. However, it is not known to which extent cytoplasmic components contribute to the formation of the tubular structure. *In vivo* observations suggest that the secretions are produced in the dorsal oesophageal gland, although an analytical evidence has not yet been provided. The studies, however, proved that the feeding tubes are exchanged during the repeated feeding cycles, when the stylet is redrawn and re-inserted, thus explaining the high number of tubes found in feeding structures of advanced nematode stages.

In the early stages of development, no feeding tubes are formed. In *H. schachtii* they were found about 36 hr after syncytium induction at earliest [45], while in *M. incognita* they were first observed in adult females [18]. However, release of secretions from the stylet orifice was observed as early as 6 hr after the induction of feeding structures [41]. In conclusion, either the nature of the secretions changes during nematode development or the response of the cytoplasm of the feeding cell is dependent on the stage of differentiation. Therefore, the mechanism of feeding must be different between early and advanced stages of nematode development.

The structure and shape of feeding tubes varies in the different species of sedentary nematodes. In general, the tubes form from a small droplet of secretions released through the stylet orifice. By still unknown mechanisms a lumen is created which is surrounded by a wall of osmiophilic material except for the continuous connection to the stylet orifice and stylet lumen. The wall material is considerably different between root-knot and cyst nematodes. In root-knot nematodes it is thick and has a regular cristalline structure [18] whereas in cyst nematodes it consists of an electron-dense amorphous material [11, 41, 47].

In general, around the feeding tube aggregations of endoplasmic reticulum (ER) are observed. In root-knot nematodes no apparent contact between the lumina of the feeding tubes and the surrounding ER system could be detected (see Bleve-Zacheo and Melillo, this volume). However, on the base of electron-micrographs Sobczak [41] was able to show that feeding tubes formed by *H. schachtii* are intruded by ER tubules.

4.2. FUNCTION

Feeding tubes are attached to the stylets of sedentary endoparasitic nematodes during the periods of active feeding. The close contact of the feeding tubes with the stylet orifice of the feeding nematodes has led to the assumption that these structures are essential for the effective withdrawal of solutes from the feeding sites by the parasitic nematodes [18, 47, 48]. The function(s) of the feeding tubes, however, are still a matter of speculation. Razak and Evans [33] proposed that feeding tubes might act as cytoplasmic filters preventing the blockage of the stylet aperture by cell organelles. By means of electron and video microscopy it was shown that the cytoplasm around the feeding tubes of *H. schachtii* is modified as it lacks larger cell organelles [47, 48]. Based on these observations, the action of enzymes was proposed, which are released by the tubes and predigest the surrounding cytoplasm in order to facilitate solute flow from the cytoplasm into the feeding tubes and further into the nematode. Similar zones of modified cytoplasm around feeding tubes are reported from other cyst and root-knot nematodes [18, 35].

Feeding tubes serving as filters for the selective transfer of nutrients from the cytoplasm of the feeding site into the stylet lumen were first proposed by Grundler and Wyss [17]. Supporting evidence for this hypothesis was provided by microinjection experiments, which were mainly performed in order to obtain first information about the nature of nutrients withdrawn by *H. schachtii* from its feeding sites [4]. Microinjections of fluorescent-labelled substances into syncytia of *H. schachtii* indicated the existence of a size exclusion limit (SEL) for the uptake of solutes from the syncytia by the nematodes. Microinjected low molecular fluorochromes such as Lucifer Yellow CH (M_r 457) and fluorescent dextrans of molecular weights up to 20 kD were withdrawn by actively pumping *H. schachtii* juveniles and adult females (Figure 1), whereas larger fluorescent dextrans of 40 kD and higher molecular weights were not ingested by feeding nematodes. As the size of a dextran is modelled by its Stokes radius [31], this apparent upper size limit for the uptake of solutes from syncytia by *H. schachtii* corresponds to a Stokes radius between approximately 3.2 and 4.4 nm [4]. It has yet to be determined if this level of the SEL is also true for other organic solutes such as proteins.

The orifice of the stylet of *H. schachtii* has a diameter of approximately 100 nm [47] suggesting that all microinjected dextrans should be able to pass. Therefore, the SEL must be due to an additional sieving mechanism at the interface between the syncytial cytoplasm and the nematode stylet. A biomembrane, somehow covering the stylet orifice seems very unlikely in this respect, since all microinjected substances are membrane-impermeable and were readily taken up by the feeding nematodes. More probable is, however, that the feeding tubes serve as molecular filters, creating the SEL for the uptake of nutrients from the feeding sites by the nematodes. Possibly, the frequent replacement of the feeding tube during the feeding cycle (see 4.1.) can be attributed to a blockage of the sieve mechanism during usage.

Microinjection studies with giant cells of root-knot nematodes have as yet not been performed. Although there are obvious differences in the structure of feeding tubes of cyst and root-knot nematodes (see 4.1.) their function might be similar but varying in SEL.

Very striking is the obvious close connection of the feeding tubes of cyst and root-knot nematodes with the elaborate ER of the feeding structures (see 4.1.). Functions of the ER include intracellular communication and transport as well as the synsthesis and modification of different substances, for example membrane components and proteins [7]. With regard to the nutrient supply of the nematode the ER might provide specific solutes which are substantial components of the nematode diet. The close association of the ER with the feeding tube might facilitate a quick and directed transport of these nutrients for withdrawal by the parasite.

The SEL and the observed connections between the ER and the feeding tubes are in contradiction to each other, unless two different mechanisms contribute simultaneously to the nematode nutrient supply.

5. Nutrient and Water Supply of Feeding Sites

The feeding structures and the associated nematodes in the roots of infested plants demand a considerably high amount of nutrients and water. A typical trait of these structures is their high metabolic activity. The amount of food taken up daily by a female *H. schachtii* has been calculated by Sijmons *et al.* [39] to be equivalent to four times the volume of the syncytium. As nematodes take up all nutrients as an aqueous solution their high demand for water is obvious.

Evidence for the accumulation capacity of the feeding sites for water and solutes was provided by the analysis of the water relation parameters turgor and osmotic potential in *H. schachtii*-infested roots of *A. thaliana*. Turgor pressure differences of 5000 - 7000 hPa and differences of the osmotic potential of 8000 hPa between syncytia and surrounding non-infested root cells established that the syncytia are very strong sinks for water and solutes [3]. Obviously, under the influence of the nematode the turgor pressure and osmotic potential of the modified plant cells are raised to new equilibria without changing the viability of the cells. Due to this increased sink strength of syncytia, the feeding sites are even under conditions of water and/or nutrient stress much better supplied with water and solutes than non-infested root cells. How the nematode triggers these changes in the water relations on the cellular level remains obscure.

5.1. TRANSFER OF ASSIMILATES INTO FEEDING SITES

On the basis of the elaborate cell wall protuberances of the feeding sites of sedentary nematodes towards xylem elements in the vascular cylinder, syncytia and giant cells were described to behave functionally as xylem-related transfer cells [21]. Therefore, for a long time the xylem was regarded as the main source for the nutrients of feeding cells and associated nematodes [19, 38]. However, the concept of feeding sites functioning as xylem-related transfer cells was only based on morphological studies. Physiological transport studies to confirm this hypothesis had not been performed. Besides, due to the low concentration of organic solutes in the xylem stream it seemed questionable whether transfer processes from the xylem into feeding sites of cyst and root-knot nematodes alone can meet the high demand for assimilates.

The phloem as the main source for the translocation of nutrients into feeding sites of sedentary nematodes was first discussed by Bird and Loveys [2] and McClure [24] who observed an accumulation of radioactive assimilates in feeding cells and associated root-knot nematodes after exposure of the infested host plants to $^{14}CO_2$. The relation between the feeding sites and the phloem elements, however, remained unclear. Dorhout *et al.* [8] used the fluorochrome 5(6) carboxyfluorescein (CF) as a phloem tracer and showed a transfer of the dye into the giant cells and associated individuals of the root-knot nematode *M. incognita*. They concluded a symplastic pathway between the phloem elements and the giant cells, although the cellular pathway was not studied in detail.

Transport studies using both CF and radioactively labelled sucrose established that syncytia of the cyst nematode *H. schachtii* act as major sinks for phloem-derived solutes [5]. After loading the phloem of *H. schachtii* infested *A. thaliana* plants with CF and ^{14}C sucrose both tracers accumulated in syncytia and associated feeding nematodes. Moreover, confocal laser scanning microscopy revealed that *H. schachtii* induces anormalous phloem unloading in roots of *A. thaliana*. Phloem unloading in non-infested roots of *A. thaliana* takes place only at the root apex in the zone of elongation (Figure 2) and at developing lateral root primordia. This phloem unloading in intact roots appears to follow the symplastic pathway [27]. However, in roots of *A. thaliana* which were parasitized by *H. schachtii*, localized phloem unloading of CF was observed into syncytia, no matter where they were situated along the root system (Fig. 3). This transfer of solutes between phloem elements and syncytia was proved to be unidirectional. Microinjection of various fluorochromes into syncytia of *H. schachtii* revealed that these feeding structures are symplastically isolated within the root tissue (Fig. 4). A translocation of the microinjected fluorescence tracers out of the syncytia into neighbouring cells or elements of the vascular bundle was never observed [4].

The mechanisms for this irregular phloem unloading, which are obviously induced by *H. schachtii*, are still a matter of speculation. A symplastic pathway for the transport of solutes from phloem elements into the syncytium is very unlikely due to the totally unidirectional nature of this transfer. Moreover, ultrastructural analyses showed that functional plasmodesmata are missing in the cell walls of syncytia of *H. schachtii* [14, 42]. Localized apoplastic phloem unloading might be achieved by an inhibition of a proton pump, which promotes the retrieval of sucrose from the apoplast into the phloem elements [30]. The sucrose which diffuses passively out of the phloem could then be transported across the plasma membrane of the syncytium by a putative sucrose carrier. It is questionable, however, if the inhibition of the proton pump would lead to a local acidification in the phloem sufficient to induce the passive leakage of undissociated CF to the apoplast. How the nematode might trigger this inhibition of the proton pump is also obscure. Alternative hypotheses for the apoplastic transfer of both CF and sucrose from the phloem into syncytia include the action of putative non-specific solute pores in the plasma membranes of the phloem elements and/or the syncytium or the action of

113

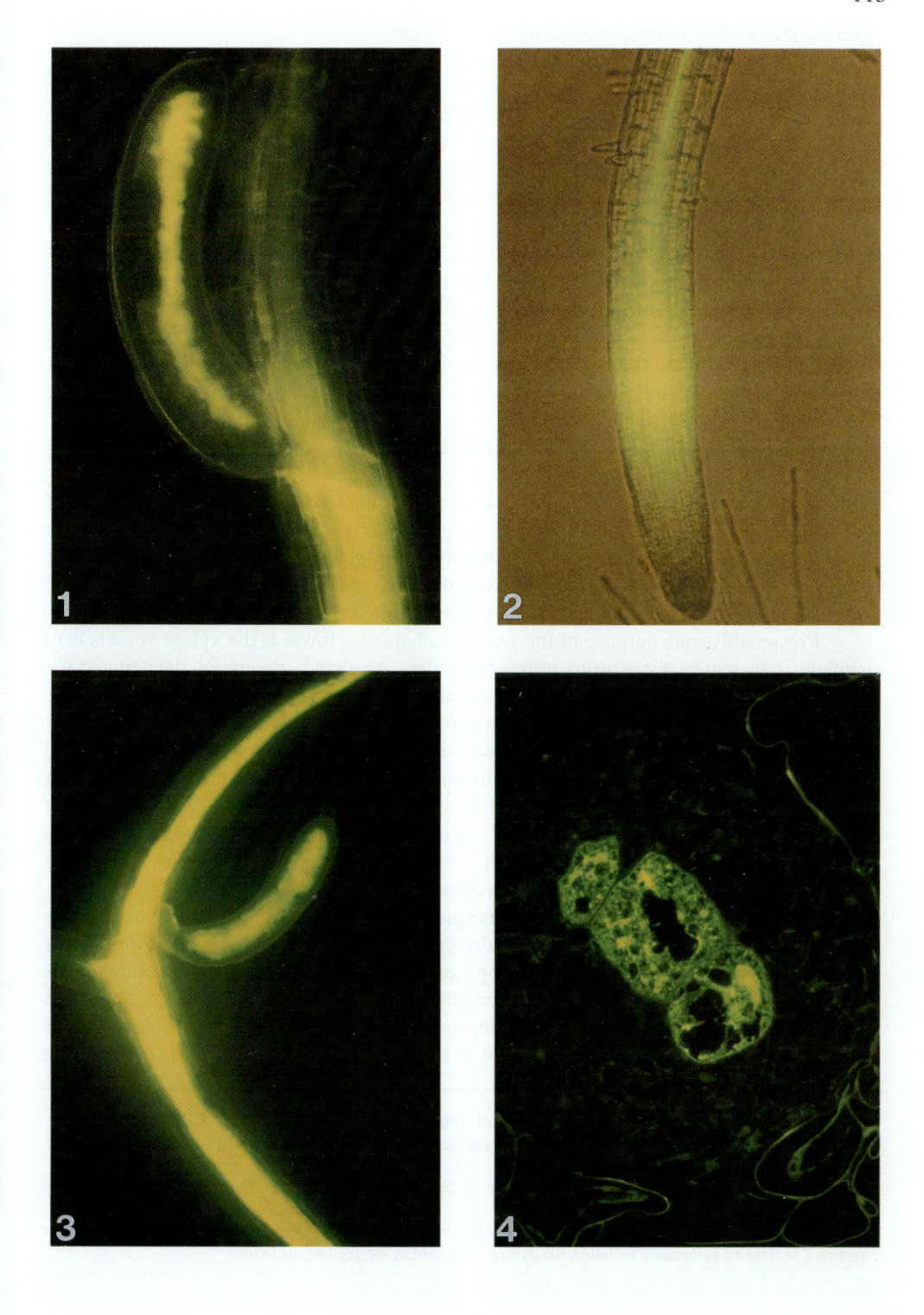

specific plasmalemma-integrated channels or carriers.

Conventional physiological analyses alone will probably not suceed in solving the question of which mechanisms are involved in the transfer of solutes between the phloem and syncytia. New possibilities are offered by combining physiological analyses with techniques of molecular biology and genetic engineering. For example, the role of sucrose carriers in the syncytial plasmalemma in the transfer process might be clarified by performing phloem translocation studies with transformed plants, which exhibit no or a stronger expression of the specific carrier gene. For this purpose antisense and sense transformations with known genes which are expressed under the control of syncytium-specific promoters must be performed. A number of sucrose carriers of *A. thaliana* are already genetically characterized [36].

Irrespective of the actual mechanism of phloem unloading at the feeding sites, the investigations indicate the important role of phloem-derived solutes for the nutrient supply of giant cells and syncytia.

5.2. TRANSFER OF WATER INTO FEEDING SITES

Wall ingrowths are usually interpreted as sites of enhanced short distance transport of solutes [21]. Although the phloem must be regarded as the main source of assimilates for feeding sites and associated nematodes the xylem might also contribute to the nutrient supply of syncytia and giant cells. However, the close contact of the feeding sites and the xylem elements indicates a high demand of the feeding cells for ions.

Presumably, one purpose of the wall protuberances towards the xylem vessels is to mediate an increased water influx into feeding sites. As already described, the demand for water of the feeding nematodes is considerably high due to the consumption of large amounts of solutes. Furthermore, syncytia are strong sinks for water which is indicated by their enhanced turgor pressure and osmotic potential (see above). Dorhout *et al.* [9] found a high transfer rate of the fluorochrome Tinopal CBS from xylem elements into the giant cells of *M. incognita*. From this observation they concluded a high influx of water and solutes into feeding sites of *M. incognita* from the xylem. To what extent the fluorescence marker Tinopal CBS with a molecular weight of 562.6 D is suited to monitor water flux was not determined in detail.

However, in infested roots, as a consequence of feeding structure formation, there is a notable reduction of the number of continuous xylem vessels [14, 46]. Therefore, water may be a limiting factor in the physiology of feeding sites and the wall protuberances might be structures that specifically compensate this limitation. As the syncytial and giant cell plasmalemma lines the wall ingrowths, it can locally be amplified up to 15 times [20]. This increase of the interface to the xylem vessels is likely to serve for the

---

Figure 1. H. schachtii J3 after uptake of a 10 kD dextran labelled with Lucifer Yellow CH.

Figure 2. Root apex of *A. thaliana* with regular symplastic phloem unloading of 5(6) carboxyfluorescein in the elongation zone.

Figure 3. Nematode-induced phloem unloading of 5(6) carboxyfluorescein into the syncytium of *H. schachtii* in a root of *A. thaliana*.

Figure 4. Cross section of a syncytium of *H. schachtii* in a root of *A. thaliana* after microinjection of Lucifer Yellow CH. Due to symplastic isolation the dye remains within the feeding structure.

supply of feeding structures and associated nematodes with water. In addition, specific water channels might be incorporated in these plasmalemma protuberances, which would lead to a considerable increase of their water conductivity. An indication for this mechanism of increased influx of water into nematode feeding cells was provided by Opperman et al. [28], who isolated a root-specific and nematode-responsive promoter that regulates the expression of a putative water channel of the plasmalemma. The expression of the corresponding gene *TobRB7* was shown to be enhanced in giant cells induced by *M. incognita* in roots of tobacco, but not by cyst nematodes [28, 40].

6. Nutritional Effects on Nematode Development

As biotrophic parasites sedentary plant-parasitic nematodes are essentially dependent on the nutrient supply provided by the host plant. As shown above the specific feeding structures obviously improve the efficacy of the nutrient supply by a number of physiological changes at the plant side. Assimilates, water and proteins are transferred and enriched for the benefit of the parasites. The concentration of proteins in the syncytia of *H. schachtii* was determined to be more than four times higher than that of the control tissue [37]. Electrophoretic separations of protein extracts from syncytia revealed that the protein composition of the feeding structures are clearly different from that of control tissue. Presumably, the parasites are able to optimize their diet by upregulating benefitial proteins and downregulating redundant proteins [37].

On the other hand the nematodes themselves have developed adaptations to variable nutritional conditions and environmental factors that influence the metabolism of their host plants. Under adverse conditions the growth of the nematodes can be reduced leading to a retarded growth rate and smaller mature animals [22]. A specific adaptation is the environmentally controlled sex determination that has been observed in a number of species in root-knot and cyst nematodes [32, 44]. The relation between females and males is obviously influenced by factors that limit the nematode nutrient supply. Under unfavourable conditions male development is triggered whereas under favourable conditions female development is supported. This relation reflects the different demands for food which has been calculated in *H. schachtii* to be about 40 times lower in males than in females [26]. Accordingly, the volume of the induced syncytia is smaller [6, 22] and the anatomical structure of the syncytia is different [42].

Male development can often be triggered by heavy nematode infestation of the plant so that the parasitising juveniles are exposed to a high level of intraspecific competition. Under these conditions the proportion of males in the nematode population dominates in root-knot and cyst nematodes. Factors that reduce plant growth have a similar effect. Under conditions of limited nutrient supply, treatment with herbicides or anti-metabolites (reviewed in [44]), and manipulations of the plants such as removal of the shoot or just advanced plant age [15, 41] the number of females decreases whereas the number of males increases.

A number of experimental data support the concept that nutritional factors are decisive about the development of the nematode juveniles´ sex. The unsexed infective juvenile enters the root, induces a feeding structure and is then able to either follow female or male development. This mechanism is obviously restricted to the early development of the parasitising juveniles. In *H. schachtii* under optimal conditions more than 90 percent of the invaded juveniles were observed to develop to females [16]. When the nutrient medium of the plants was changed from optimal to adverse conditions about 5 days after syncytium induction, the specific adaptation seemed to be no longer possible and most of the female juveniles died. Root-knot nematodes seem to possess a certain flexibility as a reversal of the sexual organs from female to male gonads could be observed following the change to unfavourable conditions [29]. As yet, the determinative factors in the environmental control of sex determination are not known. With regard to nutrition it has still to be clarified whether qualitative or quantitative factors are important. Betka *et al.* [1] measured in syncytia of stagnating females an increased relative concentration of certain free amino acids (lysine, methionine, phenylalanine, tryptophan) suggesting that they are syncytial components inhibiting nematode development.

Sobczak *et al.* [42] observed that under conditions supporting male development procambial cells of *A. thaliana* responded hypersensitively when selected for syncytium induction by *H. schachtii* juveniles, whereas pericycle cells allowed syncytium induction. They raised the question whether the particular response of the root tissue selected for the induction of a feeding structure determines male or female development. It might well be that induced or natural plant physiological conditions support, limit or prevent the early development of feeding structures.

In resistant plants almost only males develop to the adult stage (e.g. [25]). The analysis of the resistance mechanisms revealed that feeding structures are induced but degrade in the course of differentiation in earlier or later stages [10, 13, 47]. Also under these conditions the mechanism of sex determination leads to the development of those juveniles that have a reasonable chance to complete their life cycle depending on the speed of degradation of the feeding structure.

Future research should aim to establish exactly controlled experimental conditions that allow to determine the role of single nutritional factors in the course of nematode development.

7. Conclusions

In order to conceive the interaction between nematodes and their host plants the physiology of the complex system of transport, metabolism and consumption has to be analysed in detail. The extremely small size of the feeding structures in relation to the whole plant and the obligat mode of parasitism of sedentary nematodes make direct analyses very difficult. Molecular biological methods are limited to the characterization of genes, while physiological analyses alone come up against limiting factors due to the complexity of the system. Therefore, a comprehensive approach combining conventional physiological techniques with molecular biological methods and genetic engineering seems to be most promising to develop theoretical models that explain the function and regulation of single elements of the system - an essential basis to develop new concepts in nematode control.

ACKNOWLEDGEMENTS

The authors thank Dr. Jens Aumann for critically reading the manuscript and Petra Sedlag for preparing the figures.

References

1. Betka, M., Grundler, F.M.W., and Wyss, U. (1991) Influence of changes in the nurse cell system (syncytium) on the development of the cyst nematode *Heterodera schachtii*, *Phytopathology* **81**, 75-79.
2. Bird, A.F. and Loveys, B.R. (1975) The incorporation of photosynthates by *Meloidogyne javanica*. *J. Nematol.* **7**, 112-113.
3. Böckenhoff, A. (1995) Untersuchungen zur Physiologie der Nährstoffversorgung des Rübenzystennematoden *Heterodera schachtii* und der von ihm induzierten Nährzellen in Wurzeln von *Arabidopsis thaliana* unter Verwendung einer speziell adaptierten *in situ* Mikroinjektionstechnik. PhD Thesis, Institut für Phytopathologie der Christian-Albrechts-Universität Kiel, Germany.
4. Böckenhoff, A. and Grundler, F.M.W. (1994) Studies on the nutrient uptake of the beet cyst nematode *H. schachtii* by *in situ* microinjection of fluorescent probes into the feeding structures in *Arabidopsis thaliana*, *Parasitology* **109**, 249-254.
5. Böckenhoff, A., Prior, D.A.M., Grundler, F.M.W., and Oparka, K.J. (1996) Induction of phloem unloading in *Arabidopsis* roots by the parasitic nematode *Heterodera schachtii*, *Plant Physiol.*, Fehler.
6. Caswell-Chen, E.P. and Thomason, I.J. (1993) Root volumes occupied by different stages of *Heterodera schachtii* in sugarbeet, *Beta vulgaris*, *Fund. Appl. Nematol.* **16**, 39-42.
7. Chrispeels, M.J. (1980) The endoplasmic reticulum, in N.E. Tolbert (ed.), *The Biochemistry of Plants*, Volume 1, Academic Press, New York, pp. 390-412.
8. Dorhout, R., Gommers, F.J., and Kollöffel, C. (1993) Phloem transport of carboxyfluorescein through tomato roots infected with *Meloidogyne incognita*, *Physiol. Mol. Plant Pathol.* **43**, 1-10.
9. Dorhout, R., Kollöffel, C., and Gommers, F.J. (1988) Transport of an apoplastic fluorescent dye to feeding sites induced in tomato roots by *Meloidogyne incognita*, *Physiol. Mol. Plant Pathol.* **78**, 1421-1424.
10. Dropkin, V.H., Davis, D.W., and Webb, R.E. (1967) Resistance of tomato to *Meloidogyne incognita acrita* and to *M. hapla* (root-knot nematodes) as determined by a new technique, *Proc. Amer. Soc. Horticult. Sci.* **90**, 316-323.
11. Endo, B.Y. (1987) Ultrastructure of esophageal gland secretory granules in juveniles of *Heterodera glycines*, *J. Nematol.* **19**, 469-483.
12. Endo, B.Y. (1991) Ultrastructure of initial response of susceptible and resistant soybean roots to infection by *Heterodera glycines*, *Revue Nématol.* **14**, 73-94.
13. Golinowski, W. and Magnusson, C. (1991) Tissue response induced by *Heterodera schachtii* (Nematoda) in susceptible and resistant white mustard cultivars, *Can. J. Bot.* **69**, 53-62.
14. Golinowski, W., Grundler, F.M.W., and Sobczak, M. (1996) Changes in the structure of *Arabidopsis thaliana* induced during development of females of the plant parasitic nematode *Heterodera schachtii*, *Protoplasma*, **194**, 103-116.
15. Grundler, F.M.W. (1989) Untersuchungen zur Geschlechtsdetermination des Rübenzystennematoden *Heterodera schachtii* Schmidt, *PhD Thesis*, Institut für Phytopathologie der Christian-Albrechts-Universität Kiel, Germany.
16. Grundler, F.M.W., Betka, M., and Wyss, U. (1991) Influence changes in the nurse cell system (syncytium) on sex determination and development of the cyst nematode *Heterodera schachtii:* Total amounts of proteins and amino acids, *Phytopathology* **81**,70-74.
17. Grundler, F.M.W. and Wyss, U. (1994) Strategies of root parasitism by sedentary plant parasitic nematodes, in U.S. Singh, R.P. Singh and K. Kohmoto (eds.), *Pathogenesis and Host specificity in Plant diseases*, Vol. 2 Eukaryotes, Pergamon Press, Oxford, pp. 309-319.
18. Hussey, R.S. and Mims, C.W. (1991) Ultrastructure of feeding tubes in giant-cells induced in plants by the root-knot nematode *Meloidogyne incognita*, *Protoplasma* **162**, 99-107.
19. Jones, M.G.K. (1981) The development and function of plant cells parasitized by endoparasitic

nematodes, in B.M. Zuckerman and R.A. Rohde (eds.), *Plant Parasitic Nematodes*, Vol. 3, Academic Press, London, pp. 255-279.
20. Jones, M.G.K. and Dropkin, V.H. (1976) Scanning electron microscopy of nematode-induced giant transfer cells, *Cytobios* **15**, 149-161.
21. Jones, M.G.K. and Northcote, D.H. (1972) Nematode-induced syncytium - A multinucleate transfer cell, *J. Cell Sci.* **10**, 789-809.
22. Kerstan, U. (1969) Die Beinflussung des Geschlechterverhältnisses in der Gattung *Heterodera* II. Minimallebensraum - Selektive Absterberate der Geschlechter - Geschlechterverhältnis *(Heterodera schachtii)*, *Nematologica* **15**, 210-228.
23. Linford, M.B. (1937) The feeding of the root-knot nematode in root tissue and nutrient solution, *Phytopathology* **27**, 824-834.
24. McClure, M.A. (1977) *Meloidogyne incognita*: A metabolic sink, *J. Nematol.* **9**, 89-90.
25. Müller, J. (1985) Der Einfluß der Wirtspflanze auf die Geschlechtsdeterminierung bei *Heterodera schachtii*, *Beitr. Nematodenforsch.* **226**, 46-63.
26. Müller, J., Rehbock, K., and Wyss, U. (1981) Growth of *Heterodera schachtii* with remarks on amounts of food consumed, *Revue Nématol.* **4**, 227-234.
27. Oparka, K.J., Duckett, C.M., Prior, D.A.M., and Fisher, D.B. (1994) Real-time imaging of phloem unloading in the root tip of *Arabidopsis*, *Plant J.* **6**, 759-766.
28. Opperman, C.H., Taylor, C.G., and Conkling, M.A. (1994) Root-knot nematode-directed expression of a plant root-specific gene, *Science* **263**, 221-223.
29. Papadopoulou, J. and Triantaphyllou, A.C. (1982) Sex differentiation in *Meloidogyne incognita* and anatomical evidence of sex reversal, *J. Nematol.* **14**, 549-566.
30. Patrick, J.W. (1990) Sieve element unloading: cellular pathway, mechanism and control, *Physiol. Plant.* **78**, 298-308.
31. Peters, R. (1986) Fluorescence microphotolysis to measure nucleo-cytoplasmic transport and intracellular mobility, *Biochim. Biophys. Acta* **864**, 305-359.
32. Poinar, G.O. and Hansen, E. (1983) Sex and reproductive modifications in nematodes, *Helminthol. Abstr.* **B 52**, 145-163.
33. Razak, A.R. and Evans, A.A.F. (1976) An intracellular tube associated with feeding by *Rotylenchulus reniformis* on cowpea root, *Nematologica* **22**, 182-189.
34. Rebois, R.V. (1980) Ultrastructure of a feeding peg and tube associated with *Rotylenchulus reniformis* in cotton, *Nematologica* **26**, 396-405.
35. Rumpenhorst, H.J. (1984) Intracellular feeding tubes associated with sedentary plant parasitic nematodes, *Nematologica* **30**, 77-85.
36. Sauer, N., Baier, K., Gahrtz, M., Stadler, R., Stolz, J., and Truernit, E. (1994) Sugar transport across the plasma membrane of higher plants, *Plant Mol. Biol.* **26**, 1671-1679.
37. Schmidt, K.-P. (1995) Proteinanalytische Charakterisierung pathogenspezifischer Vorgänge im Wurzelgewebe von *Arabidopsis thaliana* nach Infektion mit dem Rübenzystennematoden *Heterodera schachtii*, PhD Thesis, Institut für Phytopathologie der Christian-Albrechts-Universität Kiel, Germany.
38. Sijmons, P.C., Atkinson, H.J., and Wyss U. (1994a) Parasitic strategies of root nematodes and associated host cell responses, *Annu. Rev. Phytopathol.* **32**, 235-259.
39. Sijmons, P.C., Grundler, F.M.W., von Mende, N., Burrows, P.R., and Wyss, U. (1991) *Arabidopsis thaliana* as a new model host for plant parasitic nematodes, *Plant J.* **1**, 245-254.
40. Sijmons, P.C., von Mende, N., and Grundler, F.M.W. (1994b) Plant-parasitic nematodes, in E.M. Meyerowitz and C.R. Somerville, *Arabidopsis*, Cold Spring Harbor Laboratory Press, Cold Spring Harbor, New York, pp. 749-767.
41. Sobczak, M. (1996) Investigations on the structure of syncytia in roots of *Arabidopsis thaliana* induced by the beet cyst nematode *Heterodera schachtii* and its relevance to the sex of the nematode, *PhD Thesis*, Institut für Phytopathologie der Christian-Albrechts-Universität Kiel, Germany.
42. Sobczak, M., Grundler, F.M.W., and Golinowski, W. (1996) Changes in the structure of *Arabidopsis thaliana* induced during development of males of the plant parasitic nematode *Heterodera schachtii*, *Europ. J. Plant Pathol.*, in press.
43. Steinbach, P. (1973) Untersuchungen über das Verhalten von Larven des Kartoffelzystenälchens (*Heterodera rostochiensis* Wollenweber, 1923) an und in Wurzeln der Wirtspflanze *Lycopersicon esculentum* Mill. III. Die Nahrung von Kartoffelnematodenlarven, *Biol. Zbl.* **92**, 563-582.
44. Triantaphyllou, A.C. (1973) Environmental sex differentiation of nematodes in relation to pest

management, *Annu. Rev. Phytopathol.* **11**, 441-463.
45. Wyss, U. (1992) Observations on the feeding behaviour of *Heterodera schachtii* throughout development, including events during moulting, *Fund. Appl. Nematol.* **15,** 75-89.
46. Wyss, U., Grundler F.M.W., and Münch A. (1992) The parasitic behaviour of second-stage juveniles of *Meloidogyne incognita* in roots of *Arabidopsis thaliana, Nematologica* **38**, 98-111.
47. Wyss, U., Stender, C., and Lehmann, H. (1984) Ultrastructure of feeding sites of the cyst nematode *Heterodera schachtii* Schmidt in roots of susceptible and resistant *Raphanus sativus* L. var. *oleiformis* Pers. cultivars, *Physiol. Plant Pathol.* **25,** 21-37.
48. Wyss U. and Zunke, U. (1986) Observations on the behaviour of second stage juveniles of *Heterodera schachtii* inside host roots, *Revue Nématol.* **9**, 153-165.

CELL CYCLE REGULATION IN NEMATODE FEEDING SITES

Godelieve GHEYSEN, Janice DE ALMEIDA ENGLER, and Marc VAN MONTAGU
Laboratorium voor Genetica, Department of Genetics, Flanders Interuniversity Institute for Biotechnology (VIB), Universiteit Gent, K.L. Ledeganckstraat 35, B-9000 Gent, Belgium

Abstract

Sedentary endoparasitic nematodes induce multinucleate feeding cells in the roots of their host plants. These cells undergo multiple rounds of shortened cell cycles leading to genome amplification and hypertrophy of the cytoplasm. After explaining the specific terminology involved, this chapter reviews the cytological observations in giant cells induced by root-knot nematodes and syncytia induced by cyst nematodes. Recent molecular research into cell cycle regulation in *Arabidopsis thaliana* has been extended from normal plant development to the changes induced by these sedentary nematodes. In this way it has been demonstrated that several cell cycle genes are activated in developing feeding cells. To analyze the importance of cell cycle activation for the establishment of feeding cells, cell cycle-inhibiting drugs have been used and experiments with plants producing dominant mutant proteins will be performed in the near future.

1. Introduction

Several plant-parasitic nematode genera have evolved the ability to induce morphological changes in host cells to form feeding sites. The classification of these root parasites and their life cycles have been reviewed by Sijmons *et al.* [48]. Here, we will concentrate on the cytological and especially nuclear changes that are induced by two main groups of sedentary endoparasites: root-knot and cyst nematodes. After an initial invasion and migration towards a suitable site in the plant root, these nematodes become immobile and depend completely on the successful induction and maintenance of specialized feeding cells. This intimate relationship persists for one to several months until completion of the nematode's life cycle. These parasites are therefore biotrophic; they do not kill the cells they feed from but instead modify them into efficient food sources, most probably by the injection of unknown substances originating from their oesophageal glands [25].

The mechanism of feeding site formation is different and specific for the infecting nematode, regardless of the tissue and host in which they are induced. Root-knot nematodes induce several giant cells embedded in a gall whereas cyst nematodes generate a syncytium. However, the final large and multinucleate feeding cells are functionally similar, in that they are metabolically highly active and adapted to withdraw large amounts of nutrients from the host plant in order to feed the nematode (see Grundler and Böckenhoff, this volume). This functional analogy is reflected in the ultrastructure of the feeding cells: cell wall ingrowths adjacent to vascular tissue, breakdown of the large vacuole, dense granular cytoplasm with many organelles and numerous enlarged amoeboid nuclei [2, 27, 26; see also Golinowski et al., and Bleve-Zacheo and Melillo, this volume]. It is particularly on this last feature that we will focus in this chapter: what is the origin of the multinucleated nature of the feeding cells, how does the nuclear size correlate with the ploidy and the DNA content of the nucleus and how should these changes be interpreted in the light of feeding cell development?

2. Terminology of Mitosis and Genome Multiplication

The main event in mitosis is the doubling of the number of chromosomes. Then, the nuclear envelope disintegrates and a division spindle is formed to equally distribute the doubled chromosomes (chromatids) between the daughter cells. The mitotic cycle (schematically shown in Figure 1) does not always end with cell division. Two main forms of incomplete cell cycles, leading to genome multiplication, are known [9, 6]. These are:

(1) Polyploidizing mitosis
   Mitosis is blocked at or after metaphase and the result is a polyploid cell; examples are:
   - c-mitosis (blocked at metaphase as by colchicine)
   - endomitosis (mitosis within the nucleus)
   - acytokinetic mitosis (mitosis without cell division).
(2) Endocycle, endoreduplication, or polytenization
   The cell cycle is blocked immediately after DNA replication; in fact, there is no mitosis and it results in polytene chromosomes. Several of these shortened cycles can occur, leading to a substantial genome amplification within the cell.

We will explain some of these terms in a little more detail since they are important for describing feeding cell development and they have not always been used in the current meaning. *Acytokinetic mitosis* or mitosis without cell division is caused by incomplete or no cell plate formation and results in a binucleate cell, or a multinucleate when the process is repeated. Since the nuclei of these cells generally divide synchronously, mitotic disturbances, especially spindle fusions (fusion of metaphase or anaphase plates) are not uncommon. Multinucleate cells can also be generated by cell fusion. This can be *refusion*, a process very similar to acytokinetic mitosis, since the two daughter cells from the mitosis remain connected by a cytoplasmic bridge and soon re-unite. *Cell fusion* between adjacent non-sister cells is a completely different phenomenon, totally unrelated to mitosis, but also resulting in a bi- or multinucleate cell. *Endomitosis* is mitosis within a nucleus: the chromosomes are doubled, condensed, and separated inside the nuclear envelope without a spindle being formed or the nucleolus being destroyed [14]. If this happens in a normal diploid cell, it results in a (mononucleate) cell with a polyploid nucleus. It is important to know that endomitosis has previously been used as a synonym for endoreduplication and has also been confused sometimes with acytokinetic mitosis. *Endoreduplication* or *polyteny* is the (repeated) doubling of the number of chromatids in the interphase nucleus without their subsequent spiralization and division, leaving the chromosome number unchanged. Polytene nuclei of plant cells, like those of most animal cells, differ in structure from the well known polytene nuclei of dipteran cells. Instead of the clearly visible large chromosomes with distinct bands, bundles of chromatin threads with vague bands or even a completely diffuse chromatin structure are typical of the nuclei that have accumulated huge quantities of DNA without undergoing mitosis. Therefore, in plants this phenomenon is sometimes called masked or cryptic polyteny.

When nuclei are larger than normal, often the best way to distinguish polyploid from polytene nuclei is the number and size of the chromocenters. Chromocenters are compact chromatin regions and usually their number corresponds to the number of chromosomes. In polytene nuclei only the size of the chromocenters increases, in polyploid nuclei their number is increased. It should be noted that chromocenters can sometimes fuse, resulting in fewer and larger chromocenters.

3. Nuclear Changes during Giant Cell Formation

Root-knot nematodes (*Meloidogyne* ssp) transform parenchyma cells of the differentiating vascular cylinder into large hypertrophied multinucleate giant cells. Although there were many early reports that cell wall breakdown and fusion of neighboring cells contribute to their formation (for references, see [28]), it is now generally believed that these giant cells develop by repeated mitosis without cytokinesis [23, 24, 28, 26]. Advanced electron-microscopical techniques could not demonstrate cell wall dissolution in developing giant cells, despite extensive searches. Furthermore, Jones and Payne [28] showed that cell plate vesicles initially lined up between the two daughter nuclei but then dispersed, resulting in the abortion of the new cell plate formation.

Although giant cells with as many as 150 nuclei have been reported in *Glycine max* [10], the mean number of nuclei per mature giant cell is between 30 and 60 in most studied plant hosts [49]. The increase rate of the number of nuclei for all studied plant species is greatest during the first 7 days after inoculation and no mitotic activity was observed in giant cells associated with adult nematodes [49]. In pea giant cells, it was observed that the number of nuclei doubled each day during the period of highest mitotic activity [49].

Cytogenetic studies have shown that the individual giant cell nuclei are polyploid or aneuploid. While Bird [4] used his observation that dividing sets of chromosomes in giant cells of *Vicia faba* could correlate with 2n, 4n, 6n, 8n or 10n as an argument for cell fusion, Huang and Maggenti [23] found total chromosome numbers in *Vicia faba* giant cells of 4n, 8n, 16n, 32n, and 64n, which are consistent with a repeated mitotic sequence. In contrast, the majority of pea giant cell nuclei seem to be aneuploid with

123

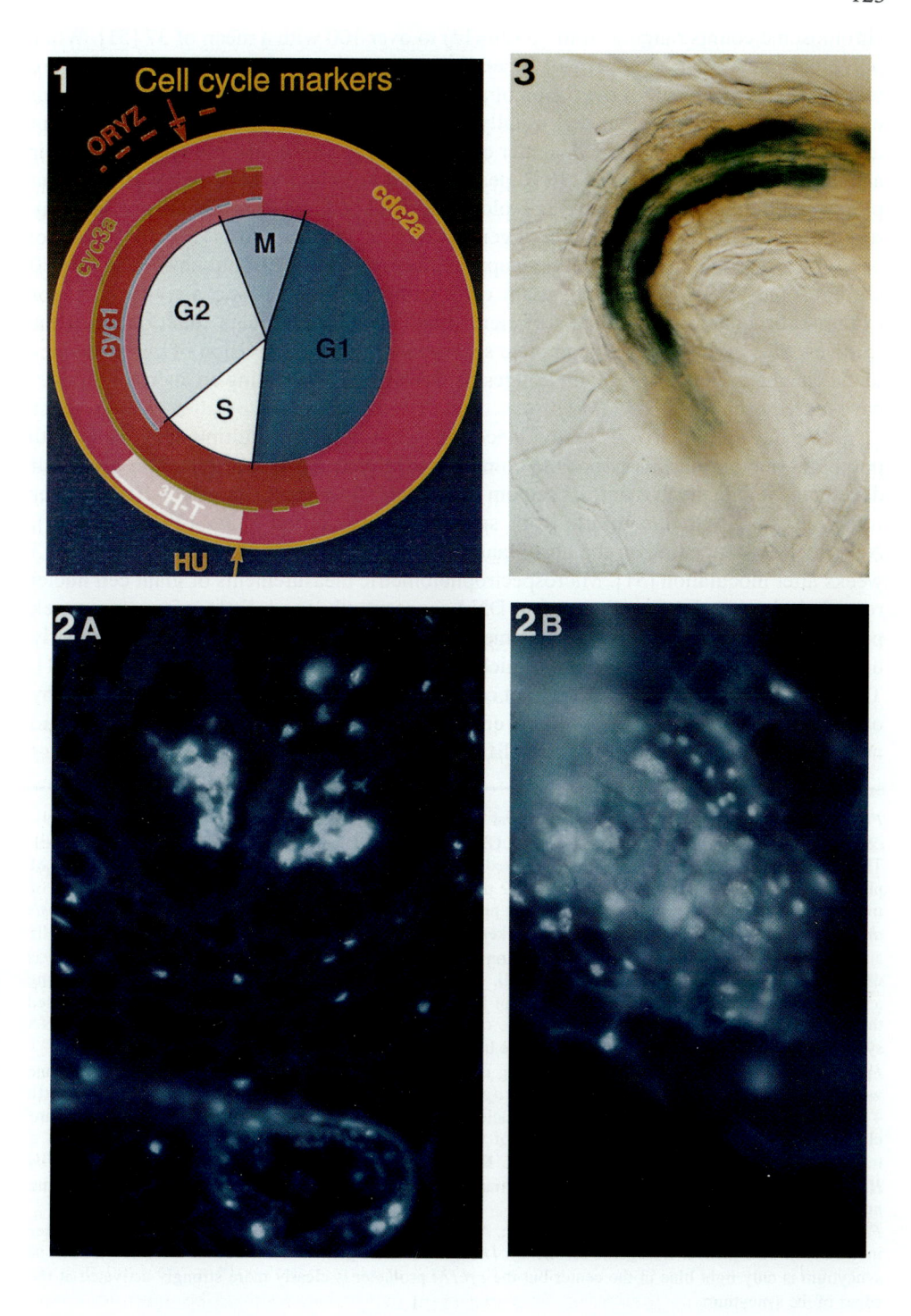

chromosome counts ranging from 15 (2n=14) to over 100 with a mean of 37 [51]. While polyploid nuclei can be generated by endomitosis (see above), in giant cells they are more probably the result of the fusion of mitotic apparatuses. Indeed, since the numerous nuclei in an individual giant cell usually divide synchronously, two or more adjacent metaphase plates can fuse, the daughter chromosome sets subsequently conducting their anaphase migrations together [23]. Repeated fusions of chromosome sets with different ploidy could explain the variable polyploidy of giant cell nuclei, while aneuploidy may result from the unequal distribution of genetic material during mitosis [51]. In the case of pea, an average mature giant cell with approximately 60 nuclei would have nearly 2,220 chromosomes [49]. Developing giant cells with this high mitotic activity may, for example, be unable to produce sufficient tubulin for the adequate functioning of the spindles, thus giving rise to incomplete separation or unequal division of chromosomes to the daughter nuclei. Chromatin bridges between nuclei, indicating an aberrant mitosis, have indeed been reported [23, 51].

We have summarized the evidence that giant cells are multinucleate cells with polyploid nuclei, but this is not the complete story yet. All observations on giant cell development indicate that mitosis seldom occurs later than approximately 5-10 days after infection, i.e., with the moult from second to third larval stage [39, 4, 51, 49]. Nevertheless, depending on the host plant, the DNA content can increase further up to 3 weeks after inoculation [51]. Microspectrophotometric measurements of giant cell nuclei revealed that in lettuce the mean DNA content changed little after one week post-inoculation. In pea, however, the mean DNA content per nucleus increased linearly up to 3 weeks after inoculation, at which stage adult females are already present [51]. This implies that at later stages of giant cell development DNA increase must happen by other methods than mitosis, such as endoreduplication (polytenization of polyploid nuclei occurs in plant cells; 6) or amplification of specific sequences. Amplification of

---

Figure 1. Scheme of the cell cycle. In the central circle, the four successive phases of a standard eukaryotic cell cycle are shown. During interphase (G1, S, G2) the cell grows continuously, during M phase it divides. The S phase is the part of the interphase devoted to DNA synthesis. G1 is the gap between the end of M phase and the beginning of the S phase, the G2 phase is between end of DNA synthesis and beginning of mitosis. In addition, some markers for different phases of the cell cycle are shown. First, three *Arabidopsis thaliana* genes are indicated. *Cdc2aAt* is a marker for the whole cell cycle, it is expressed in dividing cells and also in cells that are competent for division [19]. The *cyc1At* gene encodes a B-type cyclin and is expressed between early G2 and metaphase [13]. *Cyc3a* is another mitotic cyclin (A-type) expressed during a larger time interval [45]. DNA synthesis can be demonstrated by the incorporation of tritium-labeled thymidine and drugs can be used to block the cell cycle at specific stages. Hydroxyurea inhibits DNA synthesis and oryzalin blocks the cell cycle at the beginning of mitosis.

Figure 2. Nuclear changes in nematode feeding cells. **A**, a section through a gall of *A. thaliana* about one week after inoculation with *M. incognita*. The DNA has been stained blue with 4',6-diamidino-2-phenylindole (DAPI). The nuclei of the giant cells (in the top half of the picture) are clustered and clearly larger than the nuclei of the normal root cells. At the bottom of the picture, a transversal section of the nematode can be seen. **B**, a section through a syncytium in *A. thaliana* (infected by *H. schachtii*) approximately two days after initiation. Note the dense cytoplasm and the large size of the nuclei in this developing young syncytium.

Figure 3. Cyc1At-gus expression in a syncytium. An *A. thaliana* root infected by *H. schachtii* (1 week after inoculation). This plant is transgenic for the *cyc1At-gus* chimeric gene and has been stained for GUS. The syncytium is only light blue in the center but the *cyc1At* promoter is clearly more strongly activated at the edges of the syncytium.

four selected sequences could not be demonstrated, but the possibility of specific amplification of other sequences cannot be ruled out [52]. The linearity of DNA increase argues against a true endoreduplication, but this could be explained by the fact that it is the mean which increases linearly, while individual nuclei could be either unchanged or doubled in DNA content, or it is possible that some sequences are under-replicated during the process [51].

We have started an analysis of nuclear and cell cycle changes in *Arabidopsis thaliana*. Because of its characteristics, this small crucifer has become a model plant in molecular genetics [33]. It is also increasingly popular for plant-pathogen studies, and in case of plant-nematode interactions, one of the additional advantages is its simple transparent roots [47, 8]. As would be expected, the giant cells that develop upon root-knot nematode infection in *A. thaliana* are multinucleate, and the nuclei are polyploid (Figure 2). The size of the nuclei is about 10-fold that of the nuclei in root cortex cells. In comparison, this is similar to the relative nuclear size of giant cell nuclei in tomato as opposed to the monstrous proportions (100- to 200-fold increase) of giant cell nuclei in some other hosts [29, 38]. *Arabidopsis thaliana* is not really suitable for cytogenetic studies. The chromosomes are very small and mitotic cycles are short, making it difficult to observe the metaphase chromosomes, even in meristematic cells. However, the chromocenters in the interphase nucleus are clearly visible and in giant cells their number is increased up to ≈40 (2n = 10). Also their size is greater than normal suggesting an additional DNA increase by endoreduplication. ^{3}H-thymidine incorporation indicates DNA synthesis up to approximately 10 days, similar to what has been described for tomato and cotton giant cells [43, 42, de Almeida Engler *et al.*, in preparation).

The differences in giant cell nuclei that have been reported (poly- or aneuploidy; differences in number, size or timing) indicate that nuclear changes during giant cell formation may vary between different hosts, although the general basic mechanism is the same. It is not known what is causing the series of acytokinetic mitoses, but the signal is clearly originating from the root-knot nematode. A continuous stimulus from the nematode is needed for giant cell development [3] and often mitosis and DNA synthesis seem to be limited to only one giant cell at the time, possibly the one the nematode is feeding on at that moment [43, 2].

4. Cytological Observations during Syncytium Formation

The response of host cells to cyst nematode infection is the formation of a syncytium, a large multinucleated hypertrophied cell generated by the fusion of neighboring protoplasts after partial cell wall dissolution. Syncytia develop in vascular parenchyma or procambium cells, sometimes in pericycle cells or in the cortex [11, 27, 31, 17]. In sharp contrast to giant cells, convincing evidence for cell wall breakdown was obtained for syncytia induced in many host plants by different cyst nematodes [40, 11, 27, 26, 31]. As the syncytium grows, cell wall degradation occurs at the extremities, away from the nematodes head, indicating that the wall-digesting enzymes are of plant origin.

No mitosis was seen in syncytia [11], although Piegat and Wilski [40] reported an initial mitotic stimulation during syncytium induction by *Globodera rostochiensis* in mitoses [9]. Interestingly, similarities between tapetum and nematode feeding cells extend from cytological (such as the abundance of mitochondria and endoplasmic reticulum and the presence of many small vacuoles) to molecular characteristics (for example, the down-regulation of the *CaMV 35S* promoter in both organs) [41, 16].

One of the main consequences of polyploidy and/or polyteny is the increase in cell size. Indeed, genome multiplication relaxes restrictions on the expansion of the cytoplasmic mass and makes it possible for large cells to be formed. Since the transcriptional and translational activity increases at each doubling of the genome [5], a highly polyploid cell is functionally equivalent to hundreds or thousands of diploid cells. The small number of large cells compared to a diploid tissue of the same size and overall activity may itself be an advantage. The presence of only few cells could indeed facilitate the regulation of cellular functions in such an organ. And this might certainly be an important factor in nematode feeding cells: it is much simpler for the nematode to control the development and functioning of a few large cells than of many small ones. Furthermore, feeding then only requires careful puncturing of one or a few cells.

Another consequence of polyploidy and especially polyteny is that a shortened cell cycle allows accelerated growth of the organ. Less or no energy and time has to be spent for the production of mitosis-specific proteins and cell wall structures if mitosis and/or cell division is omitted. Not only is the development of the nutritional organ more rapid but the time that the cells are in their active stage is proportionally larger than in a dividing tissue. Polytenizing nuclei are permanently in interphase and are consequently always active. It is clear that a rapidly growing and functional feeding cell is advantageous for the development of the sedentary nematode.

Nematodes are not the only pathogens that influence the cell cycle in plants. Another example are the highly polytene cells in galls induced by the midge *Mayetiola poae* on stems of the grass *Poa memoralis* [21]. In other insect-induced galls, multinucleate cells are formed by acytokinetic and other polyploidizing mitoses [22, 46]. Some plant viruses can also reset the cell cycle. Tomato golden mosaic virus is capable of forcing quiescent leaf cells to progress through the S phase [35] creating an appropriate environment in the nucleus for viral replication. A possible mechanism for such changes is provided by the ability of the C1:C2 protein encoded by the wheat dwarf virus to interact with a retinoblastoma protein which is involved in controlling progression through the cell cycle [7]. It will be interesting to find out how the nematode secretions directly or indirectly influence the activity of cell cycle-controlling proteins in the initiation and development of the nematode feeding site.

7. Strategies to Block the Cell Cycle

Whereas it is obvious that cell cycle activation is involved in feeding site formation, the question is in how far the observed nuclear changes are essential for this cytodifferentiation. Drugs blocking the cell cycle in different stages have been known for long (Figure 1). Oryzalin inhibits plant microtubule polymerization and arrests cells at early M phase (see Figure 1) preventing mitotic separation of the chromatids [34]. Gall formation in cotton roots inoculated with root-knot nematodes was shown to be drastically inhibited by oryzalin [37]. Microscopical analysis revealed that the nematodes entered the root but failed to initiate giant cells, and vascular tissue had differentiated around their heads. This inhibition of giant cell development has been confirmed in *A. thaliana* roots on oryzalin-containing medium whereas control experiments show no or little toxicity of even higher concentrations on the nematodes themselves (de Almeida Engler *et al.*, in preparation).

If mitosis is not involved in syncytium formation, then oryzalin should have no effect on cyst nematode infections. However, syncytia in oryzalin-treated plants are significantly smaller (or develop more slowly) than in control plants (de Almeida Engler *et al.*, in preparation).This suggests that a mitotic inhibition may not affect initiation but inhibit expansion of a syncytium by preventing the surrounding cells to divide or maybe even to be activated for incorporation.

Hydroxyurea is a cytotoxic drug acting as a specific inhibitor of DNA synthesis [53]. If genome multiplication in giant cell and syncytium formation is essential, both types of feeding cells should be blocked in their development by hydroxyurea application. Incubation for 96 hours in concentrations up to 100 ppm were not lethal for the nematodes, which developed normally if transferred afterwards to untreated tomato roots [15]. However, hydroxyurea concentrations as low as 3 ppm inhibited *Meloidogyne* maturation by 70-90% without affecting root development [15]. In the presence of hydroxyurea, root-knot nematodes failed to induce normal feeding cells on tomato. The giant cells were small, their cytoplasm contained few organelles and was highly vacuolated, indicating inhibition of normal giant cell metabolic activities [50]. Furthermore, while males comprised less than 10% of the population of adults in the control roots, they comprised more than 50% in the hydroxyurea-treated tomato root cultures [15].

We have compared the effect of hydroxyurea on cyst and root-knot nematode infection of *A. thaliana* roots. When inoculation of plants was done on hydroxyurea-containing medium, no feeding cells were formed with either nematode, while a 3-day treatment of juveniles did not affect their viability (de Almeida Engler *et al.*, in preparation). A 48-hour hydroxyurea treatment at 3 days post inoculation resulted in decreased gall and syncytium development (observed 5 days after inoculation).

Since it is difficult to exclude any toxic effect of the applied drug concentrations on the nematodes, a much more elegant method to evaluate the necessity of cell cycle progression for the formation of functional feeding cells, is the use of plants producing cell cycle-related dominant mutant proteins. Based on known mutations in yeast CDKs, two mutant *cdc2aAt* genes were constructed by Hemerly *et al.* [20] and introduced into plants. Only the second mutant termed D-N, which led to a complete loss of kinase activity in yeast and to cell cycle arrest [31, 32] resulted in drastic developmental effects. *A. thaliana* plants expressing this dominant negative D-N mutation could not be regenerated, demonstrating the obligatory role of *cdc2aAt* in cell division [20]. Transformation of this mutation into tobacco led to the recovery of a few plants that developed apparently normally, but were smaller than usual. Interestingly, the number of cells building up their organs was considerably smaller than in wild-type plants, and, to compensate for this consequence of less cell divisions, the cell size in these plants was very large [20]. These preliminary results are promising enough to invest energy into these and other mutant plants. Recently, a number of novel dominant mutations in the *Schizosaccharomyces pombe CDC2* gene have been described [30]). It is of considerable interest to construct similar mutations in plant CDKs and to study their effect on cell division, plant development, and nematode infection. If expressed behind a feeding cell induced promoter — several of them are available now (Barthels *et al.*, in preparation; see also Fenoll *et al.*, this volume) — these mutants are expected to specifically inhibit cell cycle progression during nematode feeding site initiation.

8. Conclusions

Cell cycle regulation and differentiation are tightly coupled aspects of plant development. The common occurrence of shortened cell cycles in the formation of specific nutritional plant organs has been exploited by sedentary nematodes for establishing highly efficient feeding cells. It has been shown that induction of these feeding cells is correlated with activation of different cell cycle genes. These molecular data confirm and extend previous cytological observations. Further cell cycle research during the plant—nematode interaction will not only lead to more insights into the development of the feeding cells but also into the function of specific proteins in cell cycle regulation.

ACKNOWLEDGEMENTS

The authors thank Dirk Inzé and Gilbert Engler for critical reading of the manuscript, and Martine De Cock for help preparing it. This work was supported by grants from the Belgian Programme on Interuniversity Poles of Attraction (Prime Minister's Office, Science Policy Programming, No. 38) and the Vlaams Actieprogramma Biotechnologie (ETC 002). G.G. is a Post-doctoral Fellow of the National Fund for Scientific Research (Belgium).

References

1. Alberts, B., Bray, D., Lewis, J., Raff, M., Roberts, K. and Watson, J.D. (1994) *Molecular Biology of the Cell*, 3rd ed., Garland Publishing, New York.
2. Bird, A.F. (1961) The ultrastructure and histochemistry of a nematode-induced giant cell, *J. Biophys. Biochem. Cytol.* **11**, 701-715.
3. Bird, A.F. (1962) The inducement of giant cells by *Meloidogyne incognita*, *Nematologica* **8**, 1-10.
4. Bird, A.F. (1973) Observations on chromosomes and nucleoli in syncytia induced by *Meloidogyne javanica*, *Physiol. Plant Pathol.* **3**, 387-391.
5. Brodsky, V.Y. and Uryvaeva, I.V. (1977) Cell polyploidy: its relation to tissue growth and function, *Int. Rev. Cytol.* **50**, 275-332.
6. Brodsky, V.Y. and Uryvaeva, I.V. (1985) *Genome Multiplication in Growth and Development*, (Development and Cell Biology Series, Vol. 15), Cambridge University Press, Cambridge.
7. Collin, S., Fernández-Lobato, M., Gooding, P.S., Mullineaux, P.M. and Fenoll, C. (1996) The two nonstructural proteins from wheat dwarf virus involved in viral gene expression and replication are retinoblastoma-binding proteins, *Virology* **219**, 324-329.
8. Dolan, L., Janmaat, K., Willemsen, V., Linstead, P., Poethig, S., Roberts, K., and Scheres, B. (1993). Cellular organisation of the *Arabidopsis thaliana* root, *Development* **119**, 71-84.

9. D'Amato, F. (1984) Role of polyploidy in reproductive organs and tissues, in B.M. Johri (ed.), *Embryology of Angiosperms*, Springer-Verlag, Berlin, pp. 519-566.
10. Dropkin, V.H. and Nelson, P.E. (1960) The histopathology of root-knot infections in soybeans, *Phytopathology* 50, 442-447.
11. Endo, B.Y. (1964) Penetration and development of Heterodera glycines in soybean roots and related anatomical changes, *Phytopathology* 54, 79-88.
12. Endo, B.Y. (1971) Synthesis of nucleic acids at infection sites of soybean roots parasitized by Heterodera glycines, *Phytopathology* 61, 395-399.
13. Ferreira, P.C.G., Hemerly, A.S., de Almeida Engler, J., Van Montagu, M., Engler, G. and Inzé, D. (1994) Developmental expression of the Arabidopsis cyclin gene *cyc1At.*, *Plant Cell* 6, 1763-1774.
14. Geitler, L. (1953) Endomitose und endomitotische Polyploidisierung, *Protoplasmatologia* 6c, 1-89.
15. Glazer, I. and Orion, D. (1984) Influence of urea, hydroxyurea, and thiourea on *Meloidogyne javanica* and infected excised tomato roots in culture, *J. Nematol.* 16, 125-130.
16. Goddijn, O.J.M., Lindsey, K., van der Lee, F.M. , Klap, J.C. and Sijmons, P.C. (1993) Differential gene expression in nematode-induced feeding structures of transgenic plants harbouring promoter-*gus*A fusion constructs, *Plant Journal* 4, 863-873.
17. Golinowski, W., Grundler, F.M.W. and Sobczak, M. (1996) Changes in the structure of *Arabidopsis thaliana* during female development of the plant-parasitic nematode *Heterodera schachtii*. *Protoplasma*, 194, 103-116.
18. Grafi, G. and Larkins, B.A. (1995) Endoreduplication in maize endosperm: involvement of M phase-promoting factor inhibition and induction of S phase-related kinases, *Science* 269, 1262-1264.
19. Hemerly, A.S., Ferreira, P.C.G., de Almeida Engler, J., Van Montagu, M., Engler, G. and Inzé, D. (1993) *cdc2a* expression in *Arabidopsis thaliana* is linked with competence for cell division, *Plant Cell* 5, 1711-1723.
20. Hemerly, A., de Almeida Engler, J., Bergounioux, C., Van Montagu, M., Engler, G., Inzé, D. and Ferreira, P. (1995) Dominant negative mutants of the Cdc2 kinase uncouple cell division from iterative plant development, *EMBO J.* 14, 3925-3936.
21. Hesse, M. (1969) Anatomische und karyologische Untersuchungen an der Galle von *Mayetiola poae* auf *Poa nemoralis*, *Österr. Bot. Z.* 117, 411-425.
22. Hesse, M. (1971) Häufigkeit und Mechanismen der durch gallbildende Organismen ausgelösten somatischen Polyploidisierung, *Österr. Bot. Z.*. 119, 454-463.
23. Huang, C.S. and Maggenti, A.R. (1969a) Mitotic aberrations and nuclear changes of developing giant cells in Vicia faba caused by root-knot nematode, Meloidogyne javanica, *Phytopathology* 59, 447-455.
24. Huang, C.S. and Maggenti, A.R. (1969b) Wall modifications in developing giant cells of Vicia faba and Cucumis sativus induced by root-knot nematode, Meloidogyne javanica, *Phytopathology* 59, 931-937.
25. Hussey, R.S. (1989) Disease-inducing secretions of plant-parasitic nematodes, *Ann. Rev. Phytopathol.* 27, 123-141.
26. Jones, M.G.K. (1981) The development and function of plant cells modified by endoparasitic nematodes, in B.M. Zuckerman and R.A. Rohde (eds.), *Plant Parasitic Nematodes*, Vol. III, Academic Press, New York, pp. 255-279.
27. Jones, M.G.K. and Northcote, D.H. (1972) Nematode-induced syncytium —a multinucleate transfer cell, *J. Cell Sci.* 10, 789-809.
28. Jones, M.G.K. and Payne, H.L. (1978) Early stages of nematode-induced giant-cell formation in roots of *Impatiens balsamina*, *J. Nematol.* 10, 70-84.
29. Krusberg, L.R. and Nielsen, L.W. (1958) Pathogenesis of root-knot nematodes to the Porto Rico variety of sweetpotato, *Phytopathology* 48, 30-39.
30. Labib, K., Craven, R.A., Crawford, K. and Nurse, P. (1995) Dominant mutants identify new roles for p34^{cdc2} in mitosis, *EMBO J.* 14, 2155-2165.
31. Magnusson, C. and Golinowski, W. (1991) Ultrastructural relationships of the developing syncytium induced by *Heterodera schachtii* (Nematoda) in root tissues of rape, *Can. J. Bot.* 69, 44-52.
32. Mendenhall, M.D., Richardson, H.E. and Reed, S.I. (1988) Dominant negative protein kinase mutations that confer a G_1 arrest phenotype, *Proc. Natl. Acad. Sci. USA* 85, 4426-4430.
33. Meyerowitz, E.M. (1987) *Arabidopsis thaliana*. *Ann. Rev. Genet.* 21, 93-111.
34. Morejohn, L.C., Bureau, T.E., Molè-Bajer, J., Bajer, A.S. and Fosket, D.E. (1987) Oryzalin, a dinitroaniline herbicide, binds to plant tubulin and inhibits microtubule polymerization in vitro, *Planta* 172, 252-264.

35. Nagar, S., Pedersen, T.J., Carrick, K.M., Hanley-Bowdoin, L. and Robertson, D. (1995) A geminivirus induces expression of a host DNA synthesis protein in terminally differentiated plant cells, *Plant Cell* **10**, 1037-1043.
36. Niebel, A., de Almeida Engler, J., Hemerly, A., Ferreira, P., Van Montagu, M. and Gheysen, G. (1996) Induction of *cdc2a* and *cyc1At* expression in Arabidopsis during early phases of nematode-induced feeding cell formation, *Plant J.*, **10**, 1037-1043.
37. Orum, T.V., Bartels, P.G. and McClure, M.A. (1979) Effect of oryzalin and 1,1-dimethylpiperidinium chloride on cotton and tomato roots infected with the root-knot nematode, *Meloidogyne incognita*, *J. Nematol.* **11**, 78-83.
38. Owens, R.G. and Novotny Specht, H. (1964) Root-knot histogenesis, *Contrib. Boyce Thompson Inst.* **22**, 39-50.
39. Paulson, R.E. and Webster, J.M. (1970) Giant cell formation in tomato roots caused by *Meloidogyne incognita* and *Meloidogyne hapla* (Nematoda) infection. A light and electron microscope study, *Can. J. Bot.* **48**, 271-276.
40. Piegat, M. and Wilski, A. (1963) Changes observed in cell nuclei in roots of susceptible and resistant potato after their invasion by potato root eelworm (*Heterodera rostochiensis* Woll.) larvae, *Nematologica* **9**, 576-580.
41. Plegt, L., and Bino, R.J. (1989) ß-Glucuronidase activity during development of the male gametophyte from transgenic and non-transgenic plants, *Mol. Gen. Genet.* **216**, 321-327.
42. Rohde, R.A. and Mc Clure, M.A. (1975) Autoradiography of developing syncytia in cotton roots infected with *Meloidogyne incognita*, *J. Nematol.* **7**, 64-69.
43. Rubinstein, J.H. and Owens, R.G. (1964) Thymidine and uridine incorporation in relation to the ontogeny of root-knot syncytia, *Contrib. Boyce Thompson Inst.* **22**, 491-502.
44. Shaul, O., Van Montagu, M. and Inzé D. (1996) Regulation of cell division in *Arabidopsis*, *Crit. Rev. Plant Sci.* **15**, 97-112.
45. Shaul, O., Mironov, V., Burssens, S., Van Montagu, M. and Inzé, D. (1996) Two *Arabidopsis* cyclin promoters mediate distinctive transcriptional oscillation in synchronized tobacco BY-2 cells, *Proc. Natl. Acad. Sci. USA* **93**, 4868-4872.
46. Shorthouse, J.D. and Rohfritsch, O. (1992) *Biology of Insect-Induced Galls*, Oxford University Press, New York.
47. Sijmons, P.C., Grundler, F.M.W., von Mende, N., Burrows, P.R. and Wyss, U. (1991) *Arabidopsis thaliana* as a new model host for plant-parasitic nematodes, *Plant J.* **1**, 245-254.
48. Sijmons, P.C., Atkinson, H.J. and Wyss, U. (1994) Parasitic strategies of root nematodes and associated host cell responses, *Ann. Rev. Phytopathol.* **32**, 235-259.
49. Starr, J.L. (1993) Dynamics of the nuclear complement of giant cells induced by *Meloidogyne incognita*, *J. Nematol.* **25**, 416-421.
50. Stender, C., Glazer, I. and Orion, D. (1986) Effects of hydroxyurea on the ultrastructure of giant cells in galls induced by *Meloidogyne javanica*, *J. Nematol.* **18**, 37-43.
51. Wiggers, R.J., Starr, J.L. and Price, H.J. (1990) DNA content and variation in chromosome number in plant cells affected by *Meloidogyne incognita* and *M. arenaria*, *Phytopathology* **80**, 1391-1395.
52. Wiggers, R.J., Magill, C.W., Starr, J.L. and Price, H.J. (1991) Evidence against amplification of four genes in giant cells induced by *Meloidogyne incognita*, *J. Nematol.* **23**, 421-424.
53. Young, C.W. and Hodas, S. (1964) Hydroxyurea: inhibitory effect on DNA metabolism, *Science* **146**, 1172-1174.

Note added in proof: *Cyc1At* has been renamed as *Arath;cycB1;1* (Renaudin, J.P., Doonan, J.H., Freeman, D., Hashimoto, J., Hirt, H., Inzé, D., Jacobs, T., Kouchi, H., Rouzé, P., Sauter, M., Savoiré, A., Sorrell, D.A., Sundaresan, V., Murray, J.A.H. (1996) Plant cyclins: A unified nomenclature for plant A, B and D-type cyclins based on sequence organization, *Plant Mol. Biol.* **32**, 1003-1018.

REGULATION OF GENE EXPRESSION IN FEEDING SITES

Carmen FENOLL, Fabio A. ARISTIZABAL[1], Soledad SANZ-ALFEREZ and Francisca F. del CAMPO
Departamento de Biología, Universidad Autónoma de Madrid, Cantoblanco, 28049 Madrid, Spain
[1]*Present address: CIF, AA49490, Santa Fé de Bogotá D.C., Colombia*

Abstract

The profound structural and physiological transformations that initial root cells experience to become nematode feeding sites (NFSs) must be paralleled by molecular changes. Although it seemed obvious that these molecular changes should occur mostly through modifications in gene expression, experimental support for this idea has only been obtained in the last few years. In this Chapter we review the different approaches that have led to the identification of genes whose expression patterns are modified during the differentiation of NFSs. These approaches involve comparisons between infected and non-infected roots, including protein analysis by two-dimensional gel electrophoresis, construction and screening of cDNA libraries, differential display of transcribed sequences and the analysis of reporter genes in transgenic plants, either fused to known promoters or as part of promoter-trap constructions. The still limited information gathered through these means is clustered around two main questions: i) the role in feeding sites of those genes that respond to nematodes, and ii) the mechanisms underlying their responsiveness to the parasites, particularly at the level of transcriptional regulation. The identification of promoter elements in nematode-responsive genes is finally being addressed, thanks to the application to the study of plant-nematode interactions of techniques for the functional and structural analysis of promoters. The information is still too limited to allow the formulation of a comprehensive model, although some connections with differentiation processes in plants are starting to be unveiled.

1. Introduction

The development in a root of a feeding site induced by a sedentary nematode is a striking example of elaborated parasitism. The worm, once it becomes sedentary, relies completely on the correct functioning of this specialized structure to fulfill its needs and accomplish reproduction. It is not surprising, therefore, that feeding sites behave as very effective transfer cells, which can be kept active for a long time while the nematode withdraws water and food resources from it [23, 29, 48, 50]. How a feeding site is induced and built, and how it is maintained alive are fascinating questions that have attracted biologists from very diverse fields to the study of plant-nematode interactions.

Of course, saying that differentiation of nematode feeding sites involve changes in gene expression is an obvious statement. The first suggestions for nematode-induced changes in gene expression came from enzymatic studies [51, 9], but such changes in enzyme activity were linked to the defense reaction of the plant in response to the pathogen, rather than to the development of feeding sites. In fact, the actual demonstration of differential gene expression in feeding sites during compatible plant nematode interactions has only been achieved recently [review in 42, 43]. In the last years, and almost simultaneously, several laboratories have used molecular biology approaches to document changes in the accumulation of specific proteins or mRNAs, and in the activity of plant gene promoters during nematode infection. Although still far away from a comprehensive molecular description of feeding sites, the complementary approaches taken in the search for genes whose expression is differentially regulated in feeding sites (with respect to the surrounding cells in the root) is slowly leading to the construction of a catalogue of nematode-responsive genes and promoter sequences.

Finding the possibly subtle changes in the transcriptional machinery of the cell that undergoes the differentiation process is even further away, but the availability of differentially regulated promoters has made possible to start looking for such changes. The search for feeding site-specific transcription factors is crucial, because they must be one step closer to the enigmatic nematode signal that triggers the cascade of gene expression ultimately responsible for the development of the feeding site. As in other cell differentiation processes that occur during normal plant development, key proteins involved in the cell division cycle are likely to be the molecular adapters linking the initial (nematode) signal to the transcription factors responsible for the shift in cell fate. Our group is trying to get clues to these key regulatory proteins through the analysis of putative target sites for transcription factors in nematode-responsive promoters. The search for such promoter elements (nematode responsive elements, NREs) through their ability to confer nematode responsiveness to other promoters, using transgenic plants or transient expression systems, is only starting.

Besides its intrinsic interest in feeding site differentiation, unraveling the mechanisms that underlie differential promoter activity in feeding sites will also have a significant practical dimension. In the chapter by Ohl *et al.* there are specific examples on how to use nematode-responsive promoters to engineer nematode resistance, both in model plants and in crops. For optimal results, the use of artificial promoters with the desired specificity harboring taylored NREs, seems necessary. Cloning the proteins which interact with NREs will provide access to one or more feeding site-specific transcription factors, and might allow epistatic regulation of many genes in one step, thus diversifying and improving the engineered nematode resistance for agricultural purposes.

2. Strategies to Identify Genes Regulated by Nematodes

2.1. FROM PROTEINS TO GENES

The earliest attempts to demonstrate changes in mRNA populations during nematode infection were reported in potato [25]. It was shown that *in vitro* translated mRNA from roots infected by cyst nematodes produced protein profiles similar, but not identical, to non-infected roots, as analyzed by 2-D gel electrophoresis. Some of the differentially accumulated proteins were plant proteins, although most of them turned out to be related to the wounding produced by the nematodes during migration, rather than to the establishment of the feeding sites. More recent studies on alfalfa plants infected by root-knot nematodes also yielded mainly wound-response associated changes in gene expression [41]. These and other similar results are most probably due to the use of massively infected roots, in an attempt to obtain sufficient plant tissue for molecular studies. Under such conditions, the mechanical damage produced by the migration of the juveniles within the roots seemed to prevail over the nematode-specific effect which was intended. In fact, even a systemic induction of defense-related genes could be detected, in contrast to the apparent suppression of the defense response that has been suggested to occur in more controlled infections [20] (see below).

Very recently a reverse genetics approach has been used in sugarbeet [38] taking the due precautions to avoid the wound response. A few proteins specifically induced in resistant sugarbeet early after infection by cyst nematodes have been isolated in 2-D gels and used to produce antibodies. Microsequencing of these proteins is on the way.

2.2. cDNA LIBRARIES

Conventional cDNA libraries constructed from nematode-infected roots have been used to isolate specific clones through differential screening with probes derived from infected versus non-infected roots. This approach has been utilized in tomato plants infected by root-knot nematodes [46] and in potatoes infected with cyst nematodes [36]. In the potato study, one gene has been identified so far, corresponding to a catalase (*Cat2St*), which is induced both locally and systemically upon nematode infection. The induction, though, is not specific for NFSs, since bacterial infection, wounding or salicylic acid treatment also elicited transcript accumulation. In tomato, however, some of the genes isolated by this strategy seem to be more specifically related to feeding site development. The *Lemmi9* gene transcript (a homologue of a desiccation-related cotton gene) was shown by *in situ* hybridization to accumulate to high titers only in giant cells, although Northern analysis detected trace levels of the *Lemmi9* mRNA in aerial organs of infected and non-infected plants [46].

The limited success of strategies to identify specific proteins or mRNAs in nematode-infected plants is due in part to the difficulty of isolating sufficient amounts of infected tissue without eliciting a wound response. This problem has been overcome by several groups by the construction of PCR-based libraries. In this approach, minute amounts of highly purified infected tissue are used to obtain poly-adenylated RNA, which is reverse transcribed and PCR- amplified before producing cDNA libraries [5]. Three different groups have reported such amplified libraries. The first one was produced from cyst nematode-infected potato roots [24]. Another amplified library [30] was made 24 hours after infecting with root-knot nematodes a tomato cultivar carrying the *Mi* gene for nematode resistance [Liharska and Williamson, this volume]. Although not involved in feeding site development, and thus out of the scope of this review, several interesting sequences have been identified from this library which may play an important role during the very early resistance response to nematodes. In a compatible tomato-*Meloidogyne* system, several giant-cell specific transcripts have been cloned as cDNAs [7,49]. Sequence analysis of some 40 giant cell-specific transcripts has identified a plasmalemma proton ATPase, a putative MYB-like transcription factor and the large subunit of RNA polymerase II (Table 1) [7].

TABLE 1. Genes induced in giant cells

Gene and source	Method of identification	Putative function	Reference
Lemmi9 / Tomato	cDNA	Lea-like protein	Van der Eycken *et al.* 1995
DB#226 / Tomato	cDNA	H+ATPase	Bird and Wilson 1994
DB#280 / Tomato	cDNA	MYB-like protein	" "
Cat2 / Potato	cDNA	catalase	Niebel *et al.* 1995b
Extensin / Tobacco	mRNA	extensin	Niebel *et al.* 1993
Tob RB7 / Tobacco	GUS fusion	water channel	Opperman *et al.* 1994
cdc2aAt / Arabid.*	GUS fusion	mitotic kinase	Niebel *et al.* 1995a
cyc1aAt / Arabid.*	GUS fusion	G2 cyclin	" "
α *Tub* / maize	GUS fusion	α tubulin	Aristizábal 1996
HMG1 / Arabidopsis	GUS fusion	HMGReductase	" "
HMGR2 / Tomato	GUS fusion	defense response	Cramer *et al*. 1993
sHSP / Sunflower	GUS fusion	heat shock protein	This work
Tsw12 / Tomato	GUS fusion	lipid transfer protein	" "

* also induced in syncytia.

2.3. DIFFERENTIAL DISPLAY

Another technique which has just started to be applied to the molecular analysis of feeding sites is differential display. It is based on random PCR amplification of transcribed sequences from different tissues. Comparison of such sequences (short DNA fragments) by their electrophoretic mobility allows identification of those produced in any given combination of the examined samples (for instance, expressed in feeding sites and root tips, but not in shoot apex). The application of this technique to the Arabidopsis-*Meloidogyne* interaction has rendered some 70 fragments that are differentially amplified in the feeding sites. Sequencing of several of these bands revealed the transcription in feeding sites of various house-keeping genes and other potentially interesting genes [47]. A similar approach was taken for cyst nematodes, also in Arabidopsis [5, 6].

2.4. PROMOTER-GUS FUSIONS

All the strategies described so far are based on the identification of the actual gene products (proteins or mRNAs) in feeding sites. In spite of the intensive work devoted to these approaches by several labs in different plant-nematode systems, the number of genes added to the catalogue of "feeding site expressed genes" seems discouragingly low. A comparatively easier way of identifying nematode responsive genes was first taken by Goddijn *et al.* [20]. It makes use of transgenic plants harboring a reporter gene under the control of a known gene promoter. The reporter gene, GUS, codes for an enzyme that allows histochemical localization of individual cells in which the promoter is activated [28]. Screening for feeding site expression of these promoters is made by scoring the appearance of blue feeding sites after incubation of the infected roots with the substrate for GUS. The technique involves no sophisticated molecular manipulation, and can be made with a very low number of infected roots. Searching for nematode responsive genes by this approach is more hypothesis-oriented, rather than a blind search, since the particular genes to be tested have to be chosen from the beginning. In this way, a still small number of gene promoters have been found to be differentially expressed as compared to the adjacent root tissues, including both up (see Table 1) and down regulated promoters. One well studied promoter that illustrates the immediate practical potential of this approach is *Tob RB7*, which has been shown to retain its inducibility in tomato giant cells [40]. It is the first promoter used in giant cell ablation (see Chapter by Ohl *et al.*) and also one of the few known genes whose function might be necessary for feeding site maintenance [39].

2.5. PROMOTER TRAPS

Based in the same principle described above, a broader search for nematode-responsive promoters was started in Arabidopsis [20] by using T-DNA tagged transgenic lines which harbored a promoterless GUS gene. Those lines in which the transgene was inserted (by chance) behind a nematode-responsive promoter sequence in the host plant would show GUS regulation in feeding sites. Reverse PCR techniques allow the cloning of the plant sequences adjacent to the transgene, giving access in this way to the newly identified promoter, and to the identification of the actual plant gene to which the promoter belongs. Several labs, in a joint effort, have used a number of different tagged collections of plant lines to screen for nematode-responsive promoters, and as a result a few plant sequences have been identified, cloned and sequenced [4] (see also the chapter by Ohl *et al.*). Surprisingly, none of the sequences studied so far have allowed to find the corresponding genes which should lay adjacent to the tag in the transgenic lines, with the only exception of one line [4]. This raises the question of whether the tagged "promoters" were in fact non-promoter sequences in the plant genome which, fortuitously, were activated in feeding sites. In spite of this restriction, they are indeed useful to drive gene expression in feeding sites, and also to analyze transcriptional activation (see below). Because of that, and because new genes can still be found, the expression analysis of several Arabidopsis collections tagged with promoter traps is still in progress in several labs.

3. Promoter Analysis: *Cis* Elements and *Trans*-Acting Factors

Why are genes induced or silenced by nematodes? Obviously, any point along their expression pathway, from steps previous to promoter activation to protein subcellular location, might be a target for modification of gene function. However, the search for the mechanisms governing feeding site differentiation are mainly centered in the analysis of gene transcription. This is not only because of the immediate practical use of promoters in transgenic approaches to nematode resistance, but also because differentiation programs during normal development are known to be largely -although not exclusively- controlled at the transcriptional level. Reformulating the question in more precise terms, we can ask: why are promoters activated or repressed by nematodes? The answers to this question will have to start with the identification of the discrete DNA sequences within promoters which, often upon physical interaction with specific transcriptional factors (whose activity is, in turn, modulated by developmental or environmental cues -the nematode in our case), are responsible for turning a gene on or off. Two combined techniques are needed to correlate *cis* acting elements (DNA promoter sequences) and *trans* acting factors (transcriptional regulators) in gene promoters: functional analysis and mapping of protein-binding sites.

3.1. TRANSGENIC PLANTS

Functional analysis of nematode-responsive promoters is still incipient. Only a limited number of cases have been published involving the generation of series of transgenic plants harboring reporter gene fusions to promoter deletions, and even less to specific point mutations of putative *cis* elements. One example is the *TobRB7* promoter, a 300 bp fragment of which showing no expression elsewhere in the plant was nevertheless induced by root-knot nematodes in tomato [40]. Another example is the bi-directional *rolB-rolC* promoter from *Agrobacterium tumefaciens* in Arabidopsis [20]. A different finding relates to the Arabidopsis *HMG1* gene, which codes for a reductase acting at the starting point of isoprenoid biosynthesis [32]. In this case, 5´ promoter deletions up to 400 bp, which retain the described endogenous expression pattern (Boronat, unpublished) are sufficient for nematode inducibility [1]. Further deletion up to 200 bp, which renders the promoter totally inactive, destroys expression in giant cells as well [3], suggesting that regulatory elements may be shared by giant cells and other cell types that normally express the promoter in the non-infected plant. Other nematode-inducible promoters are currently being analyzed in transgenic plants, but the results are still preliminary. These include a number of deletion series from different geminiviral promoters [1], from a sunflower heat shock protein promoter [Montalvo and Fenoll, unpubl.] and from several Arabidopsis tagged sequences (see the Chapter by Ohl *et al*).

The most precise functional analysis of a nematode responsive promoter was performed by Sijmons *et al.* [44] on the 35S promoter, in which only 90 bp upstream from the transcriptional startpoint is sufficient to confer nematode responsiveness. Searching for the sequence responsible for the specific silencing of the promoter, they mutated the *as-1* boxes, which constitute the *cis* element known to be responsible for generalized promoter activation in root vascular tissue upon binding of TGA1, an apparently ubiquitous transcription factor. However, destruction of the *as-1* element did not alter significantly the nematode-dependent silencing. It remains to be established if such a site is able to mediate repression of transcription, both in its normal context and in heterologous promoters (see below). These experiments will determine if down regulation is due to the lack of specific transcription factors or to an active repression mechanism.

3.2. TRANSIENT EXPRESSION

The generation of a sufficient number of transgenic plants to test a collection of promoter deletions or other mutations is very time consuming, limiting in practice the functional analysis of nematode-responsive promoters. One of the alternatives to stable transformation for promoter analysis is transient expression. In this technique, many copies of a plasmid containing the modified promoter sequences fused to a reporter gene, are introduced in the target cells, where some DNA molecules reach the nucleus and are converted to transcriptionally active episomes before they are eventually degraded. The inaccessibility of feeding sites has restricted experiments of transient expression for nematode-dependent promoters. Very recently, however, microinjected reporter plasmids in Arabidopsis syncytia have transiently expressed GUS and GFP activities [1; Grundler, Ohl and Fenoll, unpublished], opening the possibility of performing detailed time-course expression analysis in a single feeding site. This technical achievement will now allow to test in a relatively fast way large numbers of mutations for syncytia-specific promoters, without the need for plant transformation. A similar system in giant cells is currently being developed. Other alternatives include biolistics and laser-mediated transformation.

3.3. IDENTIFICATION OF PROTEIN-BINDING SITES

When many promoters have to be analyzed in parallel to identify common *cis* elements, even transient expression is too laborious. If mutational saturation of even short regions within these promoters has to be performed, then microinjection renders unpracticable. An alternative *in vitro* approach is the identification of protein binding sites in such promoters, which may be targets for transcription factors related to promoter activity. Finding such putative *cis* elements will greatly reduce the length of DNA that has to be scanned by mutagenesis in the transient expression assays, besides providing a direct step towards the cloning of feeding site-specific transcription factors from their target binding sequences (for example, through the screening of expression libraries by southwestern blotting or in yeast one-hybrid protein systems).

Although the difficulty for obtaining sufficient amounts of nuclear proteins from feeding sites is a clear drawback of this strategy, it has been applied successfully to the

Lemmi9 promoter [2, 19], where a 12 bp imperfect repeat close to the TATA box has been shown to bind specifically a protein from tomato galls. Although still not analyzed in functional assays, this may be a Nematode Responsive Element, since the specific DNA-binding activity correlates well with the expression pattern of the endogenous tomato gene [19]. Functional analysis of the putative NRE in transgenic plants and cloning of the gall protein in a yeast one-hybrid system are in progress.

Another nematode responsive promoter whose NRE is being mapped by the same strategy is the 35S promoter. In this case, the functional analysis is rather exhaustive, and the silencing has been ascribed to a sequence within the 90 bp that constitutes the proximal region. In agreement with the repression of the 35S promoter, a search for the transcription factor TGA1a in feeding sites revealed the absence of the *as-1* binding activity [1], establishing that the absence of TGA1a might be the cause of nematode-dependent gene silencing for those promoters whose expression was dependent on *as-1* sites. In spite of the lack of as-1 binding activity in galls, a new, previously not described protein that bound in a specific manner to the 35S promoter has been found in infected roots, and its binding site mapped to the -90 fragment (Figure 1). The protein was also present, although very diluted, in some immature plant tissues, demonstrating the plant origin of the DNA-binding activity. It appears that such protein(s) may recognize and bind to a secondary structure around the transcription startpoint, perhaps hindering promoter activation [1, 17; Aristizábal, Sanz-Alférez and Fenoll, unpublished results]. If this proves to be the case, the first NRE identified would be associated with

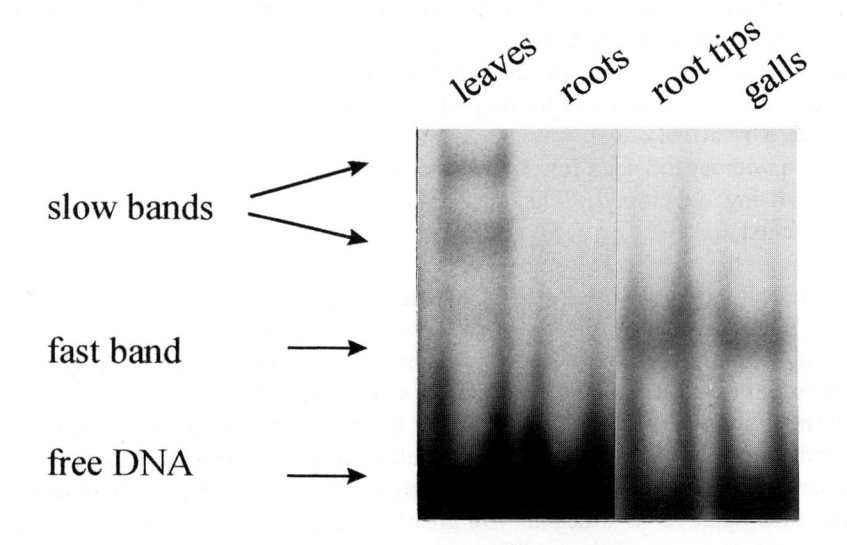

Figure 1. Interaction of nuclear proteins with the 35S promoter revealed by a bandshift assay. Nuclear protein extracts were obtained from uninfected tomato leaves, whole roots and root tips, or from *Meloidogyne incognita* induced galls. The nuclear extracts were incubated with a ^{32}P-labeled (-90 to +1) fragment from the 35S promoter, and the reaction mixture was fractionated under non-denaturing conditions by PAGE. Free DNA migrates to the bottom of the gel, while DNA molecules that form complexes with different proteins appear as retarded bands. The slow bands represent binding of TGA1a to the double *as-1* box, while the fast migrating band which appears with root tip and gall extracts is due to the binding of another protein to a sequence close to the TATA box.

gene silencing rather than with gene activation.

Other nematode-responsive promoters or tagged sequences where functional studies have limited to less than 500 bp the region necessary for induction are now being screened with powerful computer software to identify common targets for known transcription factors [1]. Testing them in DNA-protein binding assays will then be feasible. To date only one putative transcription factor has been reported to be transcribed in tomato giant cells: a MYB-like protein encoded by gene which has been identified as a partial cDNA clone [7]. It would be interesting to test if the protein encoded by this clone actually binds to any of the reported nematode-inducible promoters.

4. Functions of Genes Regulated by Nematodes in Feeding Sites

Understanding how a feeding site is induced and how it works, in terms of gene expression, requires not only to find out the transcriptional mechanisms responsible for gene expression. To know if a particular gene product is necessary to make a feeding site is also relevant, and has important practical consequences as well. But identifying genes which are turned on or off in feeding sites is a relatively simple task, as compared to demonstrating the actual involvement of the gene products in relevant processes in the feeding sites. This requires showing that inactivation of the genes leads to modifications in the feedings sites. Gene inactivation in plants can presently be done by different techniques, all of them based on transgenic approaches. Antisense inactivation, co-suppression, construction of homozygous lines from insertional mutants or expression of negative dominant mutant alleles are the most obvious. Although a few of these strategies are being conducted, this field is still in its first steps. The only successful experiment reported was performed with a *TobRB7* antisense gene, which was able to reduce giant cell development in tomato [39].

Until more experiments are reported, we have to base the functional description of feeding sites in logical hypotheses based on what is known about a particular gene product. For example, it seems natural that, if multiple mitosis are required for giant cell differentiation, the mitotic kinase encoded by *cdc2aAt* be required early after infection (that is, there is a reason for its promoter to be induced at that time). The induction by root-knot nematodes of some geminiviral promoters very early during infection might also be related to the dependence of viral replication on a particular phase of the cell [12]. Expression of cell wall proteins such as extensins [37] seems also understandable for expanding cells. Likewise, the production of tubulin in a cell that has to undergo mitosis and growth seems predictable. We have found that the maize *tubα1* promoter is induced in giant cells by root-knot nematodes [1]. Besides tubulin, other microtubule-associated proteins (MAPs) such as kinesins, myosins and dyneins play pivotal roles in cell division, cell growth, membrane trafficking and other processes that polarize plant cells both functionally and structurally [15]. The importance for nematode feeding of polarized membrane traffic within the NFS as well as with the plasmodesmata connecting it to the surrounding root cells remains to be established.

TobRB7 and *Lemmi9*, both induced in giant cells but not in syncytia, seem to be related to water status in the giant cells. The former codes for an aquaporin, and the latter for a homologue of a cotton LEA (Late Embryogenesis Abundant) protein involved in ion sequestering during seed desiccation. The actual functions of these proteins in uninfected plants, though, are still undefined, and more work needs to be done before conclusions about their specific roles in giant cells are drawn. Another gene, the sunflower *sHSP17,* firstly identified as a heat-shock gene and also inducible by water stress and ABA treatment [11], is also highly induced in giant cells and in untreated seeds, but silent in untreated roots (r. Montalvo and C. Fenoll, unpublished). We have also seen that giant cells show induction of *Tsw12,* a tomato gene induced in leaves, but not in roots, by high salt and by heat shock treatments [45]. None of these genes is induced in syncytia, what makes it improbable that their function be related to a high nutrient content in NFSs. However, in growing plant cells increased osmolarity often precede rises in turgor pressure, the necessary driving force for cell expansion; since only giant cells experience active growth, while syncytia increase their size merely by recruiting neighbouring cells, a specific role in the process of giant cell expansion may be shared by all the four said genes. An alternative explanation may be different nutrient accumulation profiles in NFSs associated with different feeding strategies by cyst and root-knot nematodes.

Catalase Cat2 induction has been correlated with the suppresion of the defense response [36]. This interpretation is in agreement with the reported silencing of several PRP genes [20] and the 35S promoter from CAMV, which are normally induced by salicylic acid during pathogen attack. Hence, interference of nematodes with SA-mediated responses might be responsible for the suppression of the defense mechanisms elicited by parasitic nematodes [44.; Ohl *et al.,* this volume] through an active gene silencing mechanism.

For most genes that exhibit differential regulation in feeding sites, a function is not so obvious. This is well illustrated by the particularly complex case of the enzyme HMGRase, that acts at the starting point of the isoprenoid biosynthetic pathway leading to the production of an array of different molecules, from hormones to carotenoids or membrane lipids [10, 18]. The enzyme is encoded by gene families in different plant species. The first reported induction was that of the tomato gene *HMGR2,* which was activated in transgenic tobacco galls; since this particular gene is pathogen- and wound-inducible, a defensive function was proposed [13]. In transgenic tobacco plants, we have shown that the Arabidopsis gene *HMG1* [32] is expressed very specifically in giant cells all throughout nematode development [3]. In this case, a defense role seems improbable, because Arabidopsis does not make isoprenoid phytoalexins and because *HMG1* is not induced by pathogens or by wounding [10, 32]. Since plant parasitic nematodes are thought to be auxotrophs for sterols [14] a part in massive sterol supply to nematodes has been suggested for the HMGRase [43], although the overproduction of the enzyme appears to be specific for giant cells and not for syncytia. We have also proposed a non-exclusive role in vesicle traffic and in membrane and cell wall biogenesis, in close connection to tubulin and kinesin production in giant cells [1, 3, 8, 9; see also the Chapter by Bleve-Zacheo and Melillo]. This case illustrates well the difficulty of drawing conclusions as to which particular role may a particular protein be playing in NFSs. It also exemplifies, like the induction of water-stress related genes do, how the apparent similarity between syncytia and giant cells may hide profound differences that are often overseen.

Some promoters whose gene products are predicted to be necessary for NFS functioning, appear, nevertheless, not to be induced by nematodes. This is what we have found for a barley Sucrose Synthase, involved in sucrose unloading in sink organs (P. Carbonero, personal communication), and for a specific RNase related to phospate mobilization in metabolically active Arabidopsis cells (J. Paz-Ares, personal communication). Similarly, Arabidopsis genes that code for homologues of LEA proteins, functionally related to *Lemmi9* and the other stress inducible genes, are not induced by root-knot nematodes in our hands. The list of non-inducible genes might be much longer, since it is probable that several genes have been tested by other groups but the negative results were never reported.

5. Feeding Sites and Cell Differentiation Processes: Making Models to Explain Changes in Gene Expression

The formation of a NFS in a root is a developmental process. As such, it is not a discrete event, but rather it extends in time as the NFS goes through a series of different stages while nematode growth and development proceeds [21, 22, 33, 34]. Although it seems evident that a NFS is not a definition for a cell type, this fact is often ignored when changes in gene expression are studied. Molecular basis for this statement comes from the fact that genes induced at a given stage are turned off at other stages. This has been described for a number of Arabidopsis tagged sequences, which show a variety of expression patterns when time courses are carefully followed [4]. Possible explanations for the kinetics of induction for the promoters of a few known genes can easily be offered. For example, cell cycle genes such as *cdc2aAt* or *cyc1aAt* seem to be induced shortly after initiation of NFS formation and are turned off in a few days, thus matching the previously described times in which nuclear division and/or DNA synthesis in the developing NFS are occurring [35]. Other highly expressed genes are induced later and remain on until completion of the nematode reproductive cycle, like *HMG1*, possibly in connection with the continuous requirement for membrane biogenesis and/or sterol synthesis [3, 8]. The reasons and implications for the particular accumulation kinetics in most other cases remain to be explained.

Another point that deserves further attention if models for NFS differentiation have to be proposed is the comparison of giant cells *vs.* syncytia. Clearly described as different both in structure and in ontogeny (see the Chapters by Bleve-Zacheo and Melillo and by Golinowski *et al.*) they do have an equivalent function in nematode feeding. Because of this, a similar physiology is expected and, at least some of the differentially regulated genes must be common. Since such common genes would be most useful in integrated transgenic approaches to construct nematode resistant plants, molecular biologists have often highlighted the similarities, rather than the differences amongst them. Therefore, comprehensive studies comparing gene expression in syncytia and giant cells are lacking, and only scattered data on differential behavior for both NFSs are available (see Table 1 and Figure 2).

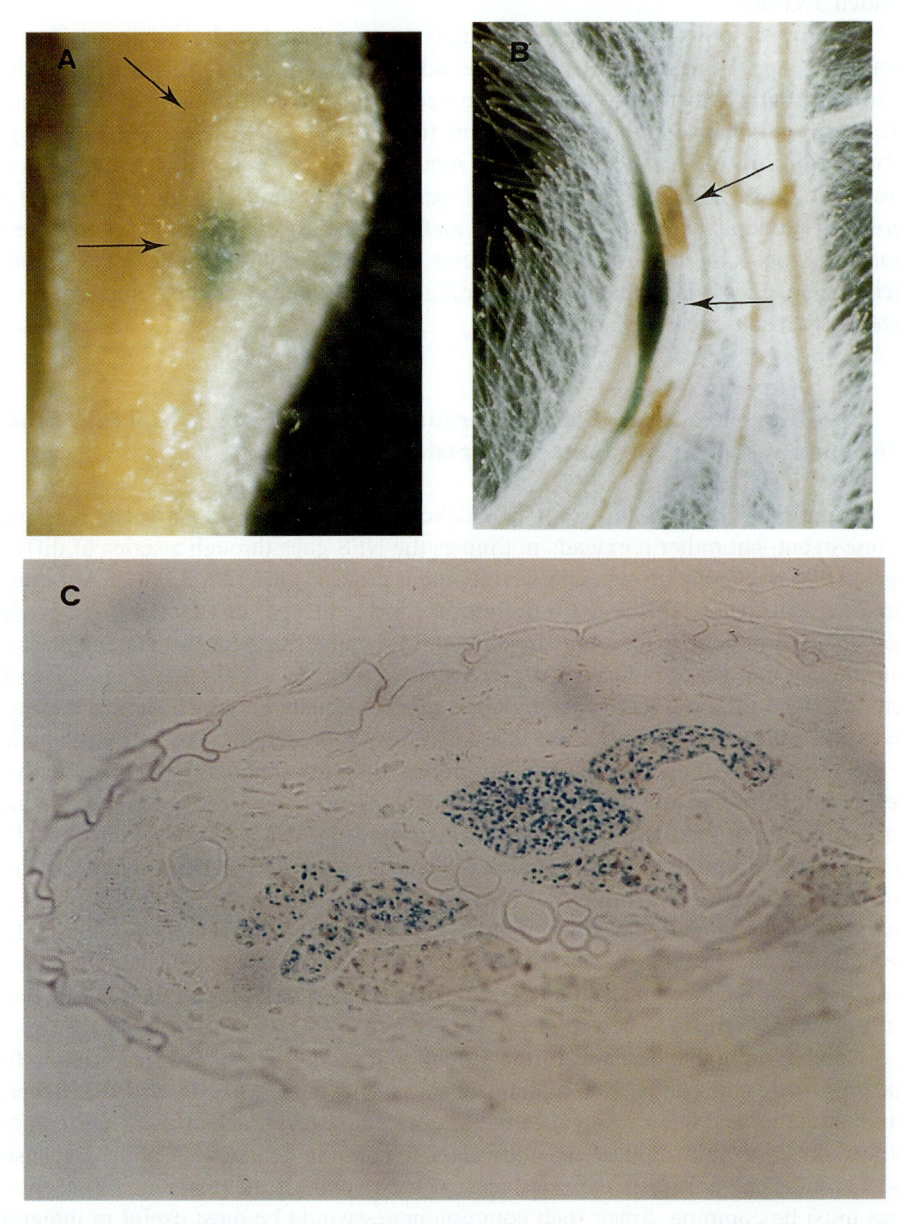

Figure 2. Expression of GUS under the control of nematode-inducible promoters. GUS activity in the NFSs is revealed by the blue color. **A**, closeview of a root from a transgenic tobacco plant harboring a GUS fusion to the Arabidopsis *HMG1* promoter, 15 days after infection with *Meloidogyne incognita*. Arrows show the white female nematode and the giant cells, which appear stained in deep blue. **B**, an Arabidopsis T-DNA tagged line shows GUS activity in a mature syncytium; the nematode (*Heterodera schachtii*) head is clearly associated with the NFSs (photograph provided by S. Ohl). **C**, cross section of a 8 days old Arabidopsis syncytium which has been microinjected with a plasmid containing the promoter- GUS fusion shown in B. Staining for GUS activity was performed 24 hours after injection.

In view of all these considerations, making a model that accounts for NFS development with the little information on gene expression that we have is a difficult task. We can establish, though, that the model would have to address several crucial points. For example, stating which cell types within the root are chosen by nematodes, and comparing their normal fate with that newly imposed as NFS, should now be possible in Arabidopsis thanks to the availability of a number of root mutants and tagged-promoter lines which are markers for specific cell types in the root (see Scheres et al., this volume). The use of antibodies against nematode secretions [26; see also the Chapter by Jones and Robertson], plus a number of well designed experiments, might help to identify the primary target(s) within the cell for the nematode signal(s). For instance, CANNTG boxes (like those present in the *lemmi9*, the *HMG1* and the *TobRB7* promoters) are recognized in animals by proteins that belong to the superfamily of the basic helix-loop-helix (bHLH) transcriptional regulators, some of which play crucial roles in cell differentiation processes such as myogenesis [31]. Until recently, plant bHLHs had only been reported in maize and *Antirrinum*, where they regulate the expression of genes involved in the anthocyanin biosynthetic pathway during endosperm and petal epidermis differentiation, respectively [31]. The first Arabidopsis bHLH protein has been identified as encoded by a water-stress related root-specific gene [52]. It is possible that root-knot nematodes make use of these preformed pathways to differentiate giant cells. In this respect, it has already been noticed [42] that *Trichinella spiralis*, a parasitic nematode of mammalian muscle which induces a nurse cell system resembling the root giant cells, seems to induce the nurse cells by shifting the normal balance of the b-HLH activator and repressor proteins that control myofiber differentiation [27]. Any clues about how the molecules that regulate this balance of transcriptional activators and repressors work might allow to formulate testable models for plant-parasitic nematodes. In this respect, protein-protein interactions are known to often precede cascades of changes in groups of genes which share common *cis* elements, and are, therefore, regulated as a block by the same transcription factors [16]. The study of the *cis* elements that control changes in promoter activity at the onset of NFS differentiation is in this respect a promising field: promoters for cell cycle specific genes (see the Chapter by Gheysen et al.) that are induced very early after nematode infection can be used to study the process. Recently, we have found that geminiviral promoters, proposed to be specific for G1/S phases, respond to nematodes (Fenoll et al., in preparation), what provides a conceptual framework to formulate new hypotheses on the mechanisms that trigger the changes in NFS gene expression. These viruses encode proteins which, by interacting very precisely with cellular cell cycle regulatory proteins, are thought to drive the cells towards a state of competence for viral replication and gene expression based, amongst other features, on an increased availability of particular families of transcription factors. Whether nematode secretions contain functionaly equivalent "transforming" proteins to trigger the differentiation of feeding sites is an attractive hypothesis.

6. Conclusions

The molecular description of NFSs in terms of gene expression, started only a few years ago, is growing steadily thanks to the combined efforts of several laboratories across the world. Some genes of identified functions are now known to be induced in NFSs, in a more or less specific manner. Several promoters (both genuine and putative pseudopromoters identified in tagged Arabidopsis lines) have been recognized as responsive to root-knot and/or cyst nematodes. For some of them, time courses of the induction upon nematode infection have been performed. In a few cases, the minimal promoter regions required for the response have been found, and these are being scanned for the presence of specific nematode-responsive sequences, as DNA fragments required both in functional assays (transgenic plants and transient expression) and for specific binding of nuclear proteins from NFSs. The identification of the first putative transcription factors that interact with such promoter elements will hopefully shed light on the possible differentiation pathways that, after being triggered by the nematode, guide the initial cell to become a specialized NFS. It will also open the study of post-transcriptional regulatory mechanisms perhaps crucial for NFS differentiation which might never be disclosed by conventional gene expression studies. Having access to such transcription factors may also allow epistatic regulation of many nematode-responsive promoters in one step, what will prove to be a powerful tool for NFS disruption by transgenic multicomponent approaches.

ACKNOWLEDGMENTS

We thank Peter Sijmons for the critical reading of this manuscript. Work in the laboratory was funded by grant number BIO95-0164 from the Spanish Interministery Comission for Science and Technology (CICYT) to CF. FAA was supported by a predoctoral fellowship from Colciencias (Colombia).

References

1. Aristizábal, F. (1996) Identificación y análisis de la expresión en plantas de genes regulados por nemátodos endoparasíticos, Ph. D. Thesis. Universidad Autónoma de Madrid. Spain.
2. Aristizábal, F., Serna, L., Sanz-Alférez, S., Escobar, C., Del Campo, F.F., Grundler, F. and Fenoll, C. (1996) Molecular analysis of nematode-inducible plant genes, *8th International Congress on Plant-Microbe Interactions*, Knoxville, USA. B-29
3. Aristizábal, F., Lumbreras, V., Boronat, A., Bleve-Zacheo, T., Arro, M, Ferrer, A. and Fenoll, C. (1995) Sobreexpresión de la enzima HMG-CoA reductasa inducida por nemátodos en raíces de tabaco y Arabidopsis, *III Reunión de Biología Molecular de Plantas*, Sevilla, Spain. pp. 123-124.
4. Barthels, N., Van der Lee, F., Klap, J., Goddijn, O.J.M., Olh, S.A., Puzio, P., Grundler, F.M.W., Karimi, M., Robertson, L., Robertson, W.M., Lindsey, K., Van Montagu, M., Gheysen, G. and Sijmons, P.C. (1996) Nematode-inducible *Arabidopsis thaliana* regulatory sequences driving reporter gene expression in nematode feeding structures, in preparation.
5. Bertioli, D.J., Smoker, M., Brown, A.C.P., Jones, M.G.K. and Burrows, P.R. (1994) A method based on PCR for the construction of cDNA libraries and probes from small amounts of tissue, *Biotechniques* **16**, 1054-1058.
6. Bertioli, D.J., Schlilchter U.A.H., Adams, M.J., Burrows, P.R., Steinbia H.H., Antoniw, J.F. (1995) An analysis of differential display shows a strong bias towards high copy number mRNAs, *Nucl. Acid Res.* **23**, 4520-4523.
7. Bird, D.McK. and Wilson, M.A. (1994) DNA sequence and expression analysis of root-knot nematode-elicited giant cell transcripts, *Mol. Plant-Microbe Interact.* **7**, 419-424.
8. Bleve-Zacheo, T., Melillo, T., Zacheo, G., Aristizábal., F., Serna, L., Del Campo, F.F., Ferre, A., Boronat, A.y Fenoll, C. (1996) β-tubulin and HMGR colocalize in meristems and nematode-induced giant cells in Arabidopsis roots, *XI European Congress on Electron Microscopy*, Dublin, Ireland.
9. Bleve-Zacheo, T., Melillo, T., Serna, L., Aristizábal., F., Sanz-Alférez, S., Del Campo, F.F. and Fenoll, C. (1996) Membrane traffic in giant cells induced by *Meloidogyne incognita* in Arabidopsis, 3rd Intl. Nematology Congress, Julio 7-12, Guadalupe.
10. Caelles, C., Ferrer, A., Balcells, L., Hegardt, F.G. and Boronat, A. (1989) Isolation and structural characterization of cDNA encoding *Arabidopsis thaliana* 3-hidroxy-3-methylglutaryl coenzyme A reductase, *Plant Mol. Biol.* **13**, 627-638.
11. Coca M.A., Almoguera C., Thomas T.L., Jordano J. (1996) Differential regulation of small heat-shock genes in plants: analysis of a water-stress-inducible and developmentally activated sunflower promoter, *Plant Mol. Biol.* **31**, 863-876.
12. Collin, S., Fernández-Lobato, M., Gooding, P.S., Mullineaux P.M. and Fenoll, C. (1996) The two nonstructural proteins from wheat dwarf virus involved in viral gene expression and replication are retinoblastoma-binding proteins, *Virology* **219**, 324-329.
13. Cramer, C.L., Weissenborn, D., Cottingham, C.K., Denbow, C.J., Eisenback, J.D., Radin, D.N. and Andu, X. (1993) Regulation of defense-related gene expression during plant-pathogen interactions, *J. Nematol.* **25**, 507-518.
14. Chitwood, D.J. and Lusby, W.R. (1991) Metabolism of plant sterols by nematodes, *Lipids* **26**, 619-627.
15. Cyr, R.J. and Palevitz, B.A. (1995) Organization of cortical microtubules in plant cells, *Current Opinion in Cell Biology* **7**, 65-71.
16. D'Arcangelo, G. and Curran, T. (1995) Smart transcription factors, *Nature* **326**, 292-293.
17. Del Campo, F.F., Aristizábal, F., Sanz-Alférez, S., Serna, L., Grundler, F. and Fenoll, C. (1996) Promoter analysis of plant genes whose expression is altered upon root-knot nematode infection, X FESPP Congress, Florence, Italy, *Plant Physiol. Biochem.* Special issue, pp. 294.
18. Enjuto, M., Balcells, L., Campos, N., Caelles, C., Arró, M. and Boronat, A. (1994) *Arabidopsis thaliana* contains two differentially expressed 3-hydroxy-methylglutaryl-CoA reductase genes, which encode microsomal forms of the enzyme, *Proc. Natl. Acad. Sci. USA* **91**, 927-931.
19. Escobar, C., Van der Eycken, W., Aristizábal, F.A., Sanz-Alférez, S., Del Campo, F.F., Barthels, N., Seurinck, J., Van Montagu, M., Gheysen, G. and Fenoll, C. (1996) A repeated sequence from the nematode-inducible tomato *Lemmi9* promoter specifically binds nuclear proteins from infected roots (submitted).
20. Goddijn, O.J.M., Lindsey, K.,Van der Lee, F.M., Klap, J.C. and Sijmons, P.C. (1993) Differential gene expression in nematode-induced feeding structures of transgenic plants harbouring promoter-*gusA*

fusion constructs, *Plant J.* **4**, 863-873.
21. Golinowski, W., Grundler, F.M.W. and Sobczak, M. (1996) Changes in the structure of *Arabidopsis thaliana* induced during development of females of the plant parasitic nematode *Heterodera schachtii*, *Protoplasma* **194**, 103-116.
22. Golinowski, W. and Magnusson, C. (1991) Tissue response by *Heterodera schachtii* (Nematoda) in susceptible and resistant white mustard cultivars, *Can. J. Bot.* **69**, 53-62.
23. Grundler, F.M.W., Wyss, U. and Golinowski, W. (1994) Sedentary nematodes in *Arabidopsis thaliana*, in Bowman, J. (ed), *Arabidopsis: An Atlas of Morphology and Development*, Springer Verlag, New York, pp. 418-423.
24. Gurr, S.J., McPherson, M.J., Scollan, C., Atkinson, H.J. and Bowles, D.J. (1991) Gene expression in nematode-infected plant roots, *Mol. Gen. Genet.* **226**, 361-366.
25. Hammond-Kosack, K.E., Atkinson, H.J., and Bowles, D.L. (1989) Local and systemic changes in gene expression in potato plants following root infection with the cyst nematode *Globodera. Rostochiensis*, *Physiol. Mol. Plant Pathol.* **37**, 339-354.
26. Hussey, R.S. (1989) Disease-inducing secretions of plant parasitic nematodes, *Ann. Rev. Phytopatol.* **27**, 123-141.
27. Jasmer, D.P. (1993) *Trichinella spiralis* infected skeletal muscle cells arrest in G2/M and cease muscle gene expression, *J. Cell Biol.* **121**, 786-793.
28. Jefferson, R.A., Kavanagh, T.A. and Bevan, M.W. (1987) GUS fusions: -glucuronidase as a sensitive and versatile gene fusion marker in higher plants, *EMBO J.* **6**, 3901-3907.
29. Jones, M.G.K. (1981) The development and function of plant cells modified by endoparasitic nematodes, in B.M. Zuckerman and R.A. Rohde (eds.), *Plant parasitic nematode*, Vol. 3, Academic Press, New York, pp. 255-278.
30. Lambert, K.N. and Williamson, V.M. (1993) cDNA library construction from small amounts of RNA using paramagnetic beads and PCR, *Nucleic Acids Res.* **21**, 775-776.
31. Littlewood, T.D. and Evan, G.I. (1994) Transcription factors 2: helix-loop-helix, *Prot. Prof.* **1**, 639-665.
32. Lumbreras, V. (1995) Estudio molecular de la 3-hidroxi-3metilglutaril CoA reductasa de Arabidopsis thaliana: expresión y regulación del gen HMG1, Ph. D. Thesis, Univesitat de Barcelona (Spain).
33. Magnusson, C. and Golinowski, W. (1991) Ultrastructural relationships of the developing syncytium induced by *Heterodera schachtii* (Nematoda) in root tissues of rape, *Can. J. Bot.* **69**, 44-52.
34. Melillo, M.T., Bleve-Zacheo, T. and Zacheo, G. (1990) Ultrastructural response of potato roots susceptible to cyst nematode *Globodera pallida* pathotype Pa 3, *Rev. Nématol.* **13**, 17-28.
35. Niebel, A., de Almeida-Engler, J., Hemerly, A., Ferreira, P., Van Montagu, M. and Gheysen, G. (1995) Induction of *cdc2a* and *cyc1* expression in Arabidopsis during early phases of nematode-induced feeding site formation, *Plant J.* (in the press).
36. Niebel, A., Heungens, K., Barthels, N., Inzé, D., Van Montagu, M. and Gheysen, G. (1995) Characterization of a pathogen-induced potato catalase and its systemic expression upon nematode and bacterial infection, *Mol. Plant-Microbe Interact.* **8**, 371-378.
37. Niebel, A., de Almeida Engler, J.,Tiré, C., Engler, G., Van Montagu, M. and Gheysen, G. (1993) Induction patterns of an extensin gene in tobacco upon nematode infection, *Plant Cell* **5**, 1697-1710.
38. Oberschmidt, O., Holtmann, B., Lange, S., Grundler, F.M.W. and Kleine, M. (1996) Studies of resistance mechanisms of sugarbeet against *Heterodera schachtii*: ultrastructure, 2D-analysis of proteins and differential display, *Fourth Annual Meeting of the European Union AIR-CAP on Mechanisms for Resistance against Plant Parasitic Nematodes*, Toledo (Spain), p. 13.
39. Opperman, C.H. and Conkling, M.A. (1996) Root-knot nematode induced plant gene expression and transgenic resistance strategies, *8th International Congress on Plant-Microbe Interactions*, Knoxville, USA, S-57.
40. Opperman, C.H., Taylor, C.G. and Conkling, M.A. (1994) Root-knot nematode-directed expression of a plant root-specific gene, *Science* **263**, 221-223.
41. Potenza, C., Higgins, E., Thomas, S. and Sengupta-Gopalan, C. (1994) Characterization of host plant genes important in *Meloidogyne incognita*/plant interaction. *4th International Congress of Plant Molecular Biology*, Amsterdam, The Netherlands, Abstract, 1765.
42. Sijmons, P.C (1993) Plant-nematode interactions, *Plant Mol. Biol.* **23**, 917-931.
43. Sijmons, P.C, Atkinsons, H.J. and Wyss, U. (1994) Parasitic strategies of root nematodes and associated host cell responses, *Ann. Rev. Phytopatol.* **32**, 235-259
44. Sijmons, P.C., Cardol, E.F. and Goddijn, O.J.M. (1994) Gene activities in nematode-induced feeding

structures, in M.J. Daniels, J.A. Downie and A.E. Osbourn (eds.), *Advances in Molecular Genetics of Plant-Microbe Interactions*, Kluwer Academic Publishers, The Netherlands, Vol. **3**, pp. 333-338.
45. Torres-Schumann, S., Godoy, J.A. and Pintor-Toro, J.A. (1992) A probable lipid transfer protein gene is induced by NaCl in stems of tomato plants, *Plant Mol. Biol.* **18**: 749-757.
46. Van der Eycken, W., de Almeida Engler, J., Inzé, D., Van Montagu, M. and Gheysen, G. (1996) A molecular study of root-knot nematode-induced feeding sites, *Plant J.* **9**, 45-54.
47. Vercauteren, I., Van der Schueren, E., Van Montagu, M. and Gheysen, G. (1996) Isolation of mRNA species expressed upon nematode infection by means of the differential display technique, *Fourth Annual Meeting of the European Union AIR-CAP on Resistance Mechanisms against plant-parasitic nematode*, Toledo,Spain, pp. 15 .
48. Williamson, V.M. and Hussey, R. S. (1996) Nematode Pathogenesis and Resistance in Plants, *The Plant Cell* **8**:1735-1745.
49. Wilson, M.A. and Bird, D.McK. (1993) Construction of a giant-cell-specific cDNA library, *Soc Nematol. Mol. Biol. Newsl.* **5**, 35-36.
50. Wyss, U. (1981) Ectoparasitic root nematodes: feeding behaviour and plant cell responses, in B.M. Zuckerman and R.A. Rohde (eds.), *Plant Parasitic Nematode*, New York, Academic Press. Vol. **3**, pp.325-351.
51. Zacheo, G., Bleve-Zacheo, T., Arrigoni-Liso, R., Arrigoni, O. and Lamberti, F. (1981) Changes in superoxide dismutase and peroxidase activities in pea roots infected by *Heterodera goettingiana*, *Nematologia Mediterranea.* **9**, 189-195.
52. Urao, T., Yamaguchi-Shinozaki, K., Mitsukawa, N., Shibata, D. and Shinozaki, K. (1996) Molecular cloning and characterization of a gene that encodes a MYC-related protein in Arabidopsis, *Plant Mol. Biol.* **32**, 571-576. (Added in proof)

NATURAL RESISTANCE: THE ASSESSMENT OF VARIATION IN VIRULENCE IN BIOLOGICAL AND MOLECULAR TERMS

Carolien ZIJLSTRA[1], Vivian C. BLOK[2] and Mark S. PHILLIPS[2]
[1]*Research Institute for Plant Protection (IPO-DLO), Binnenhaven 5, P.O. Box 9060, NL6700 GW, Wageningen, The Netherlands.*
[2]*Department of Nematology, Scottish Crop Research Institute, Invergowrie, Dundee, Scotland*

Abstract

The use of crop plants that are resistant and tolerant to attack by plant parasitic nematodes, is an important component of any control strategy. This chapter starts with defining resistance, tolerance and the nematode related terms virulence, pathotype and race.

Subsequently, the relationships between initial nematode numbers and both yield loss and nematode population dynamics are discussed and methods of assessment of resistance and virulence are described. A prerequisite for planning and implementing control programs against plant parasitic nematodes is unambiguous identification of the nematode which in turn requires a reliable diagnostic technique. As well as morphological techniques, of increasing importance are biochemical methods based principally on the analysis of proteins or DNA. These latter methods which enable specific or subspecific identification are reviewed. Finally the usefulness of molecular markers and gene isolation techniques are discussed.

1. Biological Assessment

1.1. DEFINITIONS

1.1.1. *In relation to the plant*

Resistance. Resistance, in nematology, is defined as the ability of a host plant to prevent multiplication of the parasite. It says nothing about mechanisms or symptomology.

Natural resistance occurs in two ways. Firstly, there are major genes, for instance the H1 gene conferring resistance to *Globodera rostochiensis* and the Mi gene which confers resistance to some populations of *Meloidogyne* spp. These genes tend to be dominant, can easily be detected with an avirulent pathotype or population and are useful to differentiate virulent and avirulent populations. Secondly, there are quantitative forms of resistance often derived from wild species of a crop such as resistance to *G. rostochiensis* and *G. pallida* derived from the diploid *Solanum vernei* which has been introduced into cultivars of the tetraploid *S. tuberosum* ssp. *tuberosum* [67].

Tolerance. In many host-nematode interactions the host is invaded by the nematode and may suffer damage whether the host is susceptible or resistant. A tolerant host is one that suffers proportionally less than another host genotype at a particular initial nematode density. Tolerance is independent of resistance and thus a resistant genotype may also be intolerant [83].

1.1.2. *In relation to the nematode*

Virulence. Virulence is the ability of a juvenile nematode, or of a population, to overcome the effect of host resistance gene(s) and to become a reproductive female and multiply. It can be used to describe the ability to overcome both quantitatively and qualitatively inherited resistance [61]. Although a truism, it is worth saying that to detect resistance it is necessary to have an avirulent population of nematodes.

Pathotypes/Race. The term pathotype is related to virulence. This should be used to define populations of a pathogen that share common and genetically understood virulence genes [1]. This definition is not always adhered to as in the pathotype scheme for *G. rostochiensis* and *G. pallida* [53]. In this case some pathotypes (e.g. Ro1 and Pa1) are defined in relation to their avirulence on potato clones with the major resistance genes H1 and H2 respectively. However, the other pathotypes within the scheme are differentiated on the basis of their ability to multiply on 'differential' clones with quantitative resistance. These differences are not genetically distinct, can be small and the assessments subject to the conditions in which assessments are done [69]. A similar situation occurs with the pathotype scheme for *Heterodera glycines*. The term race is also used to define groups of nematode populations but is usually based on differences in host range or ability to multiply on different host species. For example, for *Ditylenchus dipsaci* there are among others oat and onion races. Races of *Meloidogyne incognita* and *M. arenaria* are differentiated on the North Carolina Differential Host Test [43] whilst *M. hapla* is divided into races on the basis of their mode of reproduction.

1.2. POPULATION DYNAMICS AND YIELD LOSSES

The population dynamics of sedentary nematodes is density-dependent, generally with maximum rates of multiplication at the lowest initial population densities (Pi). As the Pi increases and invasion rates rise, so does competition between nematodes, shifting sex ratios in favour of males. At the same time more damage is done limiting the root space available for nematode reproduction. At high Pi values the damage can be so great as to cause a reduction in the population (Fig 1b). The relationship between yield loss and Pi is curvilinear with proportionally the greatest losses at lower initial population levels (Fig. 1a). A genotype with major gene resistance will reduce the density of an avirulent population but a genotype with quantitative resistance may allow low rates of multiplication. Whilst resistance and tolerance are independent they interact such that a tolerant but a partially resistant genotype may allow greater reproduction than an intolerant susceptible genotype at high Pi. This area is reviewed by McSorley & Phillips [57].

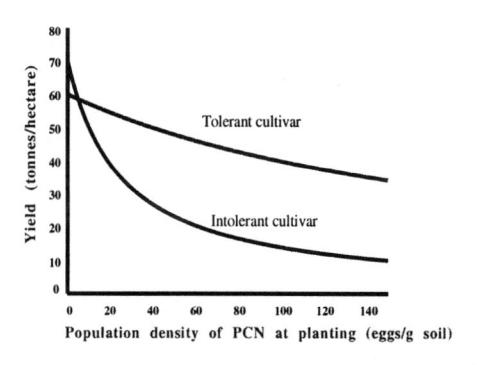

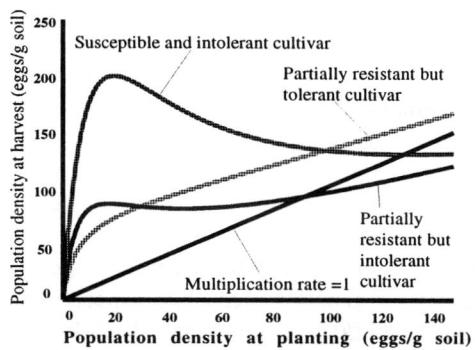

Figure 1a. Relationship between initial population density of *G. pallida* and yield.

Figure 1b. Relationship between initial population density of PCN and final population density

1.3. ASSESSMENT OF RESISTANCE AND VIRULENCE

Assessment of resistance can be done in a number of ways ranging from the use of plants from seed or tuber, culturing plants on agar in petri dishes [68], by using transformed hairy roots [51], plants grown in closed containers [70] or open containers (pots, tubes, segmented tubes) either in environment cabinets or glasshouses. Finally, assessments can be made in the field. Where testing for major gene resistance, provided an avirulent population is available, then the assay is relatively easy and the assessment is qualitative.

Nevertheless, with quantitative resistance the assays are more problematic. In such assays results can be variable and often require high levels of replication what ever system is used [52]. The rate of multiplication on such material is affected by environmental factors [69] including moisture, temperature, nutrient status of the growing medium but principally a major effect will be the initial nematode population density [67]. The problems of assessing resistance and virulence in different environments in relation to PCN and quantitative resistance are discussed by Mugniéry *et al.* [61]. Laboratory or glasshouse assays have however to be related to field results, and Phillips and Trudgill [73] showed that the measures of multiplication rates are often higher in glasshouse experiments largely due to the fact that the inoculum levels in terms of nematodes per unit of root are lower in pots where there is higher root density than in the field.

Most of the examples given here relate to potato cyst nematodes but the principles apply to a wide range of sedentary nematodes.

2. Molecular Approaches

2.1. INTERSPECIES AND INTRASPECIES VARIATION

Accurate and reliable identification of plant parasitic nematodes is fundamental to many aspects of their effective control and management. Moreover, to arrive at an effective way of studying host range, genetic variation, virulence and plant-nematode interactions in general, it is essential to work with characterized populations and to have access to a technique that can identify the species and/or pathotypes of the nematodes present in the population to be studied. For specific or subspecific identification of plant parasitic nematodes, several methods are used. Identification based on differences in morphological characters [43] is a method which requires considerable skill. Differential host range tests [43] which may identify host races, are very time consuming. Biochemical and molecular biological methods are being used more frequently to identify species and to relate the biological variation observed within species, particularly virulence differences, to the underlying genetics. This last aspect includes both studies of intraspecific genetic variation and the development of markers linked to virulence genes and ultimately the isolation of the genes involved. The various techniques which are now available to examine intraspecific relationships differ in their sensitivity in differentiating closely related isolates and, depending on the type of analysis, can group isolates in different ways due to different constraints on sequence evolution. Mode of reproduction (i.e. amphimictic such as found in most *Heterodera* or *Globodera* vs parthenogenetic such as in many *Meloidogyne, Xiphinema* or *Pratylenchus)* can influence the degree of variation within and between "species" and whether the variation is continuous or discontinuous. Because the study of plant-parasitic nematodes has concentrated primarily on isolates found parasitising crop plants, the isolates in these studies are unlikely to be representative of the total diversity in the species and if non-endemic, their geographic proximity may bear little relation to their biological similarity. Events subsequent to the introduction into a new region such as founder effects, selection, genetic drift and gene flow can have an impact on the ongoing relationships between isolates.

2.1.1. *Protein-based techniques*

Protein-based techniques including electrophoresis of total protein profiles, I.E.F., isozyme staining and 2-D electrophoresis have been used and are of particular value in genetic studies because they give allelic information which is of value for inheritance and genetic variation studies. Isozyme studies have been applied to a wide range of species including *Globodera, Meloidogyne* and *Pratylenchus* [8, 40, 47]. The similarity of species and/or isolates can be calculated based on the presence or absence of isozyme bands. Variation within species has been related to some pathotypes but variation within pathotypes has also been observed limiting the use of this technique for routine pathotype identification [37, 40]. With *Meloidogyne,* esterase and malate dehydrogenase phenotypes have been found to be a reliable method of distinguishing species and some race specific characters have been found [32, 33]. Disadvantages with isozyme analyses are that the gene products may be life-stage specific [27] and this technique is generally performed with adult females. Results can be influenced by environmental factors and the products visualised may have no relation to virulence differences.

The 2-D electrophoresis procedure is technically demanding and the gel analysis sophisticated. However, it has greater potential than isozyme studies for both revealing relationships within species and as a tool in genetic studies. It has been used to differentiate species such as *G. rostochiensis* and *G. pallida,* pathotypes of the former [4, 62] and for examining inter- and intra-population variation levels of these species [3, 28] and *H. avenae* [36]. Premachandran *et al.* [78] observed differences in the amount of protein variation in different *Meloidogyne* species and Ferris *et al.* [34] found more differences in proteins between the strict *H. avenae* isolates and the Gotland strain than they expected from morphological and pathogenicity tests. This is contrasted with later work using rDNA analysis [35, 36].

The 2-D electrophoresis technique has also been used by Castagnone-Sereno [15] in the analysis of unselected and selected lines of *M. incognita.* A protein spot was identified which distinguished these lines. It was proposed that amplification of sequences associated with the virulence phenotype could account for this observation. Further characterisation of this protein may reveal whether it is the basis for virulence differences.

A different protein-based identification system is the use of species specific monoclonal antibodies as a diagnostic tool. Such antibodies have been developed for identification of *M. incognita* [49] and *G. pallida* [5]. A disadvantage of protein-based techniques is that cellular expression of proteins often relies on developmental or environmental influences and may not be present uniformly in all individuals or nematode populations.

2.1.2. *DNA-based techniques*
DNA-based molecular diagnostics provide attractive solutions to the problems associated with protein-based identification methods. The direct examination of an organism's genotype by analysis of DNA sequence can serve as an excellent taxonomic tool. The base sequence of DNA is the primary source of biological variation and, in theory, nucleic acid analysis should provide the ultimate resolution in biochemical identification. DNA-based diagnostic methods do not rely on the expressed products of the genome, they are independent of environmental influence, independent of stage in the nematode's life cycle, and potentially extremely discriminating.

RFLPs. The first analyses conducted on nematodes concentrated on restriction fragment length polymorphisms (RFLPs) between reference populations, detecting some interspecific or even intraspecific variation. Curran *et al.* [24, 25] were the first to discriminate the species *M. arenaria* and *M. javanica* [24], and races of *M. arenaria, M. incognita* and *M. hapla* [25], by comparing RFLPs from total DNA, visualized on ethidium bromide stained gels. In this way, repetitive sequences of the genome appeared as bright bands. The same approach was used to separate *G. pallida* from *G. rostochiensis* [12, 29]. The sensitivity of this approach has been improved by using radio isotope in Southern blot hybridizations[41]. Intraspecific polymorphism has been observed using various clones as probes [17, 18, 26, 44, 71, 75, 81, 90] including labelled mtDNA [21, 71, 77]. This was of particular interest as previous work with soluble proteins and isozymes had shown little intraspecies variation. De Jong et al. [29], Schnick et al. [81] and Phillips et al. [71] have used this method to examine genetic variation between different populations of G. pallida but the relationships found did not correlate strictly with virulence and pathotype groupings. Differences in RFLP patterns obtained when total mtDNA was used as a probe was observed for different isolates of G. pallida [71] which had some relation to biological differences. RFLPs in mtDNA enabled Powers et al. [77] to discriminate between species of Meloidogyne, and upon probing with a labelled mitochondrial specific probe, Peloquin et al. [65] were able to distinguish isolates of M. hapla.

Probes. The availability of probes that can be used in simple dot blot experiments is desirable for diagnostic purposes. Besal et al. [9] isolated a species-specific mtDNA sequence from H. glycines, Curran and Webster [26] differentiated race A and B of M. hapla with an rDNA probe and Burrows and Perry [13] isolated G. pallida and G. rostochiensis specific probes by screening a genomic library of G. pallida. In similar ways specific probes were found for D. dipsaci [64], M. incognita [22] and M. arenaria [7]. Species specific satellite DNA sequences can be used as sensitive and reliable probes for direct identification of single nematodes, because the sequence is present in high copy number. Satellite DNA probes have been developed for identification of Bursaphelenchus xylophilus [84] and M. hapla [74]. However sensitive the above described techniques may be, they generally require the use of radioisotopes, or they require micrograms of DNA from several thousands of nematodes, numbers that may not always be available for routine determination.

A substantial improvement has been attained by the application of the polymerase chain reaction (PCR). This is an *in vitro* method for primer directed enzymatic amplification [31, 58] which can be used for nematode DNA amplification. It enables the exponential amplification of DNA starting from a single molecule. PCR raises the possibility of detecting minute quantities of nematode DNA.

Random sampling of genomic DNA. A group of PCR-based techniques has been introduced which randomly sample the genomic DNA. These techniques are attractive due to their high sensitivity in detecting variation, the use of random primers where there is no requirement to determine sequence information for the particular organism concerned and there is the potential to produce markers linked to virulence loci. The RAPD technique (random amplified polymorphic DNA) [88], AFLPs (amplified fragment length polymorphisms) [92], DAFs (DNA amplification fingerprinting) [6], and SSRs (simple sequence repeats) [10] are different examples of this approach.

RAPDs are generated by the amplification of random DNA segments in the target genome with single primers of approximately 10 nucleotides of arbitrary sequence. RAPDs have been employed in a number of studies to assess inter- and intra-species variation of root-knot nematodes [11, 20, 23, 89] cyst nematodes [14, 38, 54] and *Radopholus* [50]. An example of RAPD products produced with one primer from a range of widely dispersed isolates of *Meloidogyne* spp. is shown in Fig 2 which illustrates interspecies differences, the differing relationships between the species and intraspecies variation . Caswell-Chen et al. [14] compared H. cruciferae and H. schachtii populations

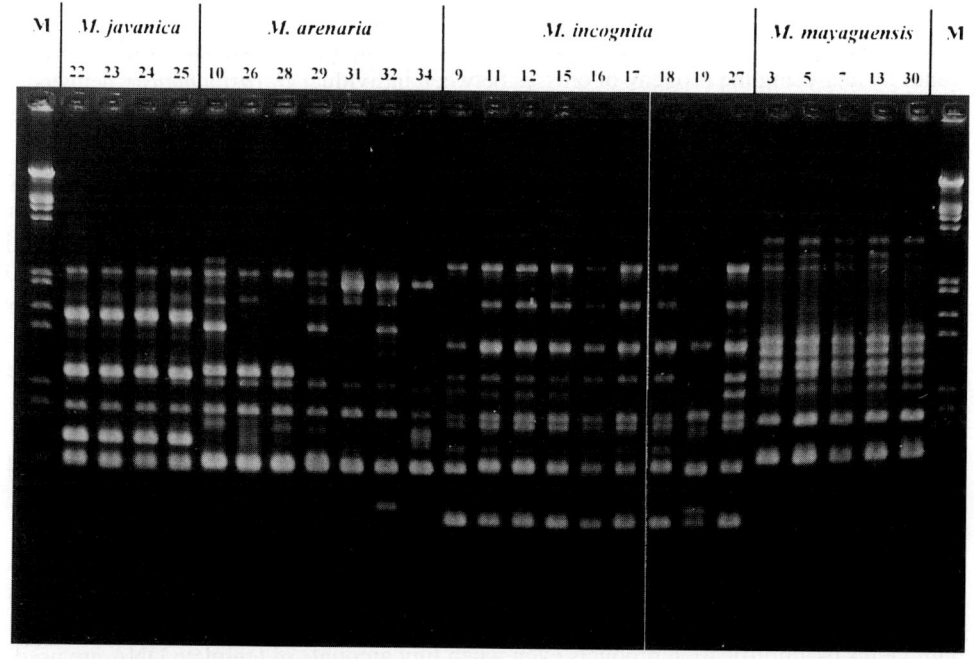

RAPDs with SC10-30

Figure 2. RAPD products from various isolates of *Meloidogyne* spp. produced with the primer 5'-CCGAAGCCCT-3' "M" is lambda-HindIII/EcoRI marker track

and examined variation within populations and found that geographic proximity did not necessarily correlate with genetic relatedness and suggested that multiple introductions into the Imperial Valley in California had occurred. They also used single cysts to compare intrapopulation variability. López-Braña *et al.* [56] have used RAPDs to differentiate the *H. avenae* Gotland strain isolates from other isolates of strict *H. avenae*. Folkertsma *et al.* [38] used RAPDs to group populations of *G. rostochiensis* and *G. pallida* from the Netherlands. Grouping according to pathotype designation was found with the first species but not with the second. However, clustering within both of these species was observed and there were differences in the intraspecific similarities. The results were interpreted in terms of gene pool similarity or "genetic signatures" which distinguish the *G. rostochiensis* pathotypes. Similar work has recently been reported using AFLPs [39].

RAPD markers that differentiate isolates which differ in virulence have been developed for soybean cyst nematodes on resistant soybean cultivars [54], *G. pallida* on the partially resistant potato cultivar Darwina [65] and *M. arenaria* with respect to the Mi gene [11]. These cases indicate a correlation of the marker fragment with a particular type of virulence. Such markers are desirable in crop managment schemes for diagnosing field isolates that share common virulence characteristics towards resistant cultivars. A simple dot blot of the RAPD-PCR product followed by hybridisation with RAPD markers linked to virulence characteristics could become a widely used approach for nematode species whose field populations exhibit different virulence patterns to resistant cultivars of their host crop. RAPDs are also being used to find markers associated with virulence genes by analysing the progeny of crosses between individual *G. rostochiensis* nematodes virulent and avirulent to the H1 gene [48, 79, and Castagnone-Sereno *et al.*, this volume]. Although tiny amounts of template DNA can be used in RAPD experiments, control of the amount of template DNA in the reaction is critical for obtaining reproducible results [59]. By increasing the primer length, nonspecific amplification is reduced and reproducibility is increased. For example, Castagnone-Sereno *et al.* [15] discriminated species and populations of *Meloidogyne* by using primers of 17-30 nucleotides. A way to improve the reproducibility in obtaining a desired RAPD marker fragment of a RAPD pattern is by specifically amplifying it. After sequencing the fragment, longer primers can be designed to complement the terminal sequences of the polymorphic DNA. These specific primers can be used to generate the sequence characterized region (SCAR). SCAR markers are less sensitive than RAPDs to varying reaction conditions. Williamson *et al.* [89] developed sets of SCAR primers that enabled identification and discrimination of *M. hapla* and *M. chitwoodi* using single juveniles. RAPDs of single individuals are too inconsistent for routine identification.

Specific amplification-based identification. Directed amplification of specific fragments that are present abundantly in the nuclear genome is very effective and ensures relatively high yields of amplification products even when tiny amounts of template DNA are used. PCR with primers derived from the cloned species specific satellite-DNA of *M. hapla* resulted in ladder patterns of monomers and multimers of the repeat unit when DNA extracted from single females, males, eggs or juveniles was used as template [16].

mtDNA. Amplification with or without subsequent restriction analysis of mtDNA sequences has been shown to be an attractive way to distinguish several *Meloidogyne* species [21, 42, 76] or within species variants including host races [46, 63]. This technique enables species identification on single juveniles or eggs [42, 76]. Several features of mtDNA, including the high copy number of mtDNA molecules per cell, the conservation in gene content, its mode of inheritance, more rapid evolution than single copy nuclear genes and lack of segregation and recombination has made this an interesting target for study. Peloquin *et al.* [65] proposed that within *M. hapla* there may be at least two distinct mitochondrial genomes which differ at a number of restriction enzyme sites.

Mitochondrial DNA analysis may well reveal evolutionary relationships within species and correlate with the presence of virulence groups provided hybridisations between different groups have not occurred, selection has not been operating or other processes such as genetic drift or founder effects but as with many of the other analyses described above, mtDNA is unlikely to be directly related to virulence genes.

rDNA. Comparative analysis of coding and non-coding regions of ribosomal DNA (rDNA) is a popular tool for species or subspecies identification of many organisms. However, the process of homogenisation [45] by which rDNA genes evolve can complicate their use as a means of assessing relationships between species. The eukaryotic rDNA repeat typically consists of three genes (18S, 28S and 5.8S), internal and external transcribed spacers, and an external non-transcribed spacer region. Direct sequence analysis of the ITS regions has been used to examine phylogenetic relationships between members of the genus *Heterodera* by Ferris *et al.* [35]. This work showed nearly as much variation between isolates of the same species as between related species and though sufficient differences were found to differentiate species, the basis for sequence differences between *H. glycines* isolates was not clear. The noncoding regions (ITS1 and 1TS2 and the IGS) are more variable than the coding regions in the rDNA cistron. This was demonstrated by Ferris *et al.* [35, 36] with *H. glycines* and *H. avenae* which illustrates that the choice of region in the rDNA can influence the amount of variability detected between isolates. Species within the *Xiphinema americanum* group [86], *Ditylenchus* [87] and *Meloidogyne* [93] have been identified by examining RFLPs in PCR products produced from amplification between the 18S and 28S genes which amplifies the intervening ITS1, 1TS2 and 5.8S gene. The ITS-RFLP approach can serve as a valuable tool to sensitively detect mixtures of species. *Dra*I, *Eco*RI and *Rsa*I restriction patterns of ITS-PCR products from mixtures of *M. hapla*, *M chitwoodi*, *M. fallax* and *M. incognita*, can detect these species when their presence in a mixture is 5% or more [94]. The ratio of the intensities of the bands of each species-specific restriction pattern observed corresponded with the ratio of the species present in the mixture, indicating that the nematodes of the populations of the different species tested each contain a similar number of ribosomal cistrons. This approach could be the basis for the development of a practical method for the determination of root-knot nematode species composition of field isolates to be used for routine analysis. The choice of restriction enzyme is critical in these experiments. Restriction of ITS-regions with some enzymes results in the appearance of minor bands in the restriction pattern. This can lead to the mis-interpretation of the result. The appearance of these minor bands can be due to variation in repeat units within an individual. This is, for instance, the case in *M. hapla* [93] and *G. pallida* (Malloch, personal communication). The work of Vahidi and Honda [85] on repeat units in the IGS region of *M. arenaria* identified another region of the rDNA which shows variation and may have potential for looking at relationships between populations.

The spliced leader RNA and 5S RNA genes also form an array of tandemly repeated genes in *G. rostochiensis* and *G. pallida* and amplification of the intervening region has been used for species discrimination. Some degree of pathotype discrimination within *G. pallida* was also observed [80]. Again these genes are attractive for PCR analysis because of their high copy number and sequence conservation. However, they are probably not directly related to sequences linked to virulence differences.

2.1.3. *Isolation of virulence genes*
The application of existing or improved molecular marker techniques such as those cited above and gene isolation strategies such as subtractive hybridisation [82], differential display [55] or cDNA/AFLP analysis [2] will no doubt become increasingly common in studies of plant-parasitic nematodes in the future. Isolation of specific loci involved in virulence and characterising their functions such as attraction to host or feeding site induction will be of immense value for understanding the biological variation observed

within species and for predicting the efficacy of different types of resistance. Access to specific nematode genes involved in plant parasitism may well have other benefits in giving rise to novel control strategies in the future. For a detailed review see Castagnone-Sereno *et al.*, this volume.

3. Conclusions

To minimise the damage that is caused by plant parasitic nematodes it is desirable to grow resistant crops. The utilisation of resistance requires an understanding of the relationship between population dynamics and yield losses as well as knowledge of the virulence characterisitics of nematode populations used in resistance assays. Qualitative resistance is relatively easy to identify when an avirulent population is available, whereas with quantitative resistance the assays are more problematic. For control and management strategies and for studying host range, genetic variation, virulence and plant-nematode interactions in general, it is important to have access to methods that enable identification of plant parasitic nematodes. Classical identification methods based on morphological characters or differential host range tests are time consuming and are now often being replaced by biochemical and molecular biological methods. These methods are also being used to study genetic variation within species, with a view to relating molecular data to biological information and virulence differences in particular. Protein-based techniques such as isozyme staining and monoclonal antibodies are being used for the identification of species of *Meloidogyne* and *Globodera* respectively and 2-D protein electrophoresis is being used to examine inter- and intra-specific variation. A disadvantage of protein based techniques is that cellular expression of proteins often relies on developmental stage of the nematode and can be subject to environmental influences and consequently may not give consistent results. DNA-based diagnostic methods, however, do not rely on the expressed products of the genome, are independent of environmental influence and developmental stage and thus have the potential of being more consistent in their discrimination. The introduction of PCR has facilitated the rapid increase in the number of identification techniques that are available which are relatively quick to use and require minute quantities of DNA. Very often, identification of an individual nematode is possible. The application of existing or improved molecular marker techniques and gene isolation strategies will no doubt become increasing common in studies of plant-parasitic nematodes in the future and offer suitable tools for control strategies.

References

1. Anderson, S. and Anderson K. (1982) Suggestions for determination and terminology of pathotypes and genes for resistance in cyst-forming nematodes, especially *Heterodera avenae, Bulletin OEPP/EPPO Bulletin* **12**, 379-386.
2. Bachem, C.W.B., van der Hoeven, R.S., de Bruijn, S.M., Vreugdenhil, D., Zabeau, M. and Visser, R.G.F. (1996) Visualisation of differential gene expression using a novel method of RNA fingerprinting based on AFLP, Analysis of gene expression during potato tuber development, *The Plant Journal* **9**, 745-753.

3. Bakker, J., Bouwman-Smits, L. and Gommers, F.J. (1992) Genetic relationships between *Globodera pallida* pathotypes in Europe assessed by using two dimensional gel electrophoresis of proteins, *Fund. appl Nematol.* **15**, 481-490.
4. Bakker, J. and Gommers, F.J. (1982) Differentiation of the potato cyst nematodes *Globodera rostochiensis* and *G. pallida* and the two *Globodera rostochiensis* pathotypes by means of two-dimensional electrophoresis, *Proc. K. Nederl. Akad. Wetensch. C* **85**, 309-314.
5. Bakker, J., Schots, A., Bouwman-Smits, L. and Gommers, F.J. (1988). Species specific and thermostable proteins from second stage larvae of *Globodera rostochiensis* and *G. pallida, Phytopathology* **78**,,300-305.
6. Baum, T.J., Gresshoff, P.M., Lewis, S.A. and Dean, R.A. (1994) Characterization and phylogenetic analysis of four root-knot nematode species using DNA amplification fingerprinting and automated polyacrylamide gel electrophoresis, *Molecular Plant-Microbe Interactions* **7**, 39-47.
7. Baum, T.J., Lewis, S.A. and Dean, R.A. (1994) Isolation, characterization, and application of DNA probes specific to *Meloidogyne arenaria. Phytopathology* **84**, 489-494.
8. Bergé, J.B. and Dalmasso, A. (1975) Characteristiques biochimiques de quelques population de *Meloidogyne hapla* et *Meloidogyne* spp,. *Cahiers de l'Office de la Recherche Scientifique et Technique OutreMer Série Biologie* **10**, 263-271.
9. Besal, E.A., Powers, T.O., Radice, A.D. and Sandall, L.J. (1988) A DNA hybridization probe for detection of soybean cyst nematode, *Phytopathology* **78**, 1136-1139.
10. Blok, V.C. and Phillips, M.S. (1995) The use of repeat sequence primers for investigating genetic diversity between populations of potato cyst nematode with differing virulence, *Fundam. appl. Nematol.* **18**, 575-582.
11. Blok, V.C., Phillips, M.S., McNicol, J.W. and Fargette, M. (1996) Genetic variation in tropical *Meloidogyne* spp. as shown by RAPDs, *Fundam. appl. Nematol.* (in press).
12. Burrows, P.R. and Boffey, S.A. (1986) A technique for the extraction and restriction endonuclease digestion of total DNA from *Globodera rostochiensis* and *Globodera pallida* second stage juveniles, *Revue Nématol.* **9**, 199-200.
13. Burrows, P.R. and Perry, R.N. (1988) Two cloned DNA fragments which differentiate *Globodera pallida* from *G.rostochiensis, Revue Nématol.* **11**, 441-445.
14. Caswell-Chen, E.P., Williamson, V.M. and Wu, F.F. (1992) Random amplified polymorphic DNA analysis of *Heterodera cruciferae* and *H. schachtii* populations,. *J. Nematol.* **24**, 343-351.
15. Castagnone-Sereno, P. (1994) Genetics of *Meloidogyne* virulence against resistance genes from Solanaceous crops. In, F. Lamberti, C. de Giorgi and D. McK. Bird (eds.) *Advances in Molecular Plant Nematology,* Plenum Press, New York, pp261-276.
16. Castagnone-Sereno, P., Esparago, G., Abad, P., Leroy, F. and Bongiovanni, M. (1995) Satellite DNA as a target for PCR specific detection of the plant parasitic nematode *Meloidogyne hapla, Curr. Genet.* **28**, 566-570.
17. Castagnone-Sereno, P., Piotte, C., Abad, P., Bongiovanni, M. and Dalmasso, A. (1991) Isolation of a repeated DNA probe showing polymorphism among *Meloidogyne incognita* populations, *J. of Nematology* **23**, 316-320.
18. Castagnone-Sereno, P., Piotte, C., Uijthof, J., Abad, P., Wajnberg, E., Vanlerberghe-Masutti, F., Bongiovanni, M. and Dalmasso, A. (1993) Phylogenetic relationships between amphimictic and parthenogenetic nematodes of the genus *Meloidogyne* as inferred from repetitive DNA analysis, *Heredity* **70**, 195-204.
19. Castagnone-Sereno, P., Vanlerberghe-Masutti, F. and Leroy, F. (1994) Genetic polymorphism between and within *Meloidogyne* species detected with RAPD markers, *Genome* **37**, 904-909.
20. Cenis, J.L. (1993) Identification of four major *Meloidogyne* spp. by random amplified polymorphic DNA (RAPD-PCR), *Phytopathology* **83**, 76-80.
21. Cenis, J.L., Opperman, C.H., and Triantaphyllou, A.C. (1992) Cytogenetic, enzymatic and restriction fragment length polymorphism variation of *Meloidogyne* spp. from Spain, *Phytopathology* **82**, 527-531.
22. Chacón, M.R., Parkhouse, R.M.E., Robinson, M.P., Burrows, P.R. and Gárate, T. (1991) A species specific oligonucleotide DNA probe for the identification of *Meloidogyne incognita, Parasitology* **103**, 315-319.
23. Chacón, M.R., Rodriguez, E., Parkhouse, R.M.E., Burrows, P.R., and Gárate, T. (1994) The differentiation of parasitic nematodes using random amplified polymorphic DNA, *J. of Helminthology* **68**, 109-113.
24. Curran, J., Baillie, D.L. and Webster, J.M. (1985) Use of genomic DNA restriction fragment length

differences to identify nematode species, *Parasitiology* **90**, 137-144.
25. Curran, J., McClure, M.A. and Webster, J.M. (1986) Genotypic differentiation of *Meloidogyne* populations by detection of restriction fragment length differences in total DNA, *J. of Nematology* **18**, 83-86.
26. Curran, J. and Webster, J.M. (1987) Identification of nematodes using restriction fragment length differences and species-specific DNA probes, *Can. J. Pl. Path.* **9**, 162-166.
27. de Boer, J.M., Overmars, H.A., Bakker, J. and Gommers, F.J. (1992) Analysis of two dimensional protein patterns from developmental stages of the potato cyst-nematode, *Globodera rostochiensis, Parasitology* **105**, 461-474.
28. de Boer, J.M., Overmars, H., Bouwman-Smits, L., de Boevere, M., Gommers F.J. and Bakker, J. (1992) Protein polymorphisms within *Globodera pallida.* assessed with mlni two dimensional gel electrophoresis of single females, *Fundam. appl. Nematol.* **15**, 495-501.
29. de Jong, A.J., Bakker, J., Roos, M. and Gommers F.J. (1989) Repetitive DNA and hybridization patterns demonstrate extensive variability between the sibling species *Globodera rostochiensis* and *G. pallid., Parasitology* **99**, 133-138.
30. Elston, D.A., Phillips, M.S. and Trudgill, D.L. (1991) The relationship between initial population density of potato cyst nematode *Globodera pallida* and the yield of partially resistant potatoes, *Revue de Nématol.* **14**, 213-219.
31. Erlich, H.A., Gelfand, D.H., and Saiki, R.K. (1988) Specific DNA amplification, *Nature* **331**, 461-462.
32. Esbenshade, P.R., and Triantaphyllou, A.C. (1990) Isozyme phenotypes for the identification of *Meloidogyne* species, *J. of Nematology* **22**, 10-15.
33. Fargette, M. (1987) Use of the esterase phenotype in the taxonomy of the genus *Meloidogyne*. 2. Esterase phenotypes observed in West African populations and their characterisation, *Revue. Nématol.* **10**, 45-56.
34. Ferris, V.R., Faghihi, J., Ireholm, A. and Ferris, J.M. (1989) Two-dimensional protein patterns of cereal cyst nematode, *Phytopathology* **79**, 927-933.
35. Ferris, V.R., Ferris, J.M. and Faghihi, J. (1993) Variation in spacer ribosomal DNA in some cyst forming species of plant parasitic nematodes, *Fundam. appl. Nematol.* **16**, 177-184.
36. Ferris, V.R., Ferris, J.M., Faghihi, J. and Ireholm, A. (1994) Comparisons of isolates of *Heterodera avenae* using 2-D PAGE protein patterns and ribosomal DNA, *J. of Nematology* **26**, 144-151.
37. Flemming, C.C. and Marks, R.J. (1983) The identification of the potato cyst nematodes *Globodera rostochiensis* and *G. pallida* by isoelectric focusing of proteins on polyacrylamide gels, *Ann. appl. Biol.* **103**, 277-281.
38. Folkertsma, R.T., Rouppe van der Voort, J.N.A.M., van Gent-Pelzer, M.P.E., de Groot, K.E., van den Bos, W.J., Schots, A. Bakker, J. and Gommers, F.J. (1994) Inter- and intraspecific variation between populations of *Globodera rostochiensis* and *G. pallida* revealed by random amplified polymorphic DNA, *Phytopathology* **84**, 807-811.
39. Folkertsma, R.T., Rouppe van der Voort, J.N.A.M., de Groot, K.E., van Zandvoort, P.M., Schots, A. Gommers, F.J., Helder, J. and Bakker, J. (1996) Gene pool similarities of potato cyst nematode populations assessed by AFLP analysis, *Molecular Plant-Microbe Interactions* **9**, 47-54.
40. Fox, P.C. and Atkinson, H.J. (1984) Isoelectric focussing of general proteins and specific enzymes from pathotypes of *Globodera rostochiensis* and *G. pallida*, *Parasitology* **88**, 131-139.
41. Gárate, T., Robinson, M.P., Chac6n, M.R. and Parkhouse, R.M.E. (1991) Characterization of species and races of the genus *Meloidogyne* by DNA restriction enzyme analysis,. *J. of Nematology* **23**, 414-420.
42. Harris, T.S., Sandall, L.J. and Powers, T.O. (1990) Identification of single *Meloidogyne* juveniles by polymerase chain reaction amplification of mitochondrial DNA, *J. of Nematology*, **22**, 518-524.
43. Hartman, K.M. and Sasser, J.N. (1985) Identification of *Meloidogyne* species on the basis of differential host tests and perineal pattern morphology. In K.R. Barker, C.C. Carter and J.N. Sasser (eds.), *An Advanced Treatise on Meloidogyne, Vol 2, Methodology*, North Carolina State University Graphics, Raleigh.
44. Hiatt, E.E., Georgi, L., Huston, S., Harsmaan, D.C., Lewis, S.A. and Abbott, A.G. (1995) Intra- and interpopulation genome variation in *Meloidogyne arenaria, J. of Nematology* **27**, 143-152.
45. Hillis, D.M. and Dixon, M.T. (1991) Ribosomal DNA, molecular evolution and phylogenetic inference, *The Quarterly Review of Biology* **66**, 411-453.
46. Hugall, A., Moritz, C., Stanton, J. and Wolstenholme, D.R. (1994) Low, but strongly structured

mitochondrial DNA diversity in root-knot nematodes (*Meloidogyne*), *Genetics* **136**, 903-912.
47. Ibrahim, S.K., Perry, R.N. and Webb, R.M. (1995) Use of isoenzyme and protein phenotypes to discriminate between six *Pratylenchus* species from Great Britian, *Ann. app. Biol.* **136**,,317-327.
48. Janssen, R., Bakker, J., and Gommers, F,J, (1991) Mendelian proof for a gene-for-gene relationship between virulence of *Globodera rostochiensis* and the Hl resistance gene in *Solanum tuberosum* spp. *andigena* CPC 1673, *Revue Nématol* **14**, 213-19.
49. Jones, J.T., Ambler, D.J. and Robinson, M.P. (1988) The application of monoclonal antibodies to the diagnosis of plant pathogens and pests. *Brighton Crop Protection Conference, Pests and Dis.*, **2**, 767-776.
50. Kaplan, D.T., Vanderspool, M.C., Garett, C., Chang, S. and Opperman, C.H. (1996) Molecular polymorphisms associated with host range in the highly conserved genomes of burrowing nematodes, *Radopholus* spp., *Molecular Plant-Microbe Interactions* **9**, 32-38.
51. Kumar, A., Forrest, J.M.S. (1990) Reproduction of *Globodera rostochiensis* on transformed roots of *Solanum tuberosum* cv. Desiree, *J. of Nematology* **22**, 395-398.
52. Kort, J., Jaspers, C.P. and Dijkstra, D.L. (1972) Testing for resistance to pathotype C of *Heterodera rostochiensis* and the practical application of *Solanum vernei* - hybrids in the Netherlands, *Ann. Appl. Biol.* **71**, 289-294.
53. Kort, J., Ross, H., Rumpenhorst, H.J. and Stone, A.R. (1978) An international scheme for identifying and classifying pathotypes of potato cyst nematodes *Globodera rostochiensis* and *G. pallida*, *Nematologica* **23**, 333-339.
54. Li, I., Faghihi, I., Ferris, J.M. and Ferris, V.R. (1996) The use of RAPD amplified DNA as markers for virulence characteristics in soybean cyst nematodes, *Fundam. Appl. Nematol.*, **19**,143-150.
55. Liang, P. and Pardee, A.B. (1992) Differential display of eukaryotic messenger RNA by means of the polymerase chain reaction, *Science*, **257**, 967-971.
56. López-Braña, I., Romero, M.D. and Delibes, A. (1996) Analysis of *Heterodera avenae* populations by the random amplified polymorphic DNA technique, *Genome* **39**, 118-122.
57. McSorley, R. and Phillips, M.S. (1993) Modelling populations dynamics and yield losses and th
58. Marx, J.L. (1988). Multiplying genes by leaps and bounds, *Science* **240**, 1408-1410.
59. Munthali, M., Ford-Lloyd, B.V. and Newburry, H.J. (1992) *PCR Methods and Applications*. Cold Spring Harbor Press, New York, pp. 274-276.
60. Mugniéry, D. and Person, F. (1976) Méthode d'élevage de quelques nématodes à kystes du genre *Heterodera*, *Sciences Agronomiques, Rennes*, 217-220.
61. Mugneiry D., Phillips, M. S., Rumpenhorst, H. J., Stone A. R., Treur, B. & Trudgill, D. L. (1989) Interim conclusions of the Eppo panel on the assessment of partial resistance to and pathotype and virulence in potato cyst nematodes, *Eppo Bulletin* **19**, 7-25.
62. Ohms, J.P. and Heinicke, D.H. (1985) Pathotypen des Kartoffelnematoden. II. Bestimmung der Rassen von *Globodera rostochiensis* durch die Mikro-2D-Elektrophorese von Einzelzysten, *J. Pt. Dis. Protect.* **92**, 225-232.
63. Okimoto, R., Chamberlin, H.M., MacFarlane, J.L. and Wolstenholme, D.R. (1991) Repeated sequence sites in mitochondrial DNA molecules of root-knot nematodes (*Meloidogyne*), nucleotide sequences, genome location and potential for host-race identification, *Nucleic Acids Res.* **19**, 1619-1626.
64. Palmer, H.M., Atkinson, H.J. and Perry, R.N. (1991) The use of DNA probes to identify *Ditylenchus dipsaci.*, *Revue Nématol.* **14**, 625-628.
65. Pastrik, K.H., Rumpemhorst, H.-J. and Burgermeister, W. (1995) Random amplified polymorphic DNA analysis of a *Globodera pallida* population selected for virulence, *Fundam. appl. Nematol.* **18**, 109-114.
66. Peloquin, J.J., Bird, D. McK., Kaloshian, I. and Matthews, W.C. (1993) Isolates of *Meloidogyne hapla* with distinct mitochondrial genomes, *J. of Nematology* **25**, 239-243.
67. Phillips, M. S. (1984) The effect of initial population density on the reproduction of *Globodera pallida* on partially resistant potato clones derived from *Solanum vernei.*, *Nematologica* **30**, 57-65.
68. Phillips, M. S. (1994) Inheritance of resistance to nematodes. In J.E. Bradshaw and G.R. Mackay (eds.), *Potato Genetics*, CAB International, Oxford, pp. 319-337.
69. Phillips, M.S., Forrest, J.M.S., and Hayter, A.M. (1979) Genotype X environment interaction for resistance to the white potato cyst nematode (*Globodera pallida*, pathotype E) in *Solanum vernei* X *S. tuberosum* hybrids, *Euphytica* **28**, 515-519.
70. Phillips, M.S., Forrest, J.M.S. and Wilson, LA. (1980) Screening for resistance to potato cyst nematode using closed containers, *Ann. appl. Biol.* **96**, 317-322.

71. Phillips, M.S., Harrower, B.E., Trudgill, D.L., Waugh, R. and Catley, M. A. (1992) Genetic variation in British populations of *Globodera pallida* as revealed by isozyme and DNA analysis, *Nematologica* **38**, 301-319.
72. Phillips, M.S. and Trudgill, D.L. (1983) Variations in the ability of *Globodera pallida* to produce females on potato clones bred from *Solanum vernei* or *S. tuberosum* ssp. *andigena* CPC 2802, *Nematologica.* **29**, 217-226.
73. Phillips, M.S. and Trudgill, D.L. (1985) Pot and field assessment of partial resistance of potato clones to different populations and densities of *Globodera pallida, Nematologica* **31**, 433-442.
74. Piotte, C. Castagnone-Sereno, P., Bongiovanni, M., Dalmasso, A. and Abad, P. (1995) Analysis of a satellite DNA from *Meloidogyne hapla* and its use as a diagnostic probe, *Phytopathology* **85**, 458-462.
75. Piotte, C., Castagnone-Sereno, P., Uijthof, J., Abad, P., Bongiovanni, M. and Dalmasso, A. (1992) Molecular characterization of species and populations of *Meloidogyne* from various geographic origins with repeated DNA homologous probes, *Fundam. appl. Nematol.* **15**, 271-276.
76. Powers, T.O. and Harris, T.S. (1993) A polymerase chain reaction method for identification of five major *Meloidogyne* species, *J. Nematol.* **25**, 1-6.
77. Powers, T.O., Platzer, E.G. and Hyman, B.C. (1986) Species-specific restriction site polymorphism in root-knot nematode mitochondrial DNA, *J. of Nematology* **18**, 288-293.
78. Premachandran, D., Bergé, J. B. and Bride, J.M. (1984) Two-dimensional electrophoresis of proteins from root-knot nematodes, *Revue Nématol.* **7**, 205-207.
79. Rouppe van der Voort, J.N.A.M., Roosien, J., van Zandvoort, R., Folkertsma, R.T., van Enckevort, E.L.J.G., Janssen, R., Gommers, F.J. and Bakker, J. (1994) Linkage mapping in potato cyst nematodes. In, F. Lamberti, C. de Giorgi and D. McK. Bird (eds.), *Advances in Molecuiar Plant Nematology*, Plenum Press, New York, pp57-63.
80. Shields, R., Flemming C.C. and Stratford, R. (1996) Identification of potato cyst nematodes using the polymerase chain reaction, *Fundam. appl. Nematol.* **19**, 167-173.
81. Schnick, D., Rumpenhorst, H.J. and Burgermeister, W. (1990) Differentiation of closely related *Globodera pallida* (Stone) populations by means of DNA restriction fragment length polymorphisms (RFLPs), *J. Phytopathology* **130**, 127-136.
82. Straus, D. and Ausubel. F.M. (1990) Genomic substraction for cloning DNA corresponding to deletion mutations, *Proc. Nati. Acad. Sci. USA* **87**, 1889-1893.
83. Trudgill, D.L. (1991) Resistance to and tolerance of plant parasitic nematodes in plants, *Annual Review of Phytopathology* **29**, 167-192.
84. Tares, S., Lemontey, J.M., de Guiran, G. and Abad, P. (1994) Use of species-specific satellite DNA from *Bursaphelenchus xylophilus* as a diagnostic probe, *Phytopathology* **84**, 294-298.
85. Vahidi, H. and Honda, B.M. (1991) Repeats and subrepeats in the intergenic spacer of rDNA from the nematode *Meloidogyne arenaria, Mol. Gen. Genet.* **227**, 334-336.
86. Vrain, T.C., Wakarchuk, DA., Levesque, AC. and Hamilton R.L. (1992) Intraspecific rDNA restriction fragment length polymorphism in the *Xiphinema americanum* group, *Fundam. appl. Nematol.* **15**, 563-573.
87. Wendt, K.R., Vrain, T.C. and Webster, J.M. (1993) Separation of 3 species of *Ditylenchus* and some host races of *D. dipsaci* by restriction fragment length polymorphism, *J. of Nematology* **25**, 555-565.
88. Williams, G.K., Kubelik, AR., Livak, K.J., Rafalski, J.A. and Tingey, S.V. (1990).DNA polymorphisms amplified by arbitrary primers are useful genetic markers, *Nucleic Acids Res.,* **18**, 6531-6535.
89. Williamson, V.M., Caswell-Chen, E.P., Westerdahl, B.B., Wu, F.F. and Caryl, G. (1996) A PCR assay to identify and distinguish single juveniles of *Meloidogyne hapla* and *M chitwoodi, J. of Nematology* (in press).
90. Xue, B., Baillie, D.L., Beckenbach, K. and Webster, J.M. (1992) DNA hybridization probes for studying the affinities of three *Meloidogyne* populations, *Fundam. appl. Nematol.* **15**, 35-41.
91. Xue, B., Balllie, D.L. and Webster, J.M. (1993) Amplified fragment length polymorphism of *Meloidogyne* spp. using oligonucleotide primers, *Fundam. appl. Nematol.* **16**, 481-487.
92. Zabeau, M. and Vos, P. (1993) Selective restriction fragment amplification, A general method for DNA fingerprinting, European Patent Application No.,92402629.7; Publication number EP 0534858 Al.
93. Zijlstra, C., Lever, A.E.M., Uenk, B.J. and van Silfhout, C.H. (1995) Differences between ITS regions of isolates of root-knot nematodes *Meloidogyne hapla* and *M chitwood, Phytopathology* **65**, 1231-1237.
94. Zijlstra, C., Uenk, B.J. and van Sillhout, C.H. (1996) A reliable, precise method to differentiate species of root-knot nematodes in mixtures on the basis of ITS-RFLPs, *Fundam. appl. Nematol.* **20**, 59-63.

GENETIC AND MOLECULAR STRATEGIES FOR THE CLONING OF (A)VIRULENCE GENES IN SEDENTARY PLANT-PARASITIC NEMATODES

Phillipe CASTAGNONE-SERENO[1], Pierre ABAD[1], Jaap BAKKER[2], Valerie M. WILLIAMSON[3], Fred J. GOMMERS[2] and Antoine DALMASSO[1]

[1]INRA, Laboratoire de Biologie des Invertébrés, BP 2078, 06606 Antibes cedex, France;
[2]Wageningen Agricultural University, Department of Nematology, PO Box 8123, 6700 ES Wageningen, The Netherlands;
[3]University of California, Department of Nematology, Davis, CA 95616, USA

Abstract

In some cases, crop resistance can be overcome by virulent nematode pathotypes, which are able to reproduce on plant carrying a resistance gene. The understanding of the molecular determinants of (a)virulence in the nematode is essential to the development of new and more durable forms of resistance. In this review will be presented the current strategies designed to identify and isolate (a)virulence genes in cyst and root-knot nematodes, in relation with the mode of reproduction of these two groups.

1. Introduction

Plant resistance to nematodes is often the most effective and environmentally safe method to control these pests. During the last decades, breeding programs have been successful in incorporating new nematode resistance into cultivated plant species. In most cases, single dominant genes were used whose phenotypic expression is characterized by a hypersensitive reaction at the infection site, This type of resistance is especially active against sedentary endoparasitic species (i.e. root-knot and cyst nematodes). However, the utility of these intensive breeding efforts have been compromised in a number of cases by the development of virulent nematode pathotypes able to overcome the resistance gene.

It is obvious that the understanding of the mechanism(s) by which the nematode is able to circumvent a plant resistance gene is essential to the development of new and more durable forms of resistance. However, relationships between sedentary nematodes and their hosts are very complex [25], and the molecular determinants of these interactions are still unknown. Therefore, no assumption can be made *a priori* about the function(s) involved in nematode (a)virulence, and it is difficult to conceive any direct approach to isolating the gene(s) related to such a phenotype. The lack of information regarding the molecular events leading to resistance in the plant further precludes a 'candidate gene' strategy. In addition, the most important sedentary endoparasitic species exhibit diverse modes of reproduction (i.e. amphimixis in *Globodera* and *Heterodera* spp., parthenogenesis in *Meloidogyne* spp.), preventing the development of a 'universal' genetic approach.

This review focuses on the current genetic and molecular strategies designed to elucidate the molecular basis of (a)virulence in cyst and root-knot nematodes in relation to the specific mode of reproduction of each of these two groups.

2. Cloning by Linkage Map Analysis and Chromosome Walking

The most damaging cyst nematode species are primarily amphimictic, allowing classical genetic experiments to be conducted (i.e. controlled single matings, segregation analysis in successive generations and reciprocal crosses). These Mendelian studies should lead to substantial information on the number of genes involved, their dominance/recessivity, the interaction or lack of interaction between them, and their possible maternal or sex-linked inheritance. Assuming that the genetic determinism of (a)virulence is not too complex, it is reasonable to consider isolating sequences related to this character by obtaining molecular markers linked to the loci of interest, followed by positional cloning. The advantage of this approach is that it requires no assumption about the function encoded by the gene to be cloned, but only that closely flanking markers are available. Two model systems for which analysis is currently in progress will be developed further.

2.1. GLOBODERA ROSTOCHIENSIS AND THE H1 RESISTANCE GENE FROM SOLANUM TUBEROSUM SSP. ANDIGENA

Currently five pathotypes are recognized in European *G. rostochiensis* populations. The pathotypes are classified as virulent or avirulent for a differential if Pf/Pi values are respectively >1 or <1. This implies that pathotypes are not necessarily fixed for alleles for virulence or avirulence. As a consequence populations classified as identical pathotypes vary in their number of virulent individuals for a given differential. For instance, as shown in Table 1, the rate of virulent phenotypes in a number of Ro3 and Ro5 field populations classified as virulent for *Solanum tuberosum* ssp. andigena CPC 1673 (containing the dominantly inherited *H1* resistance gene) varied from 1.5 to 84.6 [14].

This explains why field populations are not useful for an appropriate analysis of the inheritance of virulence. Therefore, we set out to obtain pure virulent and avirulent lines by controlled matings between one female and one male [15]. Virulent lines for the *H1* resistance gene were selected as follows. Adult females were reared in Petri dishes on roots of sprouts of the resistant cultivar 'Saturna' grown on water agar. To prevent unwanted matings only one juvenile per Petri dish was inoculated. These single crossing

TABLE 1. Percentage of virulent phenotypes for the *H1* resistance gene in a number of *Globodera rostochiensis* populations classified as Ro3 or Ro5.

Ro3						Ro5
C133	C129	C176	C150	C156	C163	Harmerz
5.5%	30.2%	1.5%	14.6%	23.4%	27.8%	84.6%

were multiplied by inoculating one cyst per pot on cultivar 'Saturna'. Multiplication was continued to the F3 or F4. Numbers of virulent phenotypes were estimated in Petri dishes by inoculating two juveniles per root tip and comparing the numbers of cysts developed on the resistant cultivar with those that were formed on the susceptible cultivar 'Eigenheimer'. Selection of lines avirulent to the *H1* resistance gene was similar to the selection of virulent lines except that for all generations the susceptible cultivar 'Eigenheimer' was used.

Mendelian proof for a gene-for-gene relationship was given by Janssen et al. [16] according to Flor's criterion [10] by crossings between virulent and avirulent lines. Actually the first indication that gene-for-gene systems may be operating in potato cyst nematodes was obtained by Jones et al. [18] in their study with crosses between rather virulent and avirulent populations of *G. pallida*. Selfing the F1 showed that virulence to the *H1* gene is controlled by a single recessive gene. Numbers of avirulent and virulent juveniles showed the expected 1:3 segregation for virulent (aa) and avirulent genotypes (AA and Aa). Reciprocal crosses did not give evidence for a sex-linked inheritance of virulence.

For a linkage map to trace markers linked to the (a)virulence gene the progeny of 300 virulent lines [1] was selected via two backcrosses with the virulent parent line (Figure 1). In the P generation virulent females were crossed with avirulent males. The virgin F1 females were back crossed with virulent males resulting in B1 individuals. These were inoculated on the resistant cultivar 'Saturna' where the virulent individuals developed into females. These were separately back crossed again with virulent males. This rather unusual second backcross with the virulent parent line is inherent to the fact that the *H1* resistance gene is not effective against males. Selfing of B1 individuals is not possible because avirulent males (AA or Aa) also do develop [16]. The B2 individuals were multiplied independently by inoculating potato plants with cysts. Thus a progeny of 300 lines can be employed for linkage studies [13]. Due to the two back crosses only markers specific for the avirulent parent line are informative. This is illustrated in Figure 1, in which the virulent (aamm) and the avirulent (AAMM) line are fixed for the alternate alleles of locus M. Marker M is defined by a fragment specific for the avirulent line which behaves as a dominant marker. In case marker M is unlinked the distribution of genotypes in the B1 will be: 25% AaMm, 25% Aamm, 25% aamm and 25% aaMm. Since only juveniles homozygous for virulence will develop into females 50% of the progeny will contain marker M. If marker M is linked to the virulence locus, the presence of M in the progeny results from a crossing-over with the virulence locus in the F1. Preliminary segregation studies were carried out with RAPD markers but are being continued with the AFLP technology [28].

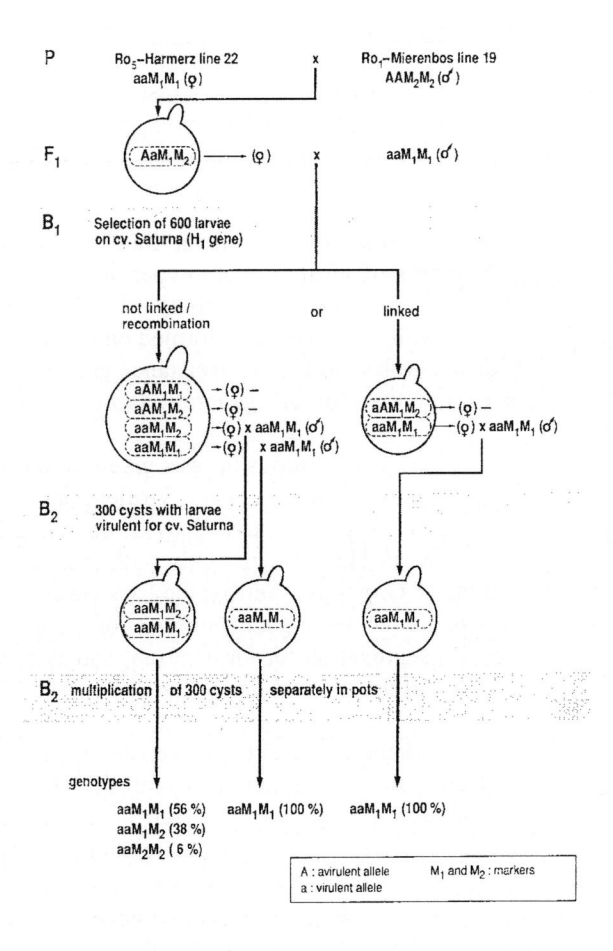

Figure 1. Backcross scheme for the identification of molecular markers linked to the (a)virulence gene in *Globodera rostochiensis* for the *H1* resistance gene in *Solanum tuberosum* ssp. *andigena*.

2.2. HETERODERA GLYCINES AND SOYBEAN

A second model system that has been extensively studied is the interaction of the soybean cyst nematode *H. glycines* with soybean, even though in some cases the genetics of this interaction does not follow the conventional gene-for-gene hypothesis. Resistance in soybean has been demonstrated to be complex, involving a number of major recessive and dominant genes, each with multiple alleles, some of which may be linked [22, 24]. Recently, the use of inbred nematode lines has allowed resolution of major resistance genes [27, 8].

Crosses have been carried out between two highly homozygous inbred lines of *H.*

glycines, each line parasitizing a different resistant soybean genotype. Segregation analysis of the F2 generation suggested the occurrence of two unlinked loci controlling virulence [23]. A search for RAPD markers linked to each locus has been initiated by bulked segregant analysis [21] using approximately 400 recombinant inbred lines derived from the initial cross. Moreover, expressed sequence tags isolated from a cDNA library constructed from *H. glycines* juveniles will be used to enhance the marker density in the genome of the nematode [23]. This approach should result in a linkage map for the parasitism loci and provide a major step toward isolating these loci.

3. Cloning by Differential Analysis

Unlike cyst nematodes, many of the most damaging *Meloidogyne* species reproduce by mitotic parthenogenesis, which precludes classical Mendelian analysis of the genetic determinism of characters of interest. To identify and isolate genes from these nematodes, other approaches must be developed. The relationship between *M. incognita* and tomato has been extensively analyzed as a model for understanding plant-nematode interactions, since resistance in tomato is reported to be governed by a single dominant locus, the *Mi* gene [11]. Previous works showed that selection of *Mi*-virulent lines from descendants of single avirulent nematodes was possible under laboratory conditions [2, 17]. The finding of variability for virulence among isofemale lines derived from a single *M. incognita* isolate and of a good correlation between the virulence of mother lines and that of their daughter lines strongly suggests that the virulence character is genetically inherited [5]. Moreover, experimental results suggested that several genes may be involved in determination of virulence [26, 5]. Considering that reproduction occurs by mitotic parthenogenesis in this species, the mechanism underlying the appearance of virulent clones from avirulent progenitors is unclear. Since the selection is initiated with a line derived from a single nematode and theoretically there is no gene segregation during parthenogenetic reproduction, the virulent and avirulent lines should be nearly isogenic apart their ability to reproduce on tomatoes carrying the *Mi* gene. Thus is is likely that any biochemical or molecular difference between them is related to this character. Such nearly isogenic nematode strain pairs have been obtained [17, 6] and constitute an ideal starting point for the cloning of (a)virulence genes by means of differential analysis.

One possible way to characterize genes involved in the (a) virulence phenotype is to identify proteins specific to nematodes with that character. For that purpose, a protein separation method with high resolution is essential. Two-dimensional gel electrophoresis (2-DGE) is currently the most efficient approach as up to seven thousand proteins were reported to have been separated in a single operation [19]. Soluble proteins from pairs of nearly-isogenic avirulent and virulent *M. incognita* lines have been compared using 2-DGE combined with a sensitive silver stain [6, 7]. Of the four hundred spots resolved on average on the gels, a very small number of differences were consistently found between avirulent and virulent nematodes. Both qualitative and quantitative differences were identified. For example, one protein, named *pda*, was present only in avirulent nematodes, while a second one, named *pdv*+, was reproducibly found to be more abundant in virulent nematodes [6]. The correlation between the presence/absence of a protein and the nematode (a)virulence in independent pairs of nearly-isogenic strains suggests that this protein may be involved in the cascade of biochemical events leading to the compatible or incompatible reaction with the plant. Such differential proteins allow the development of two complementary strategies to identify genetic determinants of (a)virulence. The first strategy consists of the purification of these proteins, production of antibodies specific to them, and use of these antibodies to screen expression libraries. The second strategy is based on the determination of partial amino acid sequence information directly from protein spots isolated from the 2-DGE gels. This sequence would be used to design degenerate synthetic oligonucleotides in order to isolate the cDNA or genomic sequences encoding the differential proteins.

Using the same biological material, a more direct cloning strategy, based on differential screening of cDNA libraries, has been recently developed [4]. A cDNA library was constructed from RNA isolated from second-stage juveniles of a nematode strain selected for virulence on plants with *Mi*. This library was screened by differential hybridization using total cDNA from avirulent and virulent lines as successive probes (Figure 2).

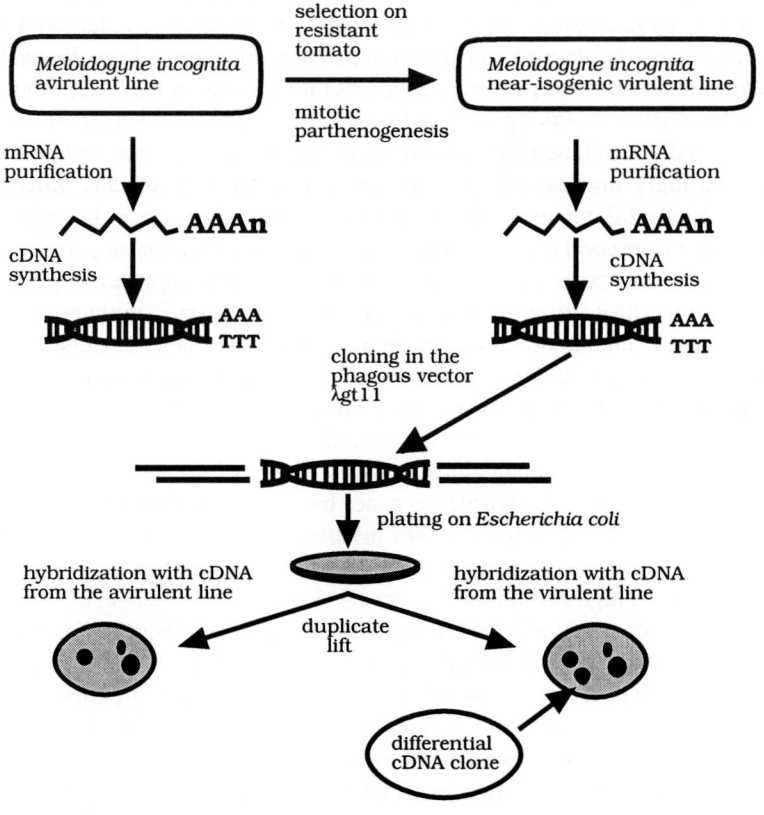

Figure 2. Schematic representation of the differential screening strategy developed for the cloning of *Meloidogyne incognita* (a)virulence genes.

By this approach, four differentially hybridizing clones have been isolated so far, all of them corresponding to genes more highly expressed in the virulent than the avirulent line. The detailed analysis of these clones is currently in progress, and should provide new information on the molecular nature of (a)virulence genes in *M. incognita*.

4. Conclusions

Given the current advances in molecular technologies, and the intensive efforts devoted to the study of plant-nematode interactions, cloning of sequences putatively involved in (a)virulence is a realistic challenge using the strategies developed above. But even where classical Mendelian genetics is possible, the only way to unambiguously demonstrate that a sequence is indeed involved in the nematode's ability to overcome a plant resistance gene is to transform the organism with a clone of the candidate DNA and test for complementation of the function. Direct transformation of the free-living nematode *Caenorhabditis elegans* by microinjection of foreign DNA into gonads is now routine [9, 20]. Very recently, a new genetic transformation system was described for the entomopathogenic nematode *Heterorhabditis bacteriophora*, using arrays of micromechanical piercing structures [12]. No transformation system is yet available for plant-parasitic nematodes, but such a technique is crucial for future studies on these organisms.

Cloning a gene involved in nematode (a)virulence will be of tremendous significance for a better understanding of the molecular determinants that regulate the compatible/ incompatible interactions between these parasites and their host plants. Study of the gene product and determination of its function should provide new insights into the complex cascade of events leading to resistance. Such investigations may also lead to an understanding of the mechanisms by which the nematode is able to overcome the resistance barriers activated by the plant. Finally, these advances in knowledge may have important consequences for the management of natural resistance genes, and, in the more long term, for the engineering of new forms of resistance.

References

1. Bakker, J., Folkertsma, R.T., Rouppe van der Voort, J.N.A.M., de Boer, J.M. and Gommers, F.J. (1993) Changing concepts and molecular approaches in the management of virulence genes in potato cyst nematodes, *Annual Review of Phytopathology* **31**, 169-190.
2. Bost, S.C. and Triantaphyllou, A.C. (1982) Genetic basis of the epidemiologic effects of resistance to *Meloidogyne incognita* in the tomato cultivar Small Fry, *Journal of Nematology* **14**, 540-544.
3. Castagnone-Sereno, P. (1994) Genetics of *Meloidogyne* virulence against resistance genes from solanaceous crops, in F. Lamberti, C. de Giorgi and D. Mck. Bird (eds.), *Advances in Molecular Plant Nematology*, Plenum Press, New york and London, pp. 261-276.
4. Castagnone-Sereno, P., Abad, P. and Dalmasso, A. (1995a) Identification of genes differentially expressed between avirulent and virulent *Meloidogyne incognita* near- isogenic lineages, EC-AIR Concerted Action Program Resistance mechanisms against plant-parasitic nematodes'. Third Annual General Meeting, 24-26 February 1995, Kiel, Germany.
5. Castagnone-Sereno, P., Wajnberg, E., Bongiovanni, M., Leroy, F. and Dalmasso, A. (1994) Genetic variation in *Meloidogyne incognita* virulence against the tomato *Mi* resistance gene: evidence from isofemale line selection studies, *Theoretical and Applied Genetics* **88**, 749-553.
6. Castagnone-Sereno, P., Rosso, M.N., Bongiovanni, M. and Dalmasso, A. (1995b) Electrophoretic analysis of near-isogenic avirulent and virulent lineages of the parthenogenetic root-knot nematode *Meloidogyne incognita*, *Physiological and Molecular Plant Pathology* **47**, 293-302.
7. Dalmasso, A., Castagnone-Sereno, P., Bongiovanni, M. and de Jong, A. (1991) Acquired virulence in the plant parasitic nematode *Meloidogyne incognita*. 2. Two- dimensional analysis of isogenic isolates, *Revue deNématologie* **14**, 305-308.
8. Faghihi, J., Vierling, R.A., Halbrendt, J.M., Ferris, V.R. and Ferris, J.M. (1995). Resistance genes in a 'Williams 82' x 'Hartwig' soybean cross to an inbred line of *Heterodera glycines*, *Journal of Nematology* **27**, 418-421.
9. Fire, A. (1986) Integrative transformation of *Caenorhabditis elegans*, *EMBO Journal* **5**, 2673-2680.
10. Flor, H.H. (1956) The complementary genetic systems in flax and flax rust, *Advances in Genetics* **8**, 29-54.
11. Gilbert, J.C. and Mac Guire, D.C. (1955) One major gene for resistance to severe galling from *Meloidogyne incognita*, *Tomato Genetics Cooperative Report* **5**, 15.
12. Hashmi, S., Ling, P., Hashmi, G., Reed, M., Gaugler, R. and Trimmer, W. (1995) Genetic transformation of nematodes using arrays of micromechanical piercing structures. *BioTechniques* **19**, 766-770.
13. Janssen, R. (1990) *Genetics of virulence in potato cyst nematodes*, PhD thesis, Agricultural University of Wageningen, The Netherlands.
14. Janssen, R., Bakker, J. and Gommers, F.J. (1990a) Assessing intraspecific variations in virulence in populations of *Globodera rostochiensis* and *G. pallida*, *Revue de Nématologie* **13**, 11-15.
15. Janssen, R., Bakker, J. and Gommers, F.J. (1990b) Selection of virulent and avirulent lines of *Globodera rostochiensis* for the *H1* resistance gene in *Solanum tuberosum* ssp. *andigena* CPC16, *Revue de Nématologie* **13**, 265-268.
16. Janssen, R., Bakker, J., and Gommers, F.J. (1991) Mendelian proof for a gene-for-gene relationship between virulence of *Globodera rostochiensis* and the *H1* resistance gene in *Solanum tuberosum* ssp. *andigena* CPC 1673, *Revue de Nématologie* **14**, 213-219.
17. Jarquin-Barberena, H., Dalmasso, A., de Guiran, G. and Cardin, M.C. (1991) Acquired virulence in the plant parasitic nematode *Meloidogyne incognita*. 1. Biological analysis of the phenomenon, *Revue de Nématologie* **14**, 299-303.
18. Jones, F.G.W., Parrot, D.M. and Perry, J.N. (1981) The gene-for-gene relationship and its significance for the potato cyst nematodes and their solanaceous hosts, in: B.M. Zuckerman and R.A. Rohde (eds.), *Plant parasitic Nematodes*, vol. III, Academic Press, New York, pp. 23-36.
19. Klose, J. (1989) Systematic analysis of the total proteins of a mammalian organism: principles, problems and implications for sequencing the human genome, *Electrophoresis* **10**, 140-152.
20. Mello, C.C., Kramer, J.M., Stinchcomb, D. and Ambros, V. (1991) Efficient gene transfer in *Caenorhabditis elegans*: extrachromosomal maintenance and integration of transforming sequences, *EMBO Journal* **10**, 3959-3970.
21. Michelmore, R.W., Paran, I. and Kesseli, R.V. (1991) Identification of markers linked to disease-

resistance genes by bulked segregant analysis: a rapid method to detect markers in specific genomic regions by using segregating populations, *Proc. Natl .Acad. Sci .U S A* **88**, 9828-32.
22. Myers, G.O. and Anand, S.C. (1991) Inheritance of resistance and genetic relationships among soybean plant introductions to races of soybean cyst nematode, *Euphytica* **55**, 197-201.
23. Opperman, C.H., Dong, K. and Chang, S. (1994) Genetic analysis of the soybean- *Heterodera glycines* interaction, in F. Lamberti, C. de Giorgi and D. Mck. Bird (eds.), *Advances in Molecular Plant nematology*, Plenum Press, New York and London, pp. 65-75.
24. Rao-Arelli, A.P., Anand, S.C and Wrather, J.A. (1992) Soybean resistance to soybean cyst nematode race 3 is conditioned by an additional dominant gene, *Crop Science* **32**, 862-864.
25. Sijmons, P.C., Atkinson, H.J. and Wyss, U. (1994) Parasitic strategies of root nematodes and associated host cell responses, *Annual Review of Phytopathology* **32**, 235-259.
26. Triantaphyllou, A.C. (1987) Genetics of nematode parasitism on plants, in A.A. Veech and D.W. Dickson (eds.), in *Vistas on Nematology*, Society of Nematologists, Hiattsville, pp. 354-363.
27. Vierling, R.A., Faghihi, J., Ferris, V.R. and Ferris, J.M. (1996) Association of RFLP markers with loci conferring broad-based resistance to the soybean cyst nematode (*Heterodera glycines*), *Theoretical and Applied Genetics* **92**, 83-86.
28. Zabeau, M. and Vos, P. (1993) *Selective restriction fragment amplification: A general method for DNA fingerprinting*, European Patent Application No: 92402629.7; Publication number EP 0534858 A1. matode chemotaxis and mechanism of host/prey recognition, *Annu. Rev. Phytophatol.* **22**, 95-113.

BREEDING FOR NEMATODE RESISTANCE IN SUGARBEET: A MOLECULAR APPROACH

Michael KLEINE[1], Daguang CAI[1], Rene M. KLEIN-LANKHORST[2], Niels N. SANDAL[3], Elma M. J. SALENTIJN[2], Hans HARLOFF[1], Sirak KIFLE[1], Kjeld A. MARCKER[3], Willem J. STIEKEMA[2], Christian JUNG[1]

[1]*Institute of Crop Science and Plant Breeding, Christian-Albrechts-University of Kiel, Olshausenstr. 40, D-24118 Kiel*
[2]*DLO-Centre for Plantbreeding and Reproduction Research (CPRO-DLO), Droevendaalsesteeg 1, NL-6700 AA Wageningen*
[3]*Department of Molecular Biology, University of Aarhus, Gustav Wieds Vej 10, DK-8000 Aarhus*

Abstract

The beet cyst nematode (BCN, *Heterodera schachtii* Schm.) is a major pest of sugarbeet (*Beta vulgaris* L.). Chemical control of the parasite poses a threat to the environment and a wide crop rotation is economically impractical. Therefore, breeding of nematode resistant sugarbeet varieties is a promising alternative. Monogenic resistance has been introduced to cultivated sugarbeet from nematode resistant wild beet of the section Procumbentes by interspecific crosses. Different nematode resistant cytogenetic mutants have been selected from the offspring. However, the agronomic performance of these lines is poor and the transmission of the resistance reduced due to meiotic disturbances. Molecular markers have been developed suitable for marker-assisted selection (MAS) of nematode resistant plants in the breeding process and for the cloning of the BCN resistance gene. The novel positional cloning strategy described here makes use of repetitive wild beet specific DNA elements that are tightly linked to the resistance gene. The next step of the cloning procedure comprises the construction of representative yeast artificial chromosome (YAC) libraries and the subsequent screening with molecular markers to extract specific YAC clones encompassing the resistance locus. Candidate genes have been selected from root specific cDNA libraries using the YACs as probes. To identify the resistance gene a root specific complementation system has been established using *Agrobacterium rhizogenes* mediated transformation. Finally, transgenic hairy roots can be analysed for the presence of the BCN resistance gene by *in vitro* nematode testing.

1. Introduction

1.1. ECONOMIC RELEVANCE

The beet cyst nematode *(Heterodera schachtii* Schm.) causes severe damage in sugarbeet *(Beta vulgaris* L.) cultivation. In general, more than 80% of the *Chenopodiaceae* and *Brassicaceae* species are hosts of *H. schachtii* [52, 58] including economically important crops like sugarbeet (*Beta vulgaris*), spinach (*Spinacea oleracea*) and rape seed (*Brassica napus*).

Chemical control is difficult because the nematode juveniles are well protected in the cysts. Also, *H. schachtii* has a high multiplication rate as one cyst can contain more than 300 eggs and several generations can be produced in one year. Typical symptoms of infected sugarbeet are wilting leaves, a reduced beet formation and vigorous growth of secondary roots. Besides chemical control, wide interval crop rotation and the use of nematode resistant catch crops are the only way to fight the parasite. Therefore, growing of nematode resistant sugarbeet varieties would be a favourable means to control this pest.

1.2. NATURAL RESOURCES FOR RESISTANCE GENES

The search for nematode resistance has been performed in all four sections of the genus *Beta* (Vulgares, Corollinae, Nanae and Procumbentes). A recessive polygenic resistance in sugarbeet leading to the reduction of cyst numbers was postulated by Curtis [7] and Heijbroek [19]. Its value for practical plant breeding, however, remained to be proven. While true resistance could not be found within *B. vulgaris,* some wilting tolerant genotypes were selected by Heijbroek *et al.* [21]. Furthermore, some tolerance against infection with *H. schachtii* was found in *B. vulgaris* by Price [43], Jorgensen and Smith [28] and Doney and Whitney (1973). A complete resistance against *H. schachtii* was found only in the three species of the Procumbentes section, *B. procumbens*, *B. webbiana*, and *B. patellaris* [24, 58, 16, 50, 53, 59]. Even under extreme infection pressure, no multiplication was observed on roots of these three species. Between 93% and 98% of the Procumbentes plants investigated by Yu [60] had no cysts, the remaining plants had a minimum number of cysts. The offspring of plants with a single cyst was not different from the parental generation indicating that there is no genetic variability for nematode resistance in these three species.

The incompatible reaction between Procumbentes species and *H. schachtii* has been studied at the cellular level. The hatching stimulance of the wild species is very efficient and is as high as for *B. vulgaris* [16]. The roots of the Procumbentes species are invaded by *H. schachtii* juveniles as well [16] but most of the nematodes die due to necrosis of cortical cells. A small proportion of juveniles is able to proceed to the vascular cylinder to induce the formation of syncytia. However, syncytia do not develop regularly as in the compatible reaction but form vacuoles causing their degeneration and consequently the death of the J2/J3 juveniles (B. Holtmann, personal communication). The wild beets are tolerant to nematode invasion and grown in infested soil their dry weight is not significantly reduced compared to controls grown in sterilized soil.

1.3. BREEDING FOR NEMATODE RESISTANT SUGARBEET

Currently there are three alternatives for breeding nematode resistant sugarbeet. One is to clone a nematode resistance gene from wild species by means of molecular genetic techniques and transfer it to susceptible beets. No such beets exist so far because the resistance gene has not been identified yet. Alternatively transgenic beets can be produced in which the resistance is caused by a hypersensitive like action of feeding cells. This system will make use of syncytium specific promotors in combination with suicide genes such as RNase (see Ohl et al., this volume). The traditional alternative is to transfer the beet cyst nematode (BCN) resistance genes from Procumbentes species to cultivated beet by species hybridization and backcrossing. In principle this strategy has been followed by three groups at the USDA, the Institute of Applied Genetics, Hanover, Germany and at the Institute for rationale Suikerproductie together with the CPRO/DLO, Wageningen, The Netherlands.

Numerous reports have been published on hybrids between *B. vulgaris* and Procumbentes species [5]. Unfortunately, there are many deleterious traits associated with the wild beets, such as poor growth, beet development, annual behaviour and others causing the need for repeated backcrossing. When using diploid *B. vulgaris* as seed parents the hybrids were totally sterile. In most cases they did not even flower because they were not able to survive. This problem could be overcome by using tetraploid seed parents and by grafting the hybrid plants [27]. A lack of homology as revealed by pairing mismatch in metaphase I of meiosis [47] and the corresponding misdivision in anaphase I led to unbalanced gametes which were the reason for sterility. Therefore, a transfer of the resistance gene by crossover events seemed to be unlikely.

The reduction of the number of wild beet chromosomes has been achieved by backcrossing with *B. vulgaris*. The resulting monosomic addition lines with only one wild beet chromosome added (2n=18+1) were as resistant as the wild beet itself proving that the resistance was inherited as a dominant gene [48, 49, 51, 22, 37, 9]. Surprisingly, three different monosomic addition lines were developed from *B. procumbens* and *B. webbiana* hybrids as revealed by phenotypical and isozyme characters [31]. Only one chromosome seems to contribute to resistance in *B. patellaris*. Later the resistance genes were designated according to the chromosome on which they are located ([36], Table1).

Additional evidence for the different nature of resistance genes came from screenings with a virulent pathotype of *H. schachtii* which is able to overcome the resistance from chromosome 1 of *B. procumbens* but not from chromosome 7 [36]. Latest evidence for the presence of at least three resistance genes in the two wild species came from molecular marker analysis.

Meanwhile different cytogenetic mutant lines carrying wild beet chromosome fragments from *B. procumbens* or *B. patellaris* have been selected [8, 2, 45]. Metaphase analysis (Fig. 1) of the chromosome fragments carrying the gene for nematode resistance led to a rough size estimation ranging from a complete chromosome arm to a small fragment [45].

TABLE 1. Nematode resistance genes in wild beets of the Procumbentes section.

Chromosome No.	Isozyme gene	Resistance gene	Growth type
B. procumbens			
pro-1	ICD, EST	$Hs1^{pro-1}$	a
pro-7		$Hs2^{pro-7}$	b
pro-8	SOD, ACO1, ACO2, 6PGDH slow band	$Hs3^{pro-8}$	c
B. webbiana			
web-1	ICD, GOT	$Hs1^{web-1}$	a
web-7	6PGDH fast band	$Hs2^{web-7}$	b
web-8	SOD, ACO, 6PGDH slow band	$Hs3^{web-8}$	c
B. patellaris			
pat-1	ICD	Hs^{pat-1}	a

1.4. SELECTION OF DIPLOID NEMATODE RESISTANT LINES

Since aneuploid lines are not useful for plant breeding diploid material (2n=18) was developed. These plants appear at very low frequencies from the offspring of monosomic addition lines and can either result from rare crossover events between sugarbeet and wild beet chromosomes or from translocations between chromosomes of both species. Successful selection of diploid resistant genotypes in progenies of monosomic addition lines strongly depends on the efficiency of the selection system. To avoid time consuming cytological analysis isozyme and molecular markers have been successfully employed in routine screening. Irradiation of monosomic plants was often used to raise the frequency of translocations. Selection of diploid nematode resistant beets in the progenies of monosomic addition lines was described by Yu [61], Jung and Wricke [32] and Heijbroek *et al.* [22] proving that resistance genes from the Procumbentes species can be successfully transferred to sugarbeet in order to breed varieties resistant against *H. schachtii*.

Recently, four of the resistance carrying translocations in sugarbeet have been mapped using the current RFLP map of *Beta* [23]. Surprisingly all translocations originating from different addition lines were mapped to the same locus distal to linkage group IV suggesting that a translocation hotspot is present on this chromosome facilitating the gene transfer.

Although significant progress has been made by classical breeding there are still several reasons for cloning the gene by molecular techniques. The lines obtained suffer from meiotic instability and consequently a reduced transmission of nematode resistance was observed [49, 32]. Further selections among diploids are necessary due to the appearance of tumors and their poor agronomical performance. The molecular identification of the mode of action of the resistance would contribute significantly to the progress in science. Although some resistance genes have been cloned from plants during the past two years the molecular nature of nematode resistance genes still lies in the dark.

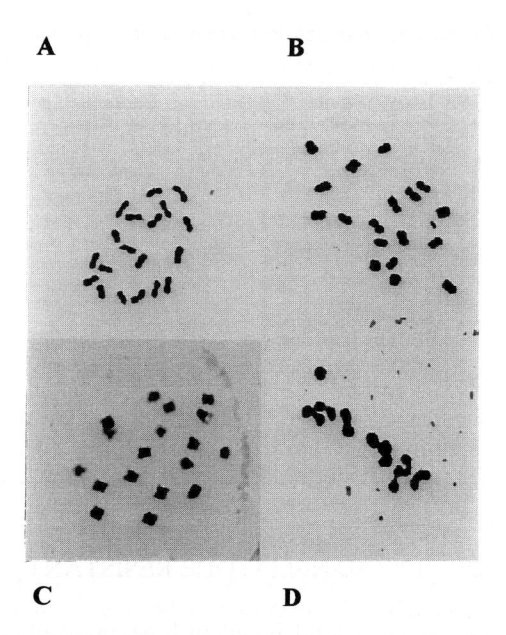

Figure 1. Cytological analysis of different nematode resistant sugarbeets: (A) Alien monosomic addition line (2n=19). (B) Fragment addition line (2n=19). (C) Translocation line (2n=18). (D) Monosomic addition line, meiotic metaphase *I*.

The scope of this article is to describe the cloning strategy, which is significantly different from commonly used map based cloning approaches since it relies on mapping of chromosomal breakage points and not on recombination mapping. As outlined above recombination within the introduced piece of wild beet chromatin in sugarbeet is a rare event making recombination mapping impossible. Therefore, markers have been selected which are specifically hybridizing wild beet DNA. These markers have been used as probes for screening a YAC library of one of the diploid translocation lines. Several YACs have been isolated in this way and arranged into contigs of overlapping clones. Transcriptionally active regions are identified from the YACs by cDNA screening and the cDNAs are finally transferred to susceptible sugarbeet by means of *Agrobacterium rhizogenes* transformation.

2. Molecular Marker Selection and its Application in the Breeding Procedure

An important prerequisite for the positional cloning strategy employed in this project is the isolation of molecular markers that are tightly linked to the gene of interest. These markers can then be used both for marker assisted selection (MAS) in the breeding process and for the extraction of corresponding genomic regions from YAC libraries which will finally lead to the isolation of the resistance gene.
During the past few years procedures for the selection of molecular markers have been

developed and improved. The first generation of molecular markers were shotgun cloned plasmid clones which were hybridized to genomic DNA to produce RFLP (restriction fragment length polymorphism, [1]) patterns. In the meantime more sophisticated PCR-based methods like the RAPD (random amplified polymorphic DNA, [57]) and the AFLP (amplified fragment length polymorphism, [56]) technique have been developed.

Three translocation lines (A906001, B883 and AN1-65-2) and of a set of monosomic addition and fragment addition lines both from *B. procumbens* and *B. patellaris* were used for the isolation of the markers.

2.1. REPETITIVE RFLP MARKERS

Repetitive markers closely linked to the nematode resistance gene have been isolated from *B. procumbens* chromosome 1 (pRK643, [30]) and *B. patellaris* chromosome 1 (Sat121, [46, 45]). In both cases these satellite markers were isolated from genomic plasmid libraries. The identified sequences all belong to the same family of satellite sequences of direct repeats of a 160 bp sequence. This type of sequence is absent from *B. vulgaris* and is found in interspersed clusters in the Procumbentes genomes.

The satellite sequences appeared in several copies in the translocation lines A906001, B883 and AN1-65-2. The pattern of hybridizing bands after restriction enzyme digests and genomic hybridization is identical for the three types of translocation lines, indicating that one single *B. procumbens* region has been transferred to the sugarbeet genome in three independant cases.

2.2. RAPD MARKERS

More than 500 RAPD primers [57] have been screened to select RAPD markers linked to the nematode resistance gene from *B. patellaris* [44] and *B. procumbens* (Sandal *et al.*, in preparation). One of them - RAPD marker X2.1 - has been cloned and hybridized to several restriction fragments which are closely linked to the nematode resistance gene on chromosome 1 in both *B. patellaris* and *B. procumbens* [44]. Three more RAPD markers were found by bulked segregant analysis on the translocation lines B883 and AN1-65-2. From the deduced sequence of the cloned RAPD fragments specific primers were developed and applied for the screening of the YAC libraries.

2.3. AFLP MARKERS

Recently a new PCR based technique (AFLP) for the isolation of molecular markers has been published [56]. It employs adapters that are ligated to genomic restriction fragments after digestion with a frequent cutter (*Mse*I, 4 bp recognition sequence) and a rare cutter (e.g. *Eco*RI, 6 bp recognition sequence). Using primers that contain the adapter sequence plus several additional nucleotides one can get a manageable amount of PCR products after separation on a sequencing gel (Fig 2). On a single gel it is therefore possible to analyze many bands for linkage to any gene of interest. In search for AFLP markers linked to the nematode resistance gene bulked segregant analysis has been applied. Two

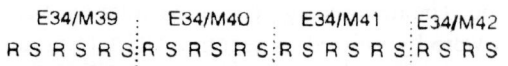

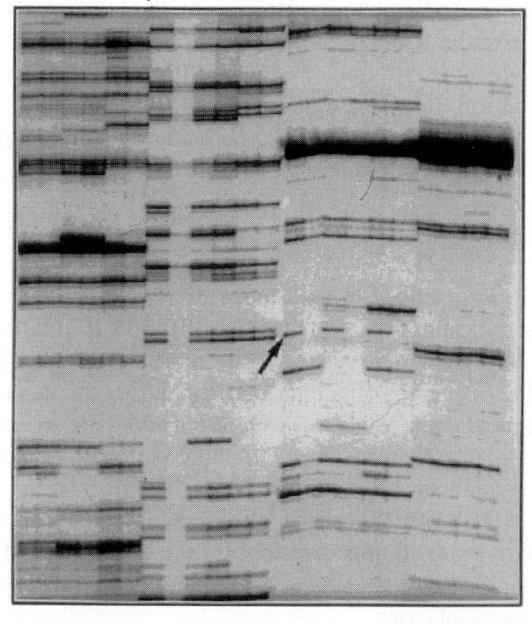

Figure 2. AFLP analysis of DNA bulks of resistant (r) and susceptible (s) sugarbeet plants. Arrow indicates specific amplification products from DNA of resistant plants.

resistant translocation line and susceptible *B. vulgaris* were used. About 25 AFLP markers have been selected cosegregating with the translocations of the three lines A906001, B883 and AN1-65-2.

2.4. MARKER ASSISTED SELECTION (MAS)

The development of the molecular marker technique led to a new application which shortens the breeding process - the marker assisted selection (MAS). Markers that are tightly linked to a trait of interest can be used to search for plants carrying this marker. Thus individual plants are already selectable within populations before the phenotype of the special trait will be expressed. This can be achieved by RFLP analysis with probes cosegregating with a specific trait or by the PCR technique employing corresponding primers. Both procedures are dependent on the isolation of DNA which in the case of the RFLP analysis is very time consuming. A third approach which has been applied successfully in this project is the squash dot test (Fig. 3). This simple but reliable method can be applied in the field by squeezing small pieces of fresh leaflets onto a nylon membrane. After alkaline denaturation of the DNA the membrane is ready for hybridisation with a specific probe [29]. The satellite sequences (pRK643 and Sat121) are suitable for squash dot analysis, as they represent repetitive sequences. A modified protocol has been developed to apply this method to the translocation lines, also because the translocated wild beet DNA contains fewer copies than e.g. monosomic addition lines.

In search of nematode resistant lines that have lost the repetitive markers but are still resistant RFLP, PCR and squash dot analysis will be used. It implies that if the size of the translocation diminishes negative traits like tumors may get lost, thus raising the breeding value of these lines. Therefore, large populations generated by backcrossing resistant translocation lines with susceptible sugarbeet will be screened for the presence or absence of the markers. These plants will be useful for both cloning resistance genes and for breeding nematode resistant sugarbeet.

Recently, the translocations of four different translocation lines have been mapped to linkage group IV using RFLP-markers from the current linkage map of sugarbeet [23]. A PCR based approach which is presently being developed using the RFLP-marker sequences will finally lead to an assay that differentiates between plants that are homozygous and heterozygous for the resistance gene within segregating F2-populations avoiding time consuming DNA isolation, restriction and Southern-analysis. This demonstrates the potential of MAS to accelerate the breeding process.

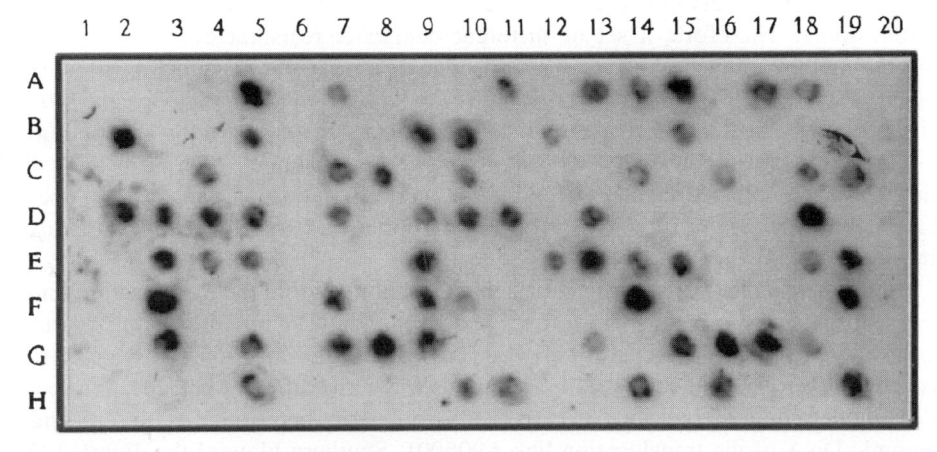

Figure 3. Autoradiography of a squash dot selection filter using leaflets of a F2 population segregating for the translocation with satellite probe Sat 121. Dark spots representing leaflets from nematode resistant individual plants.

3. Construction and Use of YAC Libraries from Nematode Resistant Sugarbeet

3.1. YAC-LIBRARIES FROM NEMATODE RESISTANT BEET

The isolation of molecular markers closely linked to the BCN resistance gene marked the beginning of the second phase of the resistance gene cloning project, the construction and screening of representative YAC-libraries. The YAC technology [3] has gained significant relevance for the analysis of complex eukaryotic genomes since it provides a tool to isolate relatively large DNA fragments in the size range between 100 and more than 1,000 kb. If molecular markers are available as landmarks for screening such libraries and for arranging individual YACs into contigs of overlapping fragments, long stretches of DNA can be cloned and genes selected.

Presently, YAC-libraries from different plant species such as *Arabidopsis thaliana* [6], carrot [17], maize [11], tomato [38], sugarbeet [13], rice [54] and barley [35] have been developed. Two representative YAC libraries were constructed from different nematode-resistant sugarbeet material. The first library consisting of about 20,000 individual clones was made with DNA of a nematode resistant cytogenetic mutant line carrying a chromosome fragment from *B. patellaris* [33], the second library consisting of about 13,000 recombinant YACs was cloned from the diploid translocation line A906001 [34].

3.2. CONSTRUCTION OF A PHYSICAL MAP ACROSS THE TRANSLOCATION

No recombination mapping was possible around the resistance gene due to the lack of crossover events between the translocated wild beet segment and the sugarbeet chromosome. Therefore, a set of different nematode resistant cytogenetic mutants harbouring the same resistance gene was used to determine chromosomal breakpoints as well as overlapping regions and consequently the location of the resistance gene [30, 45]. The strategy employed here implies the construction of a contig surpassing the chromosomal breakpoints and thus determining the extension of the translocation. This has been accomplished by the isolation and characterisation of translocation specific YAC clones using the repetitive elements as pobes and by physically mapping the translocation [34].

For the isolation of specific clones the YAC library from the translocation line has been chosen as it carries the shortest wild beet segment. The repetitive *B. procumbens*-specific 643 elements are exclusively restricted to the translocation and were therefore used as markers for screening the library. In this way three different translocation specific YAC-clones could be isolated [34]. To verify colinearity between YAC-inserts and the genomic DNA of the translocation line A906001, Southern blots of the digested DNA were hybridized with probe pRK643 (Fig. 4). This fingerprint analysis gave three main results. (1) All of the repetitive elements present on the translocation are confined to the YACs. (2) There is no evidence for rearrangements within the YACs. (3) As judged from common restriction fragments overlaps between YACs could be detected.

The size of the genomic region carrying the more or less evenly distributed 643 elements has been determined by pulsed field gel electrophoresis (PFGE) analysis using rare cutter enzymes. The 643 elements could be mapped to a single 300 kb *Sal*I restriction fragment with an internal *Mlu*I site giving rise to a first physical map around the nematode resistance locus (Figure 5). From the map it was deduced that more than 70 % of the 643 element carrying region has been cloned already [34].

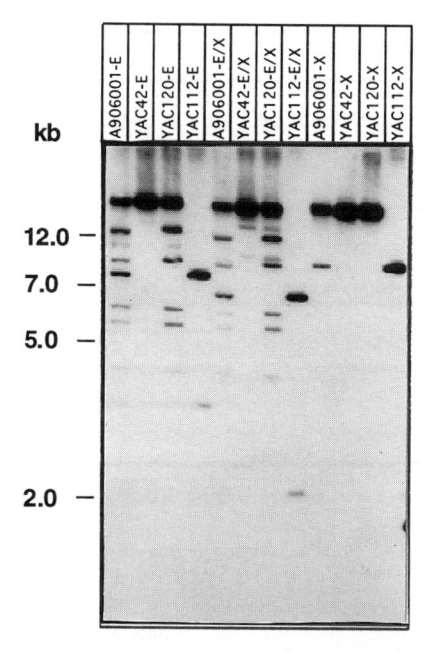

Figure 4. Southern analysis with genomic DNA of the translocation line A906001 and three translocation specific YAC-clones digested with *Eco*RI (E) and *Xba*I (X) after hybridisation with probe pRK643.

4. Isolation of Wild Beet-Specific cDNAs and Identification of Candidate Genes by Genetic Complementation Using the Hairy Root System

The third phase of this project consists of the isolation and identification of nematode resistance candidate genes with the aid of YAC clones and can be divided into four steps. (1) Screening of a cDNA library with specific YACs. (2) Cloning of the cDNA into a binary plasmid vector. (3) *Agrobacterium rhizogenes* transformation and induction of hairy roots, and (4) in vitro resistance testing with infectious *H. schachtii* juveniles.

4.1. SCREENING OF cDNA LIBRARIES WITH YAC-DNA

Two cDNA libraries from nematode infested roots of the translocation lines A906001 and B883 were constructed to isolate transcribed sequences from the YAC as candidate clones for the nematode resistance gene.

Several techniques have been developed to facilitate the isolation of genes from large genomic fragments such as YAC or cosmid contigs, e.g. restriction mapping of candidate regions using infrequently cutting restriction endonucleases [4], exon trapping [10], enrichment of coding sequences by magnetic bead capture [26], island rescue PCR [55] and cDNA hybridisation selection [15]. Also, direct identification and isolation of YAC-encoded genes can be accomplished by cDNA library screening with the entire

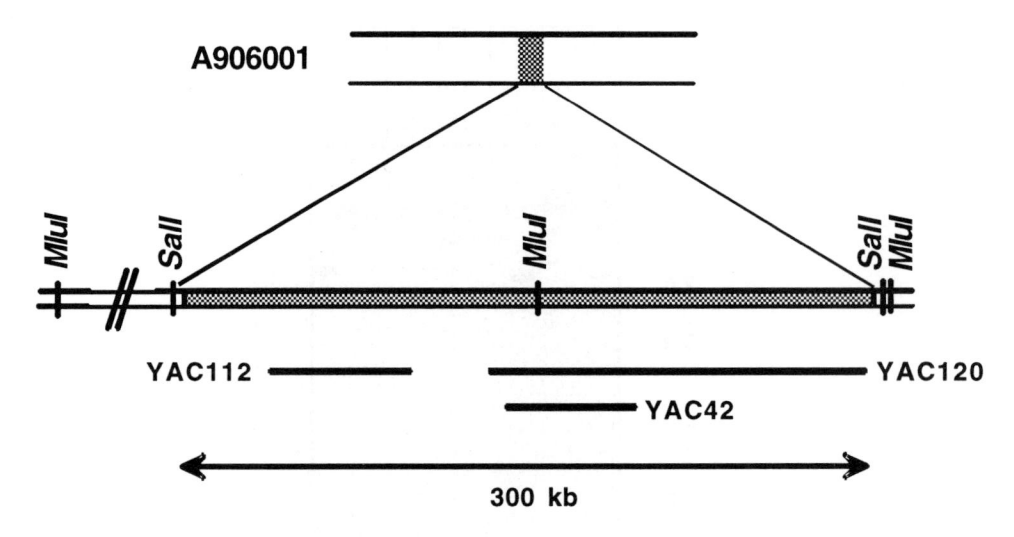

Figure 5: Physical map of the 643 element carrying region of the nematode resistant translocation line A906001 and the location of three YACs (YAC112, YAC120 and YAC42).

YAC as probe [39, 40].

This approach has been proven to be feasible despite of the large size of the genomic fragment used as a hybridisation probe and the complexity of the plant genome. However, three main obstacles leading to high background signals in the screening procedure have to be avoided. (1) Crosshybridisation with cDNA vector sequences as a part of the YAC vector. (2) Contamination of yeast DNA comigrating with the YAC DNA after preparation from pulsed-field gels and (3) extraction of non-specific cDNA clones due to repetitive sequences present in the cDNA library.

The cDNA clones have been isolated, characterized by Southern analysis (Figure 6) and used to isolate full-length transcripts suitable for genetic complementation studies.

4.2. GENETIC COMPLEMETATION IN HAIRY ROOTS

Many plant species can easily be transformed by *Agrobacterium* mediated gene transfer. In this way susceptible sugarbeet plants can be genetically complemented by introducing the resistance gene giving rise to a resistant phenotype of this plant. In contrast to many other plant species the transformation efficiency of sugarbeet with *A. tumefaciens* is low and subsequent regeneration of sugarbeet plants is difficult. The inoculation of sugarbeet leaflets with *Agrobacterium rhizogenes* induces hairy roots due to the integration of the T-DNA region from the Ri-plasmid into the plant genome [18, 42]. The hairy roots can be cultivated *in vitro* and are suitable for the study of root pathogens like fungi and nematodes [12, 41].

Hairy roots have been obtained from a susceptible line and have been inoculated with nematode juveniles in a petri dish leading to a normal development of the nematode on hairy roots. In contrast, complete resistance was observed when hairy roots of the translocation line (A906001) has been employed demonstrating that the resistance gene is

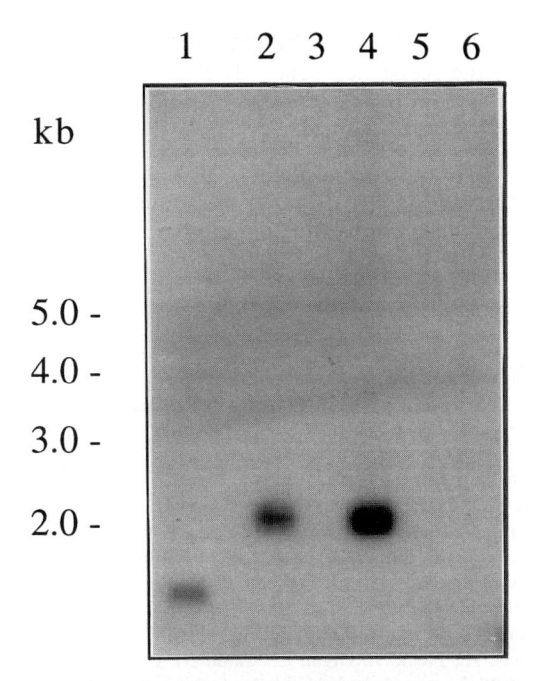

Figure 6. Southern analysis of YAC-DNA digested with *Eco*RI to verify the origin of the cDNA 1832 which was used as a probe. Lane 1 = 1 kb-ladder, 2 = resistant sugarbeet, 3 = pYAC4, 4 = YAC120 and 5 = YAC112 and lane 6 = YAC42.

functioning in hairy roots.

Next, the cDNAs were subcloned in sense and antisense orientation into a modified binary vector under the control of the 35S-promotor. An *A. rhizogenes* strain was transformed with these constructs by electroporation and transgenic hairy roots were simultaneously induced on susceptible *B. vulgaris* plants by parallel infection with *A. rhizogenes* containing the foreign gene and wild type *A. tumefaciens* using the leaf-disc or seedling-hypocotyl transformation technique. To verify the integration of the foreign gene into the plant genome Southern-, Northern- and PCR analysis have been applied. Finally, the transgenic hairy roots were inoculated with nematodes *in vitro* to test for the genetic complementation of the resistance character. Currently, more than 50 independant transformants are under investigation in search for the nematode resistance gene.

5. Conclusions

Molecular markers have been employed successfully for marker assisted selection (MAS) of resistant beets thus providing a powerful tool for the breeders for an efficient selection of nematode resistant plants in the offspring of large segregating populations.

The isolation of genes for nematode resistance is of strong economical, environmental and scientific importance. The cloning of the *Hs1^{pro-1}* gene from sugarbeet can be divided into four steps. (1) The generation of molecular markers tightly linked to the resistance gene. (2) The construction of YAC contigs encompassing large genomic regions of the translocation carrying the resistance gene. (3) The subsequent isolation of transcribed sequences from this area, and (4) the development of a fast and reliable *in vitro* system for the testing of candidate resistance genes.

The characterisation of the cDNA clones extracted from the translocation by genetic complementation will result in the isolation of the resistance gene providing the basis for studying the defence mechanisms of plants against nematodes on the molecular level.

References

1. Botstein, D., White, R.L., Skolnick, M., Davis, R.W. (1980) Construction of a genetic linkage map in man using restriction fragment length polymorphisms, *Am J Hum Genet* **32**, 314-331.
2. Brandes, A., Jung, C., Wricke, G. (1987) Nematode resistance derived from wild beet and its meiotic stability in sugarbeet, *Plant Breed* **99**, 56-64.
3. Burke, D.T., Carle, G.F., Olson, M.V. (1987) Cloning of large segments of exogenous DNA into yeast by means of artificial chromosome vectors, *Science* **236**, 806-812.
4. Chandrasekharappa, S.C., Marchuk, D.A., Collins, F.S. (1992) Analysis of yeast artificial chromosome clones. In Methodes in Molecular Biology: Pulsed-Gel Electrophoresis Techniques. L. Ulanovsky and M. Burmeister, eds., J.Walker, series ed., Humana Press, Clifton, NJ, pp. 235-257.
5. Coons, G.H. (1975) Interspecific hybrids between *Beta vulgaris* L. and the wild species of *Beta*, *Proc Am Soc Sug Beet Techn* **8**, 281-306.
6. Creusot, F,. Fouilloux, E., Dron, M., Lafleuriel, J., Picard, G., Billault, A., Lepaslier, D., Cohen, D., Chaboute, M.E., Durr, A., Fleck, J., Gigot, C., Camilleri, C., Bellini, C., Caboche, M., Bouchez, D. (1995) The CIC library: A large insert YAC library for genome mapping in *Arabidopsis thaliana*, *Plant J* **8**, 763-770.
7. Curtis, G.J. (1970) Resistance of sugar beet to the cyst-nematode *Heterodera schachtii* Schm, *Ann Appl Biol* **66**, 169-177.
8. De Jong, J.H., Speckmann, G.J., De Bock, T.S.M., Lange, W., Van Voorst, A. (1986) Alien chromosome fragments conditioning resistance to beet cyst nematode in diploid descendants from monosomic additions of *Beta procumbens* to *B. vulgaris*, *Can J Genet Cytol* **28**, 439-443.
9. De Jong, J.H., Speckmann, G.J., De Bock, T.S.M., Van Voorst, A. (1985) Monosomic additions with resistance to beet cyst nematode obtained from hybrids of *Beta vulgaris* and wild *Beta* species of the section *Patellares*. II. Comparative analysis of the alien chromosomes, *Z Pflanzenzüchtung* **95**, 84-94.
10. Duyk, G.M., Kim, S., Myers, R.M., Cox, D.R. (1990) Exon trapping: a genetic screen to idetify candidate trnscribed sequences in cloned mammalian genomic DNA, *Proc Natl Acad Sci* **87**, 8995-8999.
11. Edwards, K.J., Thompson, H., Edwards, D,. de Saizieu, A., Sparks, C., Thompson, J.A., Greenland, A.J., Eyers, M., Schuch, W. (1992) Construction and characterization of a yeast artificial chromosome library containing three haploid maize genome equivalents, *Plant Mol Biol* **19**, 299-308.
12. Ehlers, U., Commandeur, U., Frank, R., Landsmann, J., Koening, R., Burgermeister, W. (1991) Cloning of the coat protein gene from beet necrotic yellow vein virus and ist expression in sugar beet hairy roots, *Theor. Appel. Genet.* **81**, 777-782.
13. Eyers, M., Edwards, K., Schuch, W. (1992) Construction and characterisation of a yeast artificial chromosome library containing two haploid *Beta vulgaris* L. genome equivalents, *Gene* **121**, 195-201.
14. Feinberg, A.P. and Vogelstein, B. (1984) A technique for radiolabeling DNA restriction fragments to high specific activity, *Anal Biochem* **137**, 266-267.
15. Goei, V.L., Parimoo, S., Capossela, A., Chu, T.W., Gruen, J.R. (1994) Isolation of novel Non-HLA Gene Fragments from the Hemochromatosis Region (6p21.3) by cDNA Hybridization Selection, *Am J Hum Genet* **54**, 244-251.
16. Golden, A.M. (1958) Interrelationships of certain *Beta* species and *Heterodera schachtii*, the sugar-beet nematode, *Plant Dis Rep* **42**, 1157-1162
17. Guzmann, P., Ecker, J.R. (1988) Development of large DNA methods for plants: molecular cloning of large segments of *Arabidopsis* and carrot DNA into yeast, *Nucleic Acid Res* **16**, 11091-11105.
18. Hamill, J.D., Parr, A.J., Robins, R.J., Rhodes, M.J.C. (1986) Secondary product formation by cultures of

Beta vulgaris and Nicotiana rustica transformed with Agrobacterium rhizogenes, *Plant Cell Rep* **5**, 111-114.
19. Heijbroek, W. (1977) Partial resistance of sugarbeet to beet cyst eelworm (*Heterodera schachtii* Schm.), *Euphytica* **26**, 257-266.
20. Heijbroek, W. (1983) Some effects of fungal parasites on the population development of the beet cyst nematode (*Heterodera schachtii* Schm.), *Meded Fac Landbouw Gent* **48**, 433-439.
21. Heijbroek, W., McFarlane, J.S., Doney, D.L. (1977) Breeding for tolerance to beet-cyst eelworm *Heterodera schachtii* Schm. in sugar beet, *Euphytica* **26**, 557-564.
22. Heijbroek, W., Roelands, A.J., de Jong, J.H., van Hulst, C., Schoone, A.H.L., Munning, R.G. (1988) Sugar beets homozygous for resistance to beet cyst nematode (*Heterodera schachtii* Schm.) developed from monosomic additions of *Beta procumbens* to *B. vulgaris*, *Euphytica* **38**, 121-131.
23. Heller, R., Schondelmaier, J., Steinrücken, G., Jung, C. (1996) Genetic localization of four genes for nematode (*Heterodera schachtii* Sch.) resistance in sugar beet (*Beta vulgaris* L.), *Theor Appl Genet* **92**, 991-997.
24. Hijner, J.A. (1952) De gevoeligheid van wilds bieten voor het bietencystenaaltje (*Heterodera schachtii*), *Meded Inst Rat Suikerprod* **21**, 1-13.
25. Huynh, T., Young, R., Davis, R.W. (1985) Constructing and screening cDNA libraries in lambda gt10 and gt11. In DNA Cloning, A Practical Approach, Vol. 1, D.M. Glover, ed., IRL Press, Washington, DC, pp. 49-85.
26. John, G.M., Gregory, M.D., Sabrina, E.R., Linda, M.H., Lovett, M. (1992) The selection of novel cDNAs encoded by the regions surrounding the human interleukin 4 and 5 genes, *Nucleic Acids Res* **20**, 5173-5179.
27. Johnson, R.T., Wheatley, G.W. (1961) Studies on backcross generations and advanced generations of interspecific hybrids between *B. vulgaris* and *B. webbiana*, *J. Am. Soc. Sug. Beet Techn.* **11**, 429-435.
28. Jorgensen, E.C., Smith, C.H. (1966) Evaluation of selected varieties of sugarbeets for response to the sugarbeet nematode *Heterodera schachtii*, *Plant Dis. Rep.* **50**, 650-654.
29. Jung, C,. Herrmann, R.G. (1991) A DNA probe for rapid screening of sugar beet (*Beta vulgaris* L.) carrying extra chromosomes from wild beets of the *Procubentes* section, *Plant Breeding* **107**, 275-279.
30. Jung, C., Koch, R., Fischer, F., Brandes, A., Wricke, G., Herrmann, R.G. (1992) DNA markers closely linked to nematode resistance genes in sugar beet (*Beta vulgaris* L.) using chromosome additions and translocations originating from wild beets of the *Procumbentes* species, *Mol Gen Genet* **232**, 271-278.
31. Jung, C., Wehling, P., Löptien, H. (1986) Electrophoretic investigations on nematode resistant sugar beets, *Plant Breeding* **97**, 39-45.
32. Jung, C., Wricke, G. (1987) Selection of diploid nematode-resistant sugar beet from monosomic addition lines, *Plant Breeding* **98**, 205-214.
33. Klein-Lankhorst, R.M., Salentijn, E.M.J., Dirkse, W.G., Arens-de Reuver, M., Stiekema, W.J. (1994) Construction of a YAC library from *Beta vulgaris* fragment addition and isolation of a major satellite DNA cluster linked to the beet cyst nematode resistance locus *Hs1*pat-1, *Theor Appl Genet* **89**, 426-434.
34. Kleine, M., Cai, D., Eibl, C., Herrmann, R.G., Jung, C. (1995) Physical mapping and cloning of a translocation in sugar beet (*Beta vulgaris* L.) carrying a gene for nematode (*Heterodera schachtii*) resistance from *B. procumbens*, *Theor Appl Genet* **90**, 399-406.
35. Kleine, M., Michalek, W., Graner, A., Herrmann, R.G., Jung, C. (1993) Construction of a barley (*Hordeum vulgare* L.) YAC library and isolation of a *Hor 1*-specific clone, *Mol Gen Genet* **240**, 265-272.
36. Lange, W., Müller, J., De Bock, T.S.M. (1993) Virulence in the beet cyst nematode (*Heterodera schachtii*) versus some alien genes for resistance in beet, *Fund. Appl. Nematol.* **16**, 447-454.
37. Löptien, H. (1984) Breeding nematode-resistant beets. I. Development of resistant alien additions by crosses between *Beta vulgaris* L. and wild species of the section *Patellares*, *Z Pflanzenzüchtung* **92**, 208-220.
38. Martin, G., Ganal, M.W., Tanksley, S.D. (1992) Construction of a yeast artificial chromosome library of tomato and identification of cloned segments linked to two disease resistance loci, *Mol Gen Genet* **233**, 25-32.
39. Martin, G.B. , Frary, A., Wu, T., Brommonschenkel, S., Chungwongse, J., Earle, E.D., Tanksley, S.D. (1994) A Member of the Tomato Pto Gene Family Confers Sensitivity to Fenthion Resulting in Rapid Cell Death, *The Plant Cell* **6**, 1543-1552.

40. Martin, G.B., Brommenschenkel, S.H., Chunwongse, J., Frary, A., Ganal, M.W., Spivey, R., Wu, T., Earle, E.D., Tanksley, S.D. (1993) Map-based cloning of a protein kinase gene conferring disease resistance in tomato, *Science* **262**, 1432-1435.
41. Paul, H., van Deelen, J.E.M., Henken, B., de Bock, T.S.M., Lange, W., Krens, F.A. (1990) Expression *in vitro* of resistance to *Heterodera schachtii* in hairy roots of an alien monotelosomic addition plant of *Beta vulgaris*, transformed by *Agrobacterium rhizogenes*, *Euphytica* **48**, 153-157.
42. Paul, H., Zijlstra, C., Leeuwangh, J.E., Krens, F.A., Huizing, H.J. (1987) Reproduction of the beet cyst nematode *Heterodera schachtii* Schm. on transformed root cultures of Beta vulgaris L, *Plant Cell Rep.* **6**, 379-381.
43. Price, C. (1965) Breeding sugar beets for resistance to the cyst nematode *Heterodera schachtii*, *J. Am. Soc. Sug. Beet Techn.*. **13**, 397-405.
44. Salentijn, E.M.J., Arens-DeReuver, M.J.B., Lange, W., De Bock, T.S.M., Stiekema, W.J., Klein-Lankhorst, R.M. (1995) Isolation and characterization of RAPD-based markers linked to the beet cyst nematode resistance locus ($Hs1^{pat1}$) on chromosome 1 of *B. patellaris*, *Theor Appl Genet* **90**, 885-891.
45. Salentijn, E.M.J., Sandal, N.N., Klein-Lankhorst, R., Lange, W., De Bock, T.S.M., Marcker, K.A., Stiekema, W.J. (1994) Long-range organization of a satellite DNA family flanking the beet cyst nematode resistance locus (*Hs1*) on chromosome-1 of *B. patellaris* and *B. procumbens*, *Theor Appl Genet* **89**, 459-466.
46. Salentijn, E.M.J., Sandal, N.N., Lange, W., De Bock, T.S.M., Krens, F.A., Marcker, K.A., Stiekema, W.J. (1992) Isolation of DNA markers linked to a beet cyst nematode resistance locus in *Beta patellaris* and *Beta procumbens*, *Mol Gen Genet* **235**, 432-440.
47. Savitsky, H. (1960) Viable diploid triploid and tetraploid hybrids between *Beta vulgaris* and species of the section *Patellares*, *J. Am .Soc. Sug. Beet Techn.* **11**, 215-235.
48. Savitsky, H. (1973) Meiosis in hybrids between *Beta vulgaris* L. and *Beta procumbens* Chr. Sm. and transmission of sugarbeet nematode resistance, *Genetics* **74**, 241.
49. Savitsky, H. (1975) Hybridization between *Beta vulgaris* and *B. procumbens* and transmission of nematode (*Heterodera schachtii*) resistance to sugar beet, *Can J Genet Cytol* **17**, 197-209.
50. Shepherd, A.M. (1958) Experimental methods in testing for resistance to beet eelworm *Heterodera schachtii* Schmidt, *Nematologica* **111**, 127-135.
51. Speckmann, G.J., De Bock, T.S.M. (1982) The production of alien monosomic additions in *Beta vulgaris* as a source for the introgression of resistance to beet root nematode (*Heterodera schachtii*) from *Beta* species of the section *Patellares*, *Euphytica* **31**, 313-323.
52. Steele, A.E. (1965) The host range of the sugar beet nematode (*Heterodera schachtii* Schmidt), *J. Am. Soc. Sug. Beet Techn.* **13**, 573-603.
53. Steele, A.E., Savitsky, H. (1962) Susceptibility of several *Beta* species to the sugar-beet nematode (*Heterodera schachtii* Schmidt), *Nematologica* **8**, 242-243.
54. Umehara, Y., Inagaki, A., Tanoue, H., Yasukochi, Y., Nagamura, Y., Saji, S., Otsuki, Y., Fujimura, F., Kurata, N., Minobe, Y. (1995) Construction and characterization of a rice YAC library for physical mapping, *Molecular Breeding* **1**, 79-89.
55. Valdes, J.M., Tagle, D.A. and Collins, F.S. (1994) Island rescue PCR: Arapid and efficient methode for isolating transcribed sequences from yeast artificial chromosomes and cosmids, *Proc Natl Acad Sci USA* **91**, 5377-5381.
56. Vos, P., Hogers, R., Bleeker, M., Reijans, M., van de Lee, T., Hornes, M., Frijters, A., Pot, J., Peleman, J., Kuiper, M., Zabeau, M. (1995) AFLP: a new technique for DNA fingerprinting, *Nucleic Acids Research* **23**, 4407-4414.
57. Williams, J.G.K., Kubelik, A.R., Livak, K.J., Rafalski, J.A., Tingey, S.V. (1990) DNA polymorphisms amplified by arbitrary primers are useful as genetic markers, *Nucleic Acids Res* **18**, 6531-6535.
58. Winslow, R.D. (1954) Provisional list of host plants of some root eelworms (*Heterodera* spp.), *Ann Appl Biol* **41**, 591-605.
59. Yu, M.H. (1984a) Resistance to *Heterodera schachtii* in *Patellares* section of the genus *Beta*, *Euphytica* **33**, 633-640.
60. Yu, M.H. (1984b) Transmission of nematode resistance in the pedigree of homozygous resistant sugar beet, *Crop Sci* **24**, 88-91.
61. Yu, M.H., Steele, A.E. (1981) Host-parasite interaction of resistant sugarbeet and *Heterodera schachtii*, *J. Nematol* **13**, 206-212.

RESISTANCE TO ROOT-KNOT NEMATODES IN TOMATO

Towards the molecular cloning of the Mi-1 locus

Tsvetana B. LIHARSKA[1] and Valerie M. WILLIAMSON[2]
[1]Department of Molecular Biology, Wageningen Agricultural University, 6703 HA, The Netherlands
[2]Department of Nematology, University of California, Davis, CA 95616, USA

Abstract

Host resistance to three major species of root-knot nematodes, *Meloidogyne incognita*, *M. javanica* and *M. arenaria*, has been introgressed in the cultivated tomato (*Lycopersicon esculentum* Mill.) from its wild relative *L. peruvianum*. The trait is dominant and has been located at a single locus, *Mi-1*, on chromosome 6. The *Mi-1*-conferred resistance is associated with a hypersensitive response in nematode-infected tissues; it is not efficient at soil temperature higher than 28 °C, and can be overcome by some "virulent" pathotypes as well as by the species *M. hapla*. Novel resistances, that have broader specificity to root-knot nematodes and are heat-stable, are pursued and identified in the wild tomato germplasm.

Molecular cloning of the *Mi-1* locus, based on its map position, has been undertaken and reaches its final step. Some major obstacles due to the proximity of the *Mi-1* locus to the centromere had to be overcome following the positional cloning approach, namely the presence of repetitive sequences and the suppression of recombination around *Mi-1*. Narrowing down the *Mi-1* locus to less than 50 kb made complementation analysis possible. Transformation of susceptible tomato genotype with genomic and cDNA clones from the 50 kb region will reveal the sequence of *Mi-1*.

1. Introduction

Once root-knot nematodes (RKN) (*Meloidogyne* spp.) have entered the roots of a plant, like other biotrophic plant pathogens, they depend in their development upon the compatibility of the interaction with the plant genotype. The host range of *Meloidogyne* spp. is broad and comprises more than 2000 species from many plant families. Superimposed on the establishment of a compatible interaction is the potential for the plant to detect the presence of the pathogen and rapidly activate a series of defense responses leading to an incompatible reaction [6]. This ability of the plant to suppress the nematode development and reproduction, and to resist in this way infection, is under genetic control.

Naturally occurring host resistance against *Meloidogyne* spp. has been found in many crops and related wild germplasm [32]. The trait is most often dominant or incompletely dominant and is inherited as a monogenic trait [13]. Although a genetic characteristic, the resistance is conditional and probably depends on the availability and the expression in the nematode of cognate incompatibility factor(s) or avirulence (Avr) gene(s). The existence of *Avr* genes has been demonstrated in cyst nematode-potato system [21], but not yet for the parthenogenetically propagating RKN species (see chapter by Castagnone-Sereno *et al.*, this volume). However, the specificity of resistance responses as well as the occurrence of resistance-breaking races of RKN support the gene-for-gene mode of action [14] of the RKN resistance genes [9]. To elucidate the mechanism of host resistance to RKN it is essential to study the genetic factors mediating this interaction. In recent years there have been major efforts to isolate the gene(s) for resistance to RKN in tomato (*Lycopersicon esculentum* Mill.) [27; 19; 16]. Tomato is one of the first crops in which RKN resistance has been introgressed by conventional breeding programs and largely exploited [26]. Besides its economical importance, tomato has become a model in plant genetics and offers all the necessary tools to identify genes without knowledge of their phenotype.

In this chapter we describe the availability of RKN resistance in tomato and give attention to the progress in the molecular cloning of the first identified and most studied locus, *Mi-1*.

2. *L. peruvianum* Is the Origin of RKN Resistance in Tomato

In the early 1940s screening of large numbers of tomato lines for resistance to RKN revealed that the cultivated tomato is fully susceptible [4]. In the same survey some wild tomato relatives from the genus *Lycopersicon* were included and certain strains of the remote species *L. peruvianum* showed a high level of resistance. Romshe [31] suggested the use of *L. peruvianum* as a parent for the development of nematode resistant varieties. With the help of an embryo culture, a hybrid between *L.esculentum* cv.Michigan State Forcing and *L. peruvianum* accession P.I. 128657 was raised by Smith [35] (Fig.1). Backcrossing programs of the single hybrid plant were run simultaneously by two groups, one in Davis, California and the other in Hawaii, and resulted in two *L. esculentum* lines, respectively VFN8 and Anahu, homozygous for resistance to *M. incognita* [15; 18]. The symbol *Mi* was assigned for the resistance in these lines after the first letters of *Meloidogyne incognita* [17].

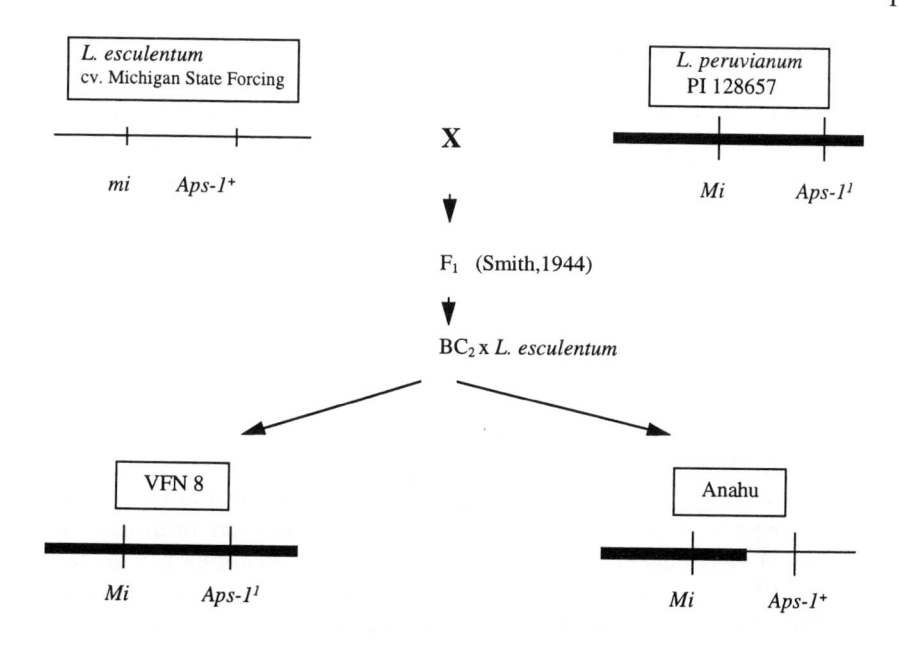

Figure 1. Breeding of *Mi*-resistance into *L. esculentum*. A single F$_1$ plant from the cross *L. esculentium x L. peruvianum* was backcrossed (BC) to *L. esculentum* as female parent. Progenies from the second backcross were further backcrossed in California and in Hawaii, resulting in two *L. esculentum* lines, VFN 8 and Anahu, respectively. Both lines carry *Mi*, along with an introgression segment from *L. peruvianum* (the black bar) on chromosome 6. The introgression in the Anahu line does not include the Aps-1 marker.

3. Properties of the Mi Locus

3.1. HYPERSENSITIVE RESPONSE

The infective stage RKN, J2, are attracted both by *Mi*-resistant and susceptible tomato cultivars. Dean and Struble [11] studied the mechanism of *Mi*-determined resistance and showed that the nematode invasion of the roots produced an extensive necrotic reaction followed by the disappearance of the J2 from the roots. Whatever is the fate of the nematode, the compatible interaction (formation of giant cells) is not established. The necrosis as a plant defense is the result of a complex of reactions very similar to those induced by other pathogens, described as hypersensitive response (HR). It is not clear though if the resistance is the result of, or coincides with, or precedes the HR. The first signs of HR can be detected as changes in cell ultrastructure in roots 8 to 12 hours after infection with root-knot nematodes [28]. Within 2-3 days the HR is localized around the nematode, which either leaves the root or dies there.

3.2. SPECTRUM OF RESISTANCE

In addition to *M. incognita*, resistance conferred by *Mi* is efficient to the prevalent pathotypes of *M. javanica* and *M. arenaria* [5], but not to *M. hapla* [3].

As any host plant resistance, the Mi-mediated resistance can be overcome by some races of nematodes. Partial or full susceptibility to *Mi*-breaking populations of *M. incognita, M. javanica* and *M. arenaria* are reported to occur naturally or arise under the selection pressure of *Mi*-genotypes in the field or in greenhouse conditions [30; Castagnone-Sereno et al., this volume].

3.3. HEAT INSTABILITY

At a soil temperature higher than 28¡C, *Mi*-containing plants show reduced necrosis and appear susceptible to RKN [20]. It has been demonstrated by Dropkin [12] that only during the first two to three days after penetration of nematodes the temperature determines the course of the interaction in *Mi* plants. This feature of the *Mi*-resistance is not unique. Temperature sensitivity of resistance against root-knot nematodes has been reported for other crops [10] and for other plant-pathogen interactions as well. Considering this, the heat instability of *Mi* resistance could be either an intrinsic feature of the host resistance pathway or a property of the *Mi*-product or /and its interaction with the RKN. Besides, factors that would ensure a heat stable HR of *Mi* may not be present in the *L. esculentum* background neither in the source *L . peruvianum* P.I. 128657 of this resistance locus [2]. Interestingly, at high (32°C) temperature *M. incognita* produce significantly more eggs on susceptible tomato cultivars than at 25°C [2].

3.4. GENETICS

Gilbert and McGuire [18] established that the resistance governed by *Mi* is dominant and segregates as a single major locus. The *Mi* locus was mapped on chromosome 6 at position 35 cM [17]. Some controversial data as to the number of genes involved in Mi-resistance and their relationship, have been discussed over the years [34]. Yet, the history of introgression of Mi-resistance in *L. esculentum* [26] indicates the introgression of a single dominant locus on chromosome 6. Further genetic analysis [29] showed a tight linkage between *Mi* and the acid phosphatase-1 ^{1} (*Aps-1^1*) marker. This linkage has been extensively exploited to select indirectly for RKN resistance in tomato in breeding programs, excluding lines originating from the Anahu line which has lost the *Aps-1^1* allele (Fig. 1).

3.5. DURABILITY

In the 45 years' history of exploiting the *Mi* resistance in tomato, the resistance has shown remarkable stability. The presence of a *L. peruvianum* segment carrying the *Mi* locus ensures the characteristic resistance discribed above in any tomato line background. Recombination events breaking the *L. peruvianum* segment with *Mi* , similar to what has happened in the Anahu line, have not occurred very often. Also, spontaneous mutations that abolish the function of the *Mi* (Hontelez *et al.*, in preparation) have been encountered rarely.

4. Novel Genes for Resistance to Root-knot Are Identified in Wild Tomato Germplasm

Until recently, the resistance conferred by the *Mi* locus was the only one identified and utilized in the tomato species. The need to broaden the genetic basis of resistance prompted new attempts to screen wild tomato species for resistance to RKN [3, 24, 30, 7, 33, 48, 42, 43, 44). Resistance effective at high temperatures and against *M. hapla* as well as against *Mi*-breaking pathotypes, was pursued. From all the *Lycopersicon* species tested, only some accessions from the *L. peruvianum* complex, different from the *Mi* source, were found to possess a variety of new resistances to RKN (see Table 1).

TABLE 1. Identified resistances to root-knot nematodes in "*L. peruvianum* complex"

Accessions	Resistance specificity[1]	Temperature limit	Genetics locus	map position	References
L. peruvianum P.I.128657	*M. incognita*, *M.arenaria*, *M.javanica*	28°C	*Mi-1*	chr.6	Bailey, 1941 Medina-Filho and Tanksley, 1983
P.I.270435 -clone 2R2	*M. incognita*	28°C	*Mi-1*	chr.6	Cap *et al.*, 1993
	M. incognita	heat stable	*Mi-2*	nd	
	M.i. 557R[2]	28°C	*Mi-8*	nd	Yaghoobi *et al.* 1995
	M. javanica	28°C	-	nd	Veremis and Roberts,
	M. javanica	heat stable	-	nd	1996b,c
P.I.270435-clone 3MH	*M. incognita*	heat stable	*Mi-6*	nd	Cap *et al.*, 1993
	M.i. 557R	28°C	*Mi-7*	nd	Roberts *et al.* 1990
	M. javanica	28°C	-	nd	Veremis and Roberts, 1996b,c
P.I. 129152	*M. incognita*	heat stable	-	nd	Ammati *et al.*, 1985
LA 1708-I	*M. incognita*	heat stable	*Mi-4*	nd	Veremis and Roberts, 1996a
P.I.126443-clone 1MH	*M.i.* 557R	28°C	*Mi-3*	chr.12	Cap *et al.*, 1991, 1993
	M. incognita, *M. javanica*	heat stable	*Mi-5*	chr.12	Yaghoobi *et al.*, 1995
	M. javanica	28°C	-	nd	Veremis and Roberts,
	M. javanica	heat stable	-	nd	1996b,c
	M. hapla	heat stable	-	nd	Ammati *et al.*, 1985, 1986
L.chilense LA2884	*M. javanica*	heat stable	-	nd	Veremis and Roberts, 1996c

Notes: 1) resistance to species or pathotypes RKN that has been tested and confirmed 2) *M. incognita* virulent strain 557; nd) not determined

Backcrossing to a susceptible cultivar or accession revealed the dominant nature of the newly identified traits. Test for allelism with *Mi* or the linkage with the *Aps-1¹* allele were exploited to classify the new loci and to discriminate between them and *Mi*. The genetic analysis of these resistances and their properties point out that they are determined by loci distinct from *Mi* [8, 48, 42]. To some, symbols from *Mi*-2 to *Mi*-8, have been assigned, and the original Mi locus is referred to as *Mi-1*. The relationship among the novel resistances needs to be studied further and some might turn out to be allelic (identical). To date, resistance to a virulent isolate of *M. incognita* (557R) has been mapped as a single locus *Mi*-3 on the telomeric end of chromosome 12 in tight linkage to the heat-stable resistance to *M. incognita* (*Mi*-5) [48, 42].

The mechanisms via which the novel genetic factors provide resistance against RKN have not been studied. It is not clear yet if they are associated with HR like the *Mi-1* confered resistance.

4.1. INTROGRESSION IN L. ESCULENTUM BACKGROUND

The efforts to introduce the newly identified resistances in *L. peruvianum* to *L. esculentum* have been hampered by the remoteness of the two species and the lack of cross-compatibility. Rescuing the interspecific hybrids by embryo cultures has been successful for some, but not all *L. peruvianum* genotypes [35, 7, 33]. Another approach that utilizes bridging lines, made between some "easy" crossable *L. peruvianum* genotypes and *L. esculentum*, has also not been successful so far [42].

5. Molecular Genetics of the *Mi-1* Locus

5.1. THE STRATEGY OF CHROMOSOME WALKING TO THE *Mi-1*

Molecular studies of *Mi-1* were started in the 1980's and were aimed at both isolation of DNA markers for indirect selection for resistance and cloning the sequence encoding the resistance based on its map position.

As a first step of a map-based strategy for cloning the *Mi-1*, closely linked markers were identified as possible starting points for chromosome walking [22, 27]. This included the major work of cloning of the *Aps-1^1* allele and converting it to a DNA marker [1, 46]. The map position of *Mi-1* on chromosome 6 was refined, using a number of genetic crosses [45] and the distance between the flanking markers GP79 and *Aps-1* was estimated at less than 2cM [27, 19]. This distance in tomato can be translated to approximately 1,5 Mb [38] and indeed, long-range physical mapping of the two markers, GP79 and *Aps-1*, revealed that they are at least 1,2 Mb apart [40]. Chromosome walking on such a large distance was not feasible, in spite of the availability of YAC libraries of nematode resistance genotypes [39, 25]. The expected presence of repetitive sequences in this region, due to its proximity to the centromeric heterochromatin, seemed to be a major obstacle for conventional chromosome walking. Indeed, during the last years deletion and cytogenetic mapping have positioned the *Mi-1* locus and the GP79 marker to the short arm of the chromosome (Figure 2) near the border with the heterochromatin [41; Xiaobo Zhong, in preparation] and separated from the *Aps-1* marker on the long arm by the centromere.

Screening of nematode resistant tomato varieties with the available RFLP markers has resulted in the identification of one introgression line, Motelle, which is nematode resistant and has none of the *L. peruvianum* alleles for GP79 and *Aps-1* [27, 19]. The extent of the *L. peruvianum* introgression carrying the *Mi-1* in Motelle has been estimated at approximately 650 kb (P. Vos, personal communication). Localizing *Mi-1* to this small chromosomal segment, as well as the availability of RAPD and AFLP techniques for isolating molecular markers, has directed a new strategy for positional cloning of *Mi-1* in recent years (Fig. 2). The approach aims at tagging the gene of interest by employing large numbers of molecular markers and recombinants with known phenotype (reviewed in [37]).

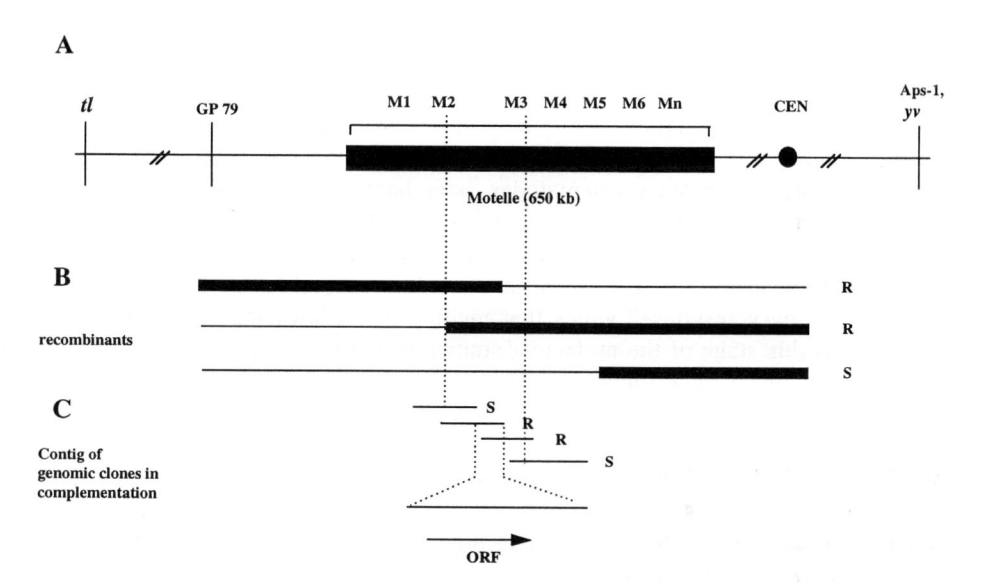

Figure 2. Chromosome landing approach for cloning the *Mi-1* locus. **A.** The line Motelle has approx. 650 kb of *L. peruvianum* introgression (the black bar), carrying the *Mi-1* locus on the short arm of chromosome 6. A considerable number of molecular markers (M1-Mn) randomly distributed within these 650 kb are needed. **B.** Recombinants from crosses between resistant (black bar) and susceptible (line) genotypes with crossover within the Motelle region and with known phenotypes (R or S). Molecular marker mapping of the crossover points reveals molecular markers (M2 and M3) in the vicinity of *Mi-1*. **C.** Complementation of a susceptible tomato genotype with selected genomic clones between the markers M2 and M3. The phenotype of the transformants will further indicate the genuine coding sequence(s) for *Mi-1*.

5.2. CHROMOSOME LANDING IN THE *Mi-1* LOCUS

Tomato lines nearly isogenic for the *Mi-1* resistance have been used to identify RFLP [19], RAPD [47] and AFLP markers (Keygene, Wageningen, unpublished) to saturate in that way the Motelle region (Fig. 2). The order of the available markers within the Motelle region was accomplished by recombinant analysis [23, 16, Williamson *et al.*, in preparation] or physical mapping (Keygene, Wageningen, personal communication).

Genetic mapping in the *Mi-1* region has been hampered by the low frequency of recombination in crosses involving the *Mi-1* locus [27, 19]. Using large populations segregating for *Mi-1* resistance with pre-selection for recombinants between the linked morphological markers tl and yv (Fig. 2) provided only a few recombinants within the Motelle region [23]. An alternative approach to obtain better resolution in the same region makes use of intraspecific *Lycopersicon peruvianum* populations [16; Williamson, in preparation]. This approach has yielded informative recombinants within the Motelle region and allowed localization of *Mi-1* to a chromosomal segment of less than 50kb.

Complementation analysis with genomic and cDNA clones corresponding to this 50 kb region as well as fine molecular mapping of the isolated recombinants is under way and will further reveal the DNA coding sequence for *Mi-1*.

6. Conclusions

During the last three years a number of dominant genes for host plant resistance against several pathogens, but not yet nematodes, have been isolated (reviewed in [36]). The remarkable similarity between the proteins encoded by many of these genes, as well as the common HR in disease resistance, suggest that plants may have a universal signal transduction mechanism to express resistance to different pathogens. The resistance genes then are "primary response" genes that mediate the expression of a host resistance pathway. At this stage of the molecular studies of the resistance against RKN we can only speculate on a similar function of *Mi-1* and of the newly identified resistance factors in tomato.

ACKNOWLEDGEMENTS

We thank John Veremis for providing in-press manuscripts and Ab van Kammen and Maarten Koorneeffor critical reading the manuscript.

References

1. Aarts, J.M.M.J.G., Hontelez, J.G.J., Fischer, P., Verkerk, R., van Kammen, A., and Zabel, P . (1991) Acid phosphatase-11, a tightly linked molecular marker for root-knot nematode resistance in tomato: from protein to gene, using PCR and degenerate primers containing deoxyinosine, *Plant Mol. Biol.* **16**, 647-61.
2. Ammati, M., Thomason, I.J., and McKinney, H.E. (1986) Retention of resistance to *Meloidogyne incognita* in *Lycopersion* genotypes at high soil temperature, *J. Nematol.* **18**, 491-495.
3. Ammati, M., Thomason, I.J., and Roberts, P.A. (1985) Screening *Lycopersion* spp. for new genes imparting resistance to root-knot nematodes (*Meloidogyne* spp.), *Plant Dis.* **69**, 112-115.
4. Bailey, D.M. (1941) The seedling test method for root-knot nematode resistance, *Proc. Am. Soc. Hortic. Sci.* **38**, 573-575.
5. Barham, W.S., and Winstead, N.N. (1957) Inheritance of resistance to root-knot nematodes, *Tomato Genet. Coop. Rep.* **7**, 3.
6. Briggs, S.P., and Johal, G.S. (1994) Genetic pattern of plant host-parasite interactions, *Trends Genet.* **10**, 12-16.
7. Cap, G.B., Roberts, P.A.,Thomason, I.J., and Murashige, T. (1991) Embryo culture of *Lycopersicon esculentum* x *L. peruvianum* hybrid genotypes possessing heat-stable resistance to *Meloidogyne*

incognita, *J.Amer Soc.Hort.Sci.* **116**, 1082-1088.
8. Cap, G.B., Roberts, P.A.and Thomason, I.J. (1993) Inheritance of heat-stable resistance to Meloidogyne incognita in L. peruvianum and its relationship to the Mi gene, *Theor. Appl..Genet* **85**, 777-783.
9. Castagnone-Sereno, P., Bongiovanni, M., Palloix, A., and Dalmasso, A. (1996) Selection for *Meloidogyne incognita* virulence against resistance genes from tomato and pepper and specificity of the virulence/resistance determinants, *European J. Plant Pathology*, in press.
10. Cook, R. (1991) Resistance in plants to cyst and root-knot nematodes, *Agricult. Zool .Rev.* **4**, 213-240.
11. Dean, J.L., and Struble, F.B. (1953) Resistance and susceptibility to root-knot nematodes in tomato and sweet potato, *Phytopathology* **43**, 290.
12. Dropkin, V.H. (1969) The necrotic reaction of tomatoes and other hosts resistant to *Meloidogyne*: reversal by temperature, *Phytopathology* **59**, 1632-1637.
13. Fassuliotis, G. (1987) Genetic basis of plant resistance to nematodes, in Veech, J.A. and Dickson, D.W.(eds.), Vistas on nematology:, Soc. of Nematologists, Inc. Hyattsville,Maryland, pp.364-371.
14. Flor, H.H. (1956) The complementary genetic systems in flax and flax rust, *Advan. Genet.* **8**, 29-54.
15. Frazier, W.A., and Dennett, R.K. (1949) Isolation of *Lycopersicon esclentum* type tomato lines essentially homozygous resistant to root-knot, *Proc. Am .Soc. Hortic. Sci* . **5**, 225-236.
16. Ganal, M.W, and Tanksley, S.D. (1996) Recombination around the Tm2a and *Mi* resistance genes in different crosses of *Lycopersicon peruvianum*, *Theor. Appl. Genet.* **92**, 101-108.
17. Gillbert, J.C. (1958) Some linkage studies with the *Mi* gene for resistance to root-knot, *Tomato Genet. Coop. Rep.* **8**, 15-17.
18. Gillbert, J.C., and McGuire, D.C. (1956) Inheritance of resistance to several root-knot from *Meloidogyne incognita* in commercial type tomatos, *Proc. Am . Soc. Hortic. Sci* **68**, 437-442.
19. Ho, J.Y., Weide, R., van Wordragen, M., Lambert, K., Koornneef, M., Zabel, P., Williamson, V.M. (1992) The root- knot nematode resistance gene (Mi) in tomato: construction of a molecular linkage map and identification of dominant cDNA markers in resistant genotypes, *Plant J.* **2**, 971-982.
20. Holzman, O.V. (1965) Effect of soil temperature on resistance of tomato to root-knot nematode (*Meloidogyne incognita*), *Phytopathology* **55**, 990-992.
21. Janssen, R., Bakker, J., and Gommers, F. (1991) Mendelian proof for a gene-for-gene relationship between virulence of *Globodera rostochiensis* and the H1 resistance gene in *Solanum tuberosum* spp. andigena CPC 1673, *Revue Nematol.* **14**, 213-219.
22. Klein-Lankhorst, R,. Rietveld, P., Machiels, B., Verkerk, R., Weide, R., Gebhardt, C., Koornneef, M., and Zabel, P. (1991) RFLP markers linked to the root-knot nematode resistance gene *Mi* in tomato, *Theor. Appl. Genet.* **81**, 661-667.
23. Liharska, T.B., Koornneef, M., van Wordragen, M., van Kammen, A., and Zabel, P. (1996) Tomato chromosome 6: effect of alien chromosomal segments on recombinant frequencies, *Genome* **39**, in press.
24. Lobo, M., Navarro, R., and Munera, G. (1988) *Meloidogyne incognita* and *Meloidogyne javanica* resistance in *Lycopersicon* species, *Tomato Genet. Coop. Rep.* **38**, 31-32.
25. Martin, G.B., Ganal, M.W., and Tanksley, S.D. (1992) Construction of yeast artificial chromosome library of tomato and identification of cloned segments linked to two disease resistance loci, *Mol. Gen. Genet.* **233**, 25- 32.
26. Medina-Filho, H., and Tanksley, S.D. (1983) Breeding for nematode resistance. In: Evans DA, Sharp WR, Ammirato PV, Yamada Y (Eds) Handbook of plant cell culture, Vol 1, Techniques for propagation and breeding, Mac Millan, New York, pp. 904-923.
27. Messenguer, R., Ganal, M., de Vicente, M.C., Young, N.D, Bolkan, H., and Tanksley, S.D. (1991) High resolution RFLP map around the root-knot nematode resistance gene (Mi) in tomato, *Theor. Appl. Genet.* **82**, 529-536.
28. Paulson, R.E., and Webster, J.M. (1972) Ultrastructure of the hypersensitive reaction in roots of tomato, *Lycopersicon esculentum L.*, to infection by root-knot nematode, *Meloidogyne incognita* , *Physiol. Plant Pathol.* **2**, 227-234.
29. Rick, C.M., and Fobes, J.F. (1974) Association of an allozyme with nematode resistance, *Tomato Genet. Coop. Rep.* **24**: 25.
30. Roberts, P.A., Dalmasso, A., Cap ,G.B., and Castagnone-Sereno, P. (1990) Resistance in *Lycopersicon peruvianum* to isolates of Mi gene-compatible Meloidogyne populations, *J .Nematol.* **22**, 585-589.
31. Romshe, F.A.(1942) Nematode resistance test of tomato, *Proc. Am . Soc . Hortic. Sci.* **40**, 423.
32. Sasser, J.N., Hartman,K.M., Carter, C.E. (1987) Summary of preliminary crop germplasm evaluation

for resistance to root-knot nematodes, NC State University/ US Agency Int. Dev. Raleigh NC, pp. 88.
33. Scott, J.W., Emmons, C.L., Overman, A.J., and Somodi, G.C. (1991) Introgression and genetics of heat stable nematode resistance from *Lycopersicon peruvianum*, *Tomato Genet. Coop. Rep.* **41**, 46.
34. Sidhu, G.S., and Webster, J.M. (1981) The genetics of plant-nematode parasitic systems, *Bot. Rev.* **47**, 387-419.
35. Smith, P.G. (1944) Embryo culture of a tomato species hybrid, Proc. Am.Soc.Hortic. Sci . **44**, 413-416.
36. Staskawicz, B.J., Ausubel, F.A., Baker, B.J., Ellis, J.G., and Jones, J.D.G. (1995) Molecular genetics of plant disease resistance, *Science* **268**, 661-667.
37. Tanksley, S.D., Ganal, M.W., and Martin, G.B.(1995) Chromosome landing: a paradigm for map-based gene cloning in plants with large genomes, *Trends Genet.* **11**, 63-68.
38. Tanksley, S.D., Ganal ,M.W., Prince, J.P., de Vincente, M.C., Bonierbale, M.W., Broun, P., Fulton, T.M., Giovanonni, J.J., Grandillo, S., Martin, G.B., Messeguer, R., Miller, J.C., Miller, L., Paterson ,A.H., Pineda, O., Roder, M.S., Wing, R.A., Wu,W., Young, N.D. (1992) High density molecular linkage maps of the tomato and potato genomes, *Genetics* **132**, 1141-1160.
39. Van Daelen, R. (1995) Towards isolation of the tomato root-knot nematode resistance gene Mi via positional cloning, PhD thesis, Wageningen Agricultural University.
40. Van Daelen, R.A.J.J., Grebens, F., van Ruissen, F., Aarts, J., Hontelez, J., and Zabel, P. (1993) Long-range physical maps of two loci (Aps-1 and GP79) flanking the root-knot nematode resistance gene (*Mi*) near the of tomato chromosome 6, *Plant. Mol. Biol.* **23**, 185-192.
41. Van Wordragen, M.F., Weide, R., Liharska, T.B., van der Steen, A., Koornneef, M., and Zabel, P. (1994) Genetic and molecular organisation of the short arm and pericentromeric region of tomato chromosome 6, *Euphytica* **79**, 169-174.
42. Veremis, J.C., and Roberts, P.A. (1996a) Differentiation of *Meloidogyne incognita* and *M. arenaria* novel resistance phenotypes in *Lycopersicon peruvianum* and derived bridge lines, *Theor. Appl. Genet.*, in press.
43. Veremis, J.C., and Roberts, P.A.(1996b) Relationships between *Meloidogyne incognita* resistance genes in *Lycopersicon peruvianum* differentiated by heat sensitivity and nematode virulence, *Theor. Appl. Genet.*, in press.
44. Veremis, J.C., and Roberts, P.A.(1996c) Identification of resistance to *Meloidogyne javanica* in the *Lycopersicon peruvianum* complex, *Theor. Appl. Genet.*, in press.
45. Weide, R., van Wordragen, M., Klein-Lankhorst, R., Verkerk, R., Hanhart, C., Liharska, T., Pap, E., Stam, P., Zabel, P., and Koornneef, M. (1993) Integration of the classical and molecular linkage maps of tomato chromosome 6, *Genetics* **135**, 1175-1186.
46. Williamson V.M., and Colwell, G. (1991) Acid phosphatase-1 from nematode resistant tomato: isolation and characterization of the gene, *Plant Physiol.* **97**, 139-146.
47. Williamson V.M., Ho, J.Y., Wu, F.F., Miller, N., and Kaloshian, I. (1994) A PCR-based marker tightly linked to the nematode resistance gene, *Mi*, in tomato, *Theor. Appl. Genet.* **87**, 757-763.
48. Yaghoobi, J., Kaloshian, I., Wen, Y., and Williamson, V. M. (1995) Mapping a new nematode resistance locus in *Lycopersicon peruvianum*, *Theor. Appl. Genet.* **91**, 457-46

BIOCHEMISTRY OF PLANT DEFENCE RESPONSES TO NEMATODE INFECTION

Giuseppe ZACHEO[1], Teresa BLEVE-ZACHEO[2] and Maria T. MELILLO[2]
[1]*Istituto di Ricerca sulle Biotecnologie Agroalimentari, CNR, Prov.le Lecce-Monteroni, 73100 Lecce*
[2]*Istituto di Nematologia Agraria, CNR, Via Amendola 165/A, 70126 Bari, Italy*

Abstract

The biochemical basis of the plant resistance response to nematode infection is discussed with special emphasis on pre-existing toxins, and protease inhibitors, pathogenesis related proteins, oxidant and antioxidant mechanisms. The relationship between oxygen radical production and induction of the hypersensitive reaction during nematode infection is explored. This review critically discusses the possible pathway for enzymatic production of oxygen species and the role of detoxifying enzymes. The pathway of induced cell wall strengthening is also briefly discussed.

1. Introduction

Plants respond to infection of pathogenic microorganisms with a co-ordinated series of changes in gene expression and cellular metabolism. The elucidation of biochemical mechanisms that underlie the ability of plants to resist nematode infection is crucial to our attempts to understand and to control plant disease. Coupled to the expression of disease resistance there are several defence mechanisms within the host. One such mechanism, the hypersensitive response (HR) involves rapid metabolic changes and localized cell necrosis in tissue at the site of contact with the nematode. The attractive idea that nematode dies because "it kills the tissue on which it depends on food" was discussed by Christie as early as 1949 [10]. Resistance as often seen in the plant-nematode interaction is associated with necrosis in which a limited number of plant cells in direct contact with the invading nematode dies rapidly. Successful resistance of the plant prevents further growth and development of the pathogen. Probably, necrosis in the plant-nematode interaction is a consequence and not the cause [22] of the plant resistance response to an incompatible parasite. The hypersensitive reaction, with its characteristic rapid cell death and subsequent necrosis, clearly constitutes one of the primary mechanisms of resistance. But not all types of genetic resistance depend upon necrotic reactions to nematodes. Associated with HR is the induction of increased products that are toxic to the pathogen, such as phytoalexins, oxydized phenolics, lipid peroxides, hydrolytic enzymes, and proteinase inhibitors. Formation of a physical barrier which prevents further penetration of the tissue by a nematode may also occur. Examples include the process of lignification and suberization. Polyphenol-oxidase and peroxidase are enzymes involved in phenol metabolism and in the production of lignin and suberin. The activities of these enzymes have been shown to increase in plant tissues that are undergoing a HR. Initiation of the HR requires the presence of many factors that may trigger or elicit rapid cell death. In recent years an exciting area of disease physiology has emerged with reports of the involvement of active oxygen species in plant pathogenesis. These active species appear to be able to affect nearly all aspects of biological systems, including proteins, lipids, polysaccharides and nucleic acids.

Several enzyme systems have been hypothesized as sources of active oxygen species, including NADPH-oxidases, peroxidases and lipoxygenases. Recently [1] it was demonstrated that exogenously supplied superoxide radicals (O^-_2) could cause death symptom similar to those of HR in leaf tissue cell. Unlike the other species reported [4] it has been demonstrated that hydrogen peroxide (H_2O_2) increase induces lignification generally via peroxidases and, depending on the timing, this could strengthen the cell wall and slow nematode ingress.

2. Pre-existing Toxins or Induced Inhibitors

As a result of frequent exposures to severe biological stresses, most plants during their evolution have acquired and developed self-defence mechanisms to protect themselves against pathogenic agents. The natural defenses that protect plants against nematodes are many and diverse, and include both passive (preformed) and active (inducible) compounds. Often plants with nematicidal and nematostatic properties were discovered because of their marked suppression of certain nematode populations in fields or glasshouse experiments. Different groups of naturally occurring plant chemicals are found as either constitutive components in various plant tissues or are synthetised in response to attacking nematodes. Gommers [17] published a comprehensive review of current knowledge on nematicidal compounds. More recently a review of naturally occurring nematicides with biological activity against plant parasitic nematodes focused on several different classes of compounds, including polythienyls, alkaloids, phenolics, polyacetylenes, fatty acids, terpenoids, and others [9]. Proof that plant resistance is regulated by one or more constitutive nematicidal compounds remains to be determined. Gommers [17] discussed the discrepancy between *in vitro* and *in vivo* effects of constitutive nematicidal compounds and concluded that their presence in plants does not necessarily prove that they confer resistance. Other authors [21] concluded that the resistance mechanism to a nematode is generally expressed after infection and that an active mechanism involves compounds produced postinfectionally rather than preformed constitutive plant products. This inducible response includes synthesis of low-molecular-weight antibiotic defence compounds, production of hydrolytic enzymes, rapid modification of existing cell walls, and others [11].

3. Induction of Generalized Resistance

The induction of a general defence response by a variety of biotic or abiotic stresses increases resistance to disease. Limited infection or treatment with certain chemical agents increases resistance to nematode [33, 34]. Such an expression of resistance is associated with the deployment of generalised defence responses such as accumulation of phytoalexins and proteins.

3.1. PHYTOALEXINS

Phytoalexins are low-molecular-weight antibiotics and their accumulation is determined by their release or the release of immediate precursors of conjugates and/or the *de novo* synthesis [26]. The involvement of phytoalexins in the plant-nematode interaction has been frequently reviewed from different perspectives [45, 51]. Early experiments and observations were reviewed by Veech [45], but it was not until the experiments of Rich *et al.* [38] that phytoalexins were identified as anti-nematode antibiotics that accumulated in host tissue. More recently other authors [20, 21] have presented evidence that resistance of soybean and potato to endoparasitic nematodes is caused by the production of glyceollin. From these data it appears that the rapidity and magnitude with which phytoalexins are produced is important in nematode resistance. With regard to glyceollin and soybean [20] it is also evident that the production of phytoalexins in a susceptible interaction is not due to the lack of defence compounds in the host but rather to the timing and intensity of intercellular migration. The inhibition of nematode development in resistant cultivars occurs at a time when levels of glyceollin are sufficient to account for the inhibition. Delayed phytoalexin accumulation is not effective in inhibiting nematode growth. Accumulation of phaseolin seems to be associated with cell death whenever the accumulation of pisatin is not associated with cell death during nematode infection. Whether phytoalexins accumulate in living or dead cells is an important factor, especially where sedentary nematodes are concerned. Phytoalexin extracts treated with an abiotic elicitor are themselves capable of eliciting phytoalexin accumulation. This suggests that nematode stylet secretions directly elicit phytoalexin accumulation or indirectly cause the release of an endogenous compound, which in turn elicits the accumulation of phytoalexins [45]. Despite the extensive studies on the role of phytoalexins in nematode resistance, current knowledge of the mode of the antibiotic action of phytoalexins is still limited. More information is required to establish the cascade of events leading *inter alia* to their biosynthesis, degradation, role in nematode resistance, regulation of genes for their synthesis, and transfer genes for their synthesis from one plant to another. A key to understanding the resistance to nematodes is the rapid activation of existing resistance mechanisms at the onset of infection.

3.2. PROTEINASE INHIBITORS

Little information is available about the nature and function of proteolytic enzymes in plant parasitic nematodes. Recent studies have demonstrated the presence of proteinases in extracts of preparasitic and parasitic nematodes [19, 25, 36] indicating their constitutive expression throughout the development of the nematode and their possible involvement during pathogenic processes. The activity detected in nematode extracts was inhibited by cysteine proteinase-inhibitors, suggesting the presence of cysteine proteinase activity [19]. Proteinase inhibitors are widely distributed in plants and probably have a function in plant protection against herbivores, fungi and bacteria. They have antinutrient effects on the digestive system [40]. Evidence that proteinase inhibitors serve as protectants against nematodes was reported by Hepler and Atkinson [19]. These authors demonstrated that cowpea trypsin inhibitor expressed constitutively as a transgene in potato plants was effective in altering the growth and fecundity in *Globodera pallida* and *Meloidogyne incognita*. Recent experiments [46, 47] described a two-step strategy to access the potential of cysteine proteinase inhibitors for nematode control. The first step estimates the degree of affinity between the target proteinase and their putative inhibitors; the second step analyzes the ability of proteinase inhibitors to recognize slightly modified proteinase variants. As shown by the differentiated effects of plant cysteine, oryzacystatin I and oryzacystatin II, on the major proteinase of *M. hapla*, it is possible to recognize proteinase variants and to predict the ability of proteinase inhibitors to inactivate the newly appeared proteinase forms during the development of resistance in target pests [47]. If these proteins are shown to be important for resistance to nematodes, then it may be possible to develop or genetically engineer plants with enhanced expression of genes encoding these proteins.

3.3. PATHOGENESIS-RELATED PROTEINS

Inoculation of tomato plants with midly virulent *Meloidogyne* species induced resistance in the plants such that the reproduction of a challenge inoculum of normally compatible *M. hapla* was highly suppressed. The increase in magnitude of induced resistance that corresponded to longer intervals between the advance and challenge inoculation was dependent on the post-infectional accumulation of nematicidal substances [33]. Limited infection or treatment with certain chemical agents increases resistance to several pathogens and induces the accumulation of a variety of proteins. A group of these proteins, collectively called pathogenesis-related (PR) proteins, is induced in plants attacked by pathogens and is associated with necrotic reactions [14]. Hammond-Kosack [18] characterized PR proteins accumulated in the apoplast of potato leaves infected with *G. rostochiensis*. Polypeptides described [24] in other plant-pathogen interactions were similar to those reported in potato leaf PR-proteins. This suggests that the signalling events arising from a range of pathogenic stimuli lead to a common expression of PR proteins. Some of the PR proteins have been identified and characterized as chitinases and ß-1,3-glucanases in leaves and roots of different potato cultivars infected with *G. pallida* [37]. The activity of chitinases increased significantly in infected roots while the activity of ß-1,3-glucanases increased significantly in the leaves. Plants produce proteolytic enzymes that may inactivate pathogens. However, the relationship of protease and nematode and the importance of protease in nematode resistance remains unknown. Utilization of an inducible mechanism of resistance would offer a valuable new option for practical nematode control as well as plant health management in general. Our understanding of the phenomenon is still primitive and needs more research in the area of

PR proteins.

4. Local Responses

The biological mechanism of resistance at the site of nematode infection in many plant species is accompanied by two main component defence responses: the hypersensitive response and the induction of a structural defence barrier.

4.1. PRODUCTION OF ACTIVE OXYGEN DURING HR

This review focuses on an active oxygen (AO) metabolism during the early stage of pathogenesis. Oxygen radicals are ubiquitously generated in many biological compartments of the cells and occur during normal metabolism in healthy plant cells. They would be toxic if allowed to accumulate. Molecular oxygen becomes dangerously reactive to the biological system when its electron structure is altered. The first one-electron reduction resulting in superoxide radical production requires a slight input of energy and it is often provided by NAD(P)H in biological systems. The reduction of superoxide that follows yields hydrogen peroxide and hydroxyl radicals, the latter being the more reactive. Although hydroxyl radicals may attack all classes of biological molecules, the polyinsaturated fatty acids of membranes are particularly vulnerable because of the lipophilic action. The report of Doke [13] indicates that plants contain a superoxide radical generating NADPH-oxidase. During the infection of plants by pathogens, the activation of NADPH-oxidase may be one of the earliest, if not the initial, biological response of the HR phenomenon.

In the incompatible plant-nematode interaction, after membrane related signals, one of the primary events in the hypersensitive reaction may be the production of oxygen free radicals. Cellular and biochemical aspects of the production of oxygen free radicals, following nematode invasion, have been reviewed previously [51]. Most recently, some authors [3] have confirmed the production of free radicals in roots of cereals during "early hypersensitive-type reacting" resulting from infection with *M. naasi*.

Biochemical perturbations caused by AO are likely to be localised. In resistant tomato and cereal roots treated with nitroblue tetrazolium (NBT), localised accumulation of formazan was found to be restricted to a short distance from the site of infection. Symptoms of AO tissue reaction were in cells not yet dead, on which the nematode had attempted to feed and in living cells immediately adjacent to the penetrating nematode. In infected tissue of a susceptible interaction no staining was visible around the nematodes or in giant cells [3, 48]. Cells that had become necrotic, as a result of nematode feeding, showed an intense dark blue pigmentation which was scattered in granular form and was distinct from the AO reaction.

Induction of hydrogen peroxide at the cell surface of infected tissue was reported by Melillo *et al.* [31]. This observation suggests that an increase in hydrogen peroxide was induced by nematode infection in the free spaces of parenchyma stelar cells adjacent to the feeding site and in the inner wall surface of xylem vessels. The granules of precipitated cesium salts were localised as a continuous layer or, more often, the layer of reaction product was discontinuous and different in amount. Frequently hydrogen peroxide precipitates accumulated along the plasma membrane and/or lined the middle lamella of the cell wall [31].

4.2. SOURCE OF ACTIVE OXYGEN

The production of AO requires the presence of chemical elicitors and enzymes. The probable sources during plant-nematode interactions could be linked to the activities of some enzymes such as NAD(P)H oxidase, peroxidase, lipoxygenase, and others. During the infection of resistant plants with nematodes, root tissues are stimulated to reduce both cytochrome c and NBT. The addition of superoxide dismutase (SOD) to the assay solution inhibits these reactions [48]. SOD-inhibitable reduction of cytochrome c or NBT are commonly used to test for the presence of free radicals. Reducing equivalents in the form of exogenous NADPH are required for significant stimulation to occur [48]. On the basis of this evidence, Doke [13] demonstrated their existence on a membrane located superoxide-generating NADPH-oxidase. The production of hydrogen peroxide seems to be related to the presence of basic peroxidases which catalyze the production of peroxide at the expense of NADH [8, 16]. With regard to HR, it would seem that the basic peroxidases may be the effect of H_2O_2 and O^-_2 accumulation.

Numerous studies have indicated that lipoxygenase (LOX) plays an important role in generating AO during the HR [43]. It has been suggested that free radical-induced-lipid peroxidation might cause the membrane damage as seen in the hypersensitive reaction, and that increased LOX activity might be a response of the plant to scavenge the free fatty acids released as a consequence. In the nematode-plant interaction, coincident with the AO production, an increased lipid peroxidation has also been found [49]. Lipoxygenase activity, measured by the oxidation level of linoleic acid, was reported to be higher in resistant than in susceptible potato roots, 2 days after *G. pallida* inoculation. Its activity was uniformly low in the 6 day-samples of all potato clones [2]. Thus, it is possible that LOX is involved in the development of the hypersensitive reaction in the plant-nematode interaction.

4.3. ANTIOXIDANT MECHANISMS

Through selective pressure and evolution, numerous defence mechanisms have emerged to protect cells against oxidative injury. The enzymatic antioxidant defenses induce enzymes capable of removing, neutralizing, or scavenging free radicals and oxy-intermediates. Ascorbate peroxidase, catalase, peroxidase, and superoxide dismutase are the most efficient antioxidant enzymes. Their combined action converts the superoxide radicals and hydrogen peroxide to water and molecular oxygen, thus avoiding cellular damage [42]. Numerous studies indicate that biotic or abiotic oxydative stress changes the activity of SOD. In investigating the resistance of tomato plant cultivars toward infection with *M. incognita*, Zacheo and Bleve-Zacheo [48] reported a fall of SOD activity in hypersensitively responding tomato plants. Similar results were obtained by Vanderspool *et al.* [44] who confirmed a decline in activity of SOD in resistant soybean, 96 hours after nematode infection. In contrast, SOD increased dramatically in galls [48] as nematodes matured and enlarged [44]. SOD might be involved in processes related to the development and maintenance of the nematode feeding site and might provide protection against superoxide-mediated damage in a compatible response. The reduced level of endogenous SOD in the resistance response leads to a preparatory action of such a defence system, allowing the formation of superoxide radicals. There is a good evidence that generation of superoxide radicals is an important feature of the local events that occur in an incompatible interaction, while an increased scavenging activity of superoxide anions by SOD, with concurrent production of H_2O_2, occurs in a compatible interaction. Clearance of H_2O_2 may be a subsequent defensive step that is induced by the action of peroxidase and catalase.

4.4. INDUCED CELL WALL STRENGTHENING

In an incompatible plant pathogen interaction, the HR often includes an oxidative burst in which H_2O_2 levels rise rapidly and dramatically. The components involved in the induction of an oxidative burst in plant cells have yet to be identified.

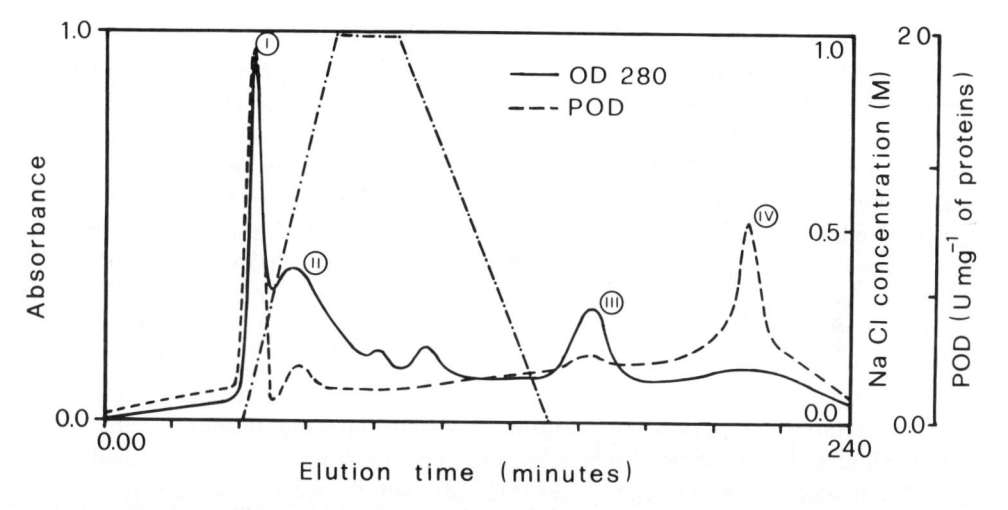

Figure 1. High performance cation-exchange chromatography of anionic and cationic peroxidases from homogenated pea roots on Ultropac TSK CM-3SW column. Symbols I, II, III, and IV indicate the peaks collected for electrophoresis analysis.

There is evidence for generation of H_2O_2 by several routes [41]. It is likely that under conditions where lignification of host cell walls is employed defensively, one or more of the pathways to H_2O_2 production proceed via O^-_2 generation. Thus, peroxidase might have a double role in both structural and physiological protection of the host [42]. As suggested by Dixon and Lamb [12], H_2O_2 is thought to be generated in cell walls by the action of an elicitor stimulated vectorial NADH-oxidase at the plasma membrane. The H_2O_2 concentration would be highest in the infected cells, thus causing cell death;

adjacent cells would receive lower levels of H_2O_2, and thus activate cellular protectant genes. The increase of H_2O_2 and peroxidase activity during pathogenesis has been demonstrated and, depending on the timing, this could strengthen the cell wall and slow pathogen ingress. There is good evidence that the action of a highly anionic peroxidase is involved in the final stages of lignin biosynthesis. Lignification is also induced in the plant pathogen response and has been correlated with local resistance and induced resistance [32]. The accumulation of peroxidases has also been correlated with nematode infection [51]. Recently, a group of fast migrating peroxidases has been chromatographically purified from resistant tomato isolines infected with *M. incognita*, and identified as anionic peroxidases [50]. Zacheo (unpublished results) later confirmed these observations by isolating four different groups of peroxidases present in resistant pea roots infected with *H. goettingiana* (Fig. 1). Electrophoresis analyses demonstrated that major changes in peroxidase isoenzymes, induced by nematode infection, were related to the fourth group of peroxidases (Zacheo, unpublished results). This group of peroxidases, as reported by other authors, has a particularly high affinity for syringaldazine. This electron donor has been used as a specific substrate for lignifying peroxidases [8], allowing a correlation between the level of anionic peroxidases and lignification [16]. Rapid induction of syringaldazine-oxidase [5] and rapid biosynthesis of lignin [30] have been detected in resistant lines of pea roots infected with *H. goettingiana*. Zacheo *et al.* [52] obtained evidence of a correlation between HR induction of anionic peroxidases and increase in lignin level. These authors pointed out that the process of lignification was partially blocked by high temperature [52]. Increase of peroxidase activity in cell walls of pea roots infected by nematodes has been found to be concentrated near the sites of the infection where lignification is greatest, and also that inhibition of lignin synthesis by cycloheximide causes inhibition of the new synthesis of peroxidases [6, 29]. Electron microscopy cytochemical studies using the 3-3'-diaminobenzidine (DAB)/H_2O_2 technique on resistant and susceptible pea roots infected with *H. goettingiana*, at acidic pH, showed that peroxidase activity was mainly detected along the plasma membrane (Fig. 2a,b) while at basic pH isoenzymes were only located in vacuoles (Fig. 2c). These results indicate a specific compartmentalization of peroxidases related to a differential and specific metabolic function. In addition, resistant infected tissues incubated in homovanillic acid, a substrate closely related to lignin monomers, strongly reacted at the level of cell wall and intercellular spaces (Fig. 2d), with the reaction being very weak in susceptible tissues (Fig. 2e). This suggests the existence of topologically distinct active sites on peroxidase molecules. The metabolic function to be assigned to each group of peroxidase isoenzymes should be in accordance with their subcellular localization. Acidic peroxidases are reported to be located in cell walls and free intercellular spaces [39] and, because of their reactivity towards cinnamyl alcohols and isoflavonoid phytoalexins, to be involved in cell wall strengthening and plant defence response [23]. The high reactivity of basic peroxidases towards alkaloids [7, 35] and phenolics, and their localization in vacuoles, supports a role for these isoenzymes in the metabolic turnover of these types of compounds [35].

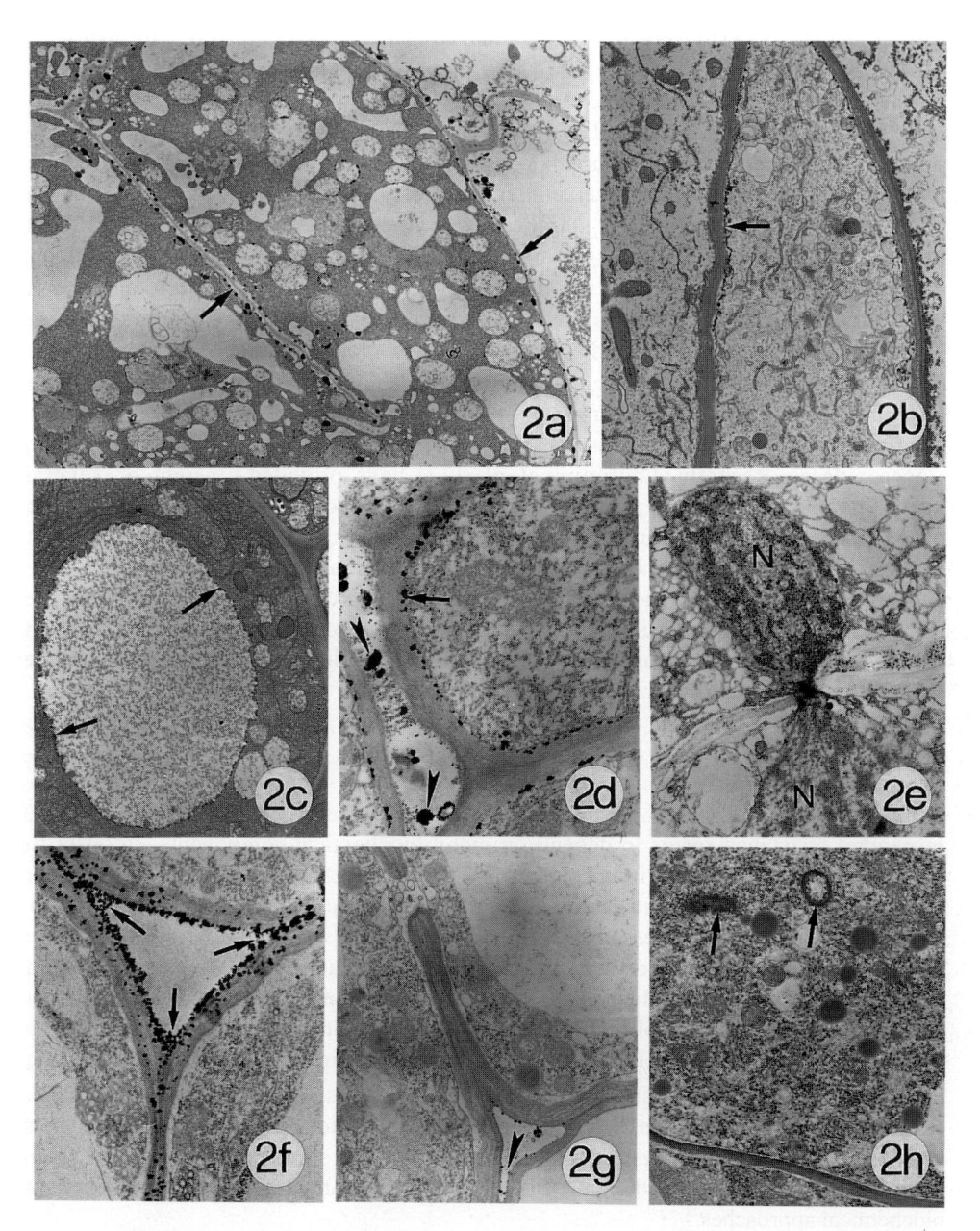

Figure 2. Subcellular cytochemical localization of peroxidases in syncytia induced by *H. goettingiana* in pea roots. a) DAB+H$_2$O$_2$ localization, at acidic pH, on the plasma membranes (->) of resistant syncytial cells. x 4,500; b) Weak reaction of the plasma membrane (->) in susceptible syncytial cells, incubated with the same medium than in a). x 3,000; c) Vacuole of a resistant syncytium, reacting to DAB+H$_2$O$_2$ at basic pH. x 4,000; d) Homovanillic oxidase along the plasma membrane (->) and in the free spaces (->) of a resistant syncytium. x 8,000; e) Very scarce homovanillic oxidase activity in susceptible syncytial cells

Elstner and Heupel [15] were able to elucidate a reaction scheme for H_2O_2-formation by cell wall peroxidases at the expense of either NADPH or NADH. Cytochemical localization of hydrogen peroxide, by using NADH as a substrate in pea roots, following *H.goettingiana* infection, showed that the reaction was consistently localized along the plasma membrane, the middle lamella of cell walls, and in the free spaces of resistant roots (Fig. 2f), and only in the intercellular space of susceptible roots (Fig. 2g); catalase added to the medium completely inhibited the reaction (Fig. 2h). It has been suggested that peroxidases involved in H_2O_2 formation and the cross-linking of cell wall polymers are attached to the primary wall, while those involved in lignification of secondary cell wall thickenings may be free in cell wall spaces [28]. Thus the accumulation of peroxidase, H_2O_2, and lignin could be causally related to development of resistance to the nematode. However, such investigations have provided little useful information about the physiological significance of changes in peroxidase activity during nematode infection. More investigations are necessary to reveal the particular functions of certain isoperoxidases. Recently, Lagrimini *et al.* [27] have provided insight on several interesting relations between peroxidase regulation and function. They created transgenic plants that overproduce a single native enzyme in many tissues that typically do not synthesize peroxidase. The altered peroxidase expression induced appearance of new phenotypes that were unpredicted and thus highlights the tremendous potential for the use of chimeric genes. By altering the expression of the peroxidase pathway, it should be possible to delineate which enzyme is rate-limiting and to determine whether this enzyme plays a role in the plant-nematode interaction.

5. Conclusions

In general resistance seems to depend on many biochemical components. In order to avoid mistaken conclusions, future studies should regard resistance as a multicomponent dynamic system, paying particular attention to time sequences and cellular specialization. Through the use of the expression of defined genes and/or their promoters in transgenic plants it should be possible to identify the precise spatial regulation of local and systemic events induced by the nematode stimulus. Much more needs to be known about the chemical enhancement of natural resistance. At the same time, we must not ignore the importance of O^-_2 and H_2O_2 generation, particularly in signal transduction, during nematode invasion and in subsequent resistance expression by the host plant cells. Research on resistance responses has already identified a wide range of proteins, enzymes and genes suitable for detailed analysis of targeting signals. Future progress in plant-nematode resistance will require the combined application of genetic, molecular and biochemical approaches.

(N=nucleus). x 8,000; f) Hydrogen peroxide detection in the free space (->) and middle lamella of resistant x 8,000 and g) susceptible syncytial cells (->). x 7,500; h) No reaction for hydrogen peroxide in susceptible cells incubated with NADH+catalase. Note the feeding tubes (->). x 6,500.

ACKNOWLEDGEMENTS

The authors are grateful to Dr. C.E. Taylor and F.J. Gommers for the critical reading of this review. The experimental work performed in our laboratory was supported by CNR, Italy, Special project RAISA, Sub-project 2, Paper n° 2816.

References

1. Adam, A.L., Bestwick, C.S., Galal, A.A., Manninger, K., Barna, B., et al. (1993) What is the putative source of free radical generation during hypersensitive response in plants?, in Mozsik G., Emerit I., Feher J., Matkovics B. and Vincze A. (eds.), *Oxygen Free Radicals and Scavengers in the Natural Sciences*, Akademial Kiado, pp.35-43.
2. Amalraj, S.F.A. (1995) Enzyme activity associated with resistance in potato to the early stages of *Globodera pallida* infection, *Nematol. medit.* 23, 199-202.
3. Balhadère, P., and Evans, A.F. (1995) Cytochemical investigations of resistance to root-knot nematode *Meloidogyne naasi* in cereals and grasses using cryosections of roots, *Fundam. Appl. Nematol.* 18, 539-547.
4. Baker, C.J., and Orlandi, E.W. (1995) Active oxygen in plant pathogenesis, *Annu. Rev. Phytopathol.* 33, 299-321.
5. Bleve-Zacheo, T., Melillo, M.T., and Zacheo, G. (1991) Syringaldazine oxidase activity in *Pisum sativum* infected with *Heterodera goettingiana, Giorn. Bot. Ital.* 125, 97-98.
6. Bleve-Zacheo, T., Melillo, M.T., and Zacheo, G. (1993) Subcellular location of benzidine and homovanillic acid-oxidase in pea roots susceptible and resistant to *Heterodera goettingiana, Giorn. Bot. Ital.* 127, 1176-1178.
7. Calderon, A.A., Zapata, J.M., Munoz, R., Ros Barcelo, A. (1993) Localization of peroxidase in grapes using nitrocellulose blotting of freezing/thawing fruits, *Hortic. Sci.* 28, 38-40.
8. Castillo, F.J. (1986) Extracellular peroxidases as marker of stress?, in Greppin H., Penel C. and Gaspar T. (eds.), *Molecular and Physiological Aspects of Plant Peroxidases*, Univ. of Geneva, Switzerland, pp. 419-426.
9. Chitwood, D.J. (1993) Naturally occurring nematicides, in Duke S.O., Menn J.J. and Plimmer J.R. (eds.), *Pest Control with Enhanced Environmental Safety*, American Chemical Society, pp. 300-315.
10. Christie, J.R. (1949) Host-parasite relationships of the root-knot nematodes, *Meloidogyne* spp. III. The nature of resistance in plants to root-knot, *Proc. Helminthol. Soc. Wash.* 16,104-108.
11. Cramer, C.L., Weissenborn, D., Cottingham, C.K., Denbow, C.J., Eisenback, J.D., Radin, D.N., and Yu, X. (1993) Regulation of defense-related gene expression during plant-pathogen interactions, *J. Nematol.* 25, 507-518.
12. Dixon, R.A., and Lamb, C.J. (1990) Molecular communication in interactions between plants and microbial pathogens, *Annu. Rev. Plant Physiol. Plant Mol. Biol.* 41, 339-367.
13. Doke, N., Chai, H.B., and Kawaguchi, A. (1987) Biochemical basis of triggering and suppression of hypersensitive cell response, in Nishimura S. (ed.), *Molecular Determinants of Plant Diseases*, Japan Sci. Soc. Press, Tokio/Springer-Verlag, Berlin, pp. 293-305.
14. Elad, Y., and Evensen, K. (1995) Physiological aspects of resistance to *Botrytis cinerea, Phytopathology* 85, 637-643.
15. Elstner, E.F., and Heupel, A. (1976) Formation of hydrogen peroxide by isolated cell walls from horseradish (*Armoracia lapathifolia* Gilib.), *Planta* 130, 175-180.
16. Goldberg, R., Le, T., and Catesson, A:M. (1985) Localization and properties of cell wall enzyme activities related to the final stages of lignin biosynthesis, *J. Experiment. Bot.* 36, 503-510.
17. Gommers, F.J. (1981) Biochemical interactions between nematodes and plants and their revelance to control, *Helminth. Abst. series B, Plant Nematol.* 50, 9-24.
18. Hammond-Kosack, K., Atkinson, H.J., and Bowles, D.J. (1989) Systemic accumulation of novel proteins in the apoplast of the leaves of potato plants following root invasion by the cyst nematode *Globodera rostochiensis, Physiol. Mol. Plant Pathol.* 35,495-506.
19. Hepher, A., and Atkinson, A.J. (1992) Nematode control with proteinase inhibitors, *Europ. patent appl.*

num., 92301890.7; Pubbl. n° 0 502 730 Al.
20. Huang, J.S., and Barker, K.R. (1991) Glyceollin I in soybean-cyst nematode interaction 1. Spatial and temporal distribution in roots of resistant and susceptible soybeans, *Plant Physiol.* **96**, 1302-1310.
21. Kaplan, D.T., and Keen, N.T. (1980) Mechanism conferring plant incompatibility to nematodes, *Rev. Nematol.* **3**, 123-134.
22. Kiràly, Z., Barna, B., and Ersek, T. (1972) Hypersensitivity as a consequence, not the cause, of plant resistance to infection, *Nature* **239**, 456-457.
23. Kolattukudy, P.E., Mohan, R., Aslam Bajar, M., and Sherf, B.A. (1992) Plant peroxidase gene expression and function, *Biochem. Soc. Trans.* **20**, 333-337.
24. Kombrink, E., Schroder, M., and Hahlbrock, K. (1988) Several "pathogenesis-related" proteins in potato are 1,3--glucanases and chitinases, *Proc. Natl. Acad. Sci. U.S.A.* **85**, 782-786.
25. Koritsas, V.M., and Atkinson, H.J. (1994) Proteinases of females of the phytoparasite *Globodera pallida* (potato cyst nematode), *Parasitology* **109**, 357-365.
26. Kuc, J. (1995) Phytoalexins, stress metabolism, and disease resistance in plants, *Annu. Rev. Phytopathol.* **33**, 275-297.
27. Lagrimini, L.M., Bradford, S., and Rothstein, S. (1990) Peroxidase-induced wilting in transgenic tobacco plants, *Plant Cell* **2**, 7-18.
28. McDougall, G.J. (1993) Covalently bound peroxidases and lignification, in Greppin H., Penel C. and Gaspar T. (eds.), *Molecular and Physiological Aspects of Plant Peroxidases*, Univ. of Geneva, Switzerland, pp. 277-282.
29. Melillo, M.T., Bleve-Zacheo, T., and Zacheo, G. (1992) Role of peroxidase and esterase isoenzymes in pea roots infected with *Heterodera goettingiana*, *Nematol. medit.* **20**, 171-179.
30. Melillo, M.T., Bleve-Zacheo, T., and Zacheo, G. (1993) Rapid lignin biosynthesis in pea roots infected with *Heterodera goettingiana*,*Giorn. Bot. Ital.* **127**, 1202-1204.
31. Melillo, M.T., Bleve-Zacheo, T., and Zacheo, G. (1995) Elicitation of H_2O_2 production on pea roots infected with nematode, *Giorn. Bot. Ital.* **129**, 915-917.
32. Nicholson, R.L., and Hammerschmidt, R. (1992) Phenolic compounds and their role in disease resistance, *Annu. Rev. Phytopathol.* **30**, 369-389.
33. Ogallo, J.L., and McClure, M.A. (1995) Induced resistance to *Meloidogyne hapla* by other *Meloidogyne* species on tomato and pyrethrum plants, *J. Nematol.* **27**, 441-447.
34. Ogallo, J.L., and McClure, M.A. (1996) Systemic acquired resistance and susceptibility to root-knot nematodes in tomato, *Phytopathology* **86**, 498-501.
35. Pedreno, M.A., Bernal, M.A., Calderon, A.A., Ferrer, M.A., Lopez-Serrano, M., Merino de Caceres, F., Munoz, R., and Ros Barcelo, A. (1993) A general pattern for peroxidase isoenzyme localization and function in vitacee, solanaceae and leguminosae, in Greppin H., Penel C. and Gaspar T. (eds.), *Molecular and Physiological Aspects of Plant Peroxidases*, Univ. of Geneva, Switzerland, pp.307-314.
36. Perry, R.N., Knox, D.P., and Beane, J. (1992) Enzymes released during hatching of *Globodera pallida* and *Meloidogyne incognita*, *Fundam. Appl. Nematol.* **15**, 283-288.
37. Rahimi, S., Perry, R.N., and Wright, D.J. (1995) Studies on pathogenesis-related proteins in potato induced by potato cyst nematode, *Globodera pallida*, and several chemicals, *Proc. 22nd Intern. Symp. E.S.N.*, Ghent, Belgium, 7-12 August, 1994, *Nematologica*, **41**, 332.
38. Rich, J.R., Keen, N.T., and Thomason, I.J. (1977) Association of coumestans with the hypersensitivity of lima bean roots to *Pratylenchus scribneri*, *Physiol. Plant Pathol.* **10**, 105-112.
39. Ros Barcelo, A. Munoz, R. and Sabater, F. (1989) Subcellular location of basic and acid soluble isoperoxidases in lupinus, *Plant Sci.* **63**, 31-38.
40. Rayan, C.A. (1990) Protease inhibitors in plants: Genes for improving defenses against insects and pathogens, *Annu. Rev. Phytopathol.* **28**, 425-449.
41. Stich, K., and Ebermann, R. (1984) Investigation of hydrogen peroxide formation in plants, *Biochemistry* **23**, 2719-2722.
42. Sutherland, M.W. (1991) The generation of oxygen radicals during host plant responses to infection, *Physiol. Mol. Plant Pathol.* **39**, 79-93.
43. Thompson, J.E., Brown, J.H., Gopinadhan, P.T., and Yao, K. (1991) Membrane phospholipid catabolism primes the production of activated oxygen, in senescing tissues, in Pell E.I. and Steffen K.L. (eds.), *Active Oxygen/Oxidative Stress and Plant Metabolism*, Am. Soc. Plant Physiol., Rockville, MD, pp. 57-66.
44. Vanderspool, M.C., Kaplan, D.T., McCollum, T.G., and Wodzinski, R.J. (1994) Partial characterization

of cytosolic superoxide dismutase activity in the interaction of *Meloidogyne incognita* with two cultivars of *Glycine max, J. Nematol.* **26**, 422-429.
45. Veech, J.A. (1981) Plant resistance nematodes, in Zuckerman B.M. and Rhode R.A. (eds.), *Plant Parasitic Nematodes*, Academic Press, New York, pp. 508-520.
46. Vrain, T., Cantin, L., and Michaud, D. (1995) Choosing proteinase inhibitors to control nematodes, *Abst. Soc. Nematol.,* 34th Annual Meeting, Little Rock, Arkansas, 5-9 August, p.523.
47. Vrain, T., Michaud, D., Cantin, L., Bottino, M.B., and Jouanin, L. (1995) Cystatin-sensitive proteinases in *Meloidogyne hapla*, *Abst. Soc. Nematol.,* 34th Annual Meeting, Little Rock, Arkansas, 5-9 August, p.523.
48. Zacheo, G., and Bleve-Zacheo, T. (1988) Involvement of superoxide dismutases and superoxide radicals in the susceptibility and resistance of tomato plants to *Meloidogyne incognita* attack, *Physiol. Mol. Plant Pathol.* **32**, 313-322.
49. Zacheo, G., Bleve-Zacheo, T., Melillo, M.T., and Perrino, P. (1990) Some biochemical properties of *Pisum sativum* susceptible and resistant to *Heterodera goettingiana, Nematol. medit.* **18**, 253-259.
50. Zacheo, G., Orlando, C., and Bleve-Zacheo, T. (1993) Characterization of anionic peroxidases in tomato isolines infected by *Meloidogyne incognita, J. Nematol.* **25**, 249-256.
51. Zacheo, G., Bleve-Zacheo, T. (1995) Plant-nematode interactions: histological, physiological and biochemical interactions, in Kohmoto K., Singh U.M. and Singh R:P. (eds.), *Pathogenesis and Host Specificity in Plant Diseases*, Elsevier Science Ltd., Oxford, pp.321-353.
52. Zacheo, G., Bleve-Zacheo, T., Pacoda, D., Orlando, C., and Durbin, R.D.(1995) The association between heat-induced susceptibility of tomato to *Meloidogyne incognita* and peroxidase activity, *Physiol. Mol. Plant Pathol.* **46**, 491-507.

C: ENGINEERED RESISTANCE

ENGINEERING RESISTANCE AGAINST PLANT PARASITIC NEMATODES USING ANTI-NEMATODE GENES

Paul R. BURROWS[1] and Dirk DE WAELE[2]
[1]IACR-Rothamsted, Harpenden Herts, AL5 2JQ. UK.
[2]Laboratory of Tropical Crop Improvement, Katholieke Universiteit Leuven Kardinaal Mercierlaan 92, B-3001 Heverlee, Belgium.

Abstract

Due to the continued reduction in the use of toxic and expensive nematicides there is now a greater than ever need for crop varieties that are resistant to plant parasitic nematodes. Genetically engineered nematode resistance is not as well developed as other engineered traits but, even so, the first genetically modified plants with enhanced nematode resistance have been produced and tested. Several classes of potential anti-nematode genes, encoding lectins, enzymes and enzyme inhibitors, are being evaluated for their ability to confer broad spectrum nematode resistance. Early indications are that these are likely to be effective for controlling nematodes. More research is required to identify other nematicidal peptides and proteins and to test these in plants. Gene pyramiding is almost certainly going to be important to increase field durability and to widen the spectrum of nematodes controlled by any one transgenic line.

This article reviews the various classes of potential anti-nematode genes and explores how they might be developed for effective and durable resistance in the field.

1. Introduction

In recent years concern over the environmental impact of some highly toxic and non-specific nematicides has led to the banning of several chemicals that were once in common use. In addition, many countries, particularly in developed temperate agriculture, have adopted policies aimed at the further reduction in use of nematicides. This has placed greater emphasis on alternative control methods, especially the development of nematode resistant crop cultivars. The value of nematode resistant crops, either alone or as part of integrated control, has long been recognised. Plant breeding programmes over the last 50 years have resulted in a few useful sources of resistance being identified and incorporated into commercial crop cultivars [55]. Nevertheless, there are still many important crops and related wild germplasm in which nematode resistance has not been found or is difficult to develop adequately.

Modern plant biotechnology is now capable of producing a much greater range of nematode resistant crops by using genetic engineering [1, 15]. Engineered nematode resistance is still in its infancy compared to other engineered traits, such as viral resistance, but promising preliminary results are being reported from a number of laboratories (see below and chapters by McPherson *et al.* and Ohl *et al.* this volume). Engineering nematode resistance has three main advantages over conventional breeding. First, it can facilitate the transfer of resistance traits across species barriers; second, resistance can be introduced into existing elite crop lines; and third, resistance mechanisms can be designed or adjusted to minimise the emergence of resistance breaking nematodes.

Research on engineering nematode resistance has so far concentrated on sedentary nematodes, mainly from the genera *Meloidogyne, Heterodera* and *Globodera* [12]. This is mainly because of their economic importance but also because the complex feeding sites of these nematodes are considered to be potentially vulnerable to disruption by genetic engineering. However, to engineer broad spectrum nematode resistance that is not only active against sedentary nematodes but also the many damaging species of migratory nematodes, we need to identify effective anti-nematode genes. An anti-nematode gene is defined here as one that produces a peptide/protein that is toxic, damaging or inhibitory to nematodes but is not deleterious (or significantly less so) to the host plant or to the animals/humans that will eventually eat the plant or its products. This sets anti-nematode genes apart from natural resistance (R) genes that in the main mediate hypersensitive/secondary metabolite based responses.

The purpose of this chapter is to review the various classes of potential anti-nematode genes and to explore how they might be developed in the field for effective and lasting resistance.

2. Anti-Nematode Genes

It is clear from the published literature that the current list of likely anti-nematode genes is not a long one. Examples of potential anti-nematode genes can be drawn from the extensive research on engineering insect resistance which shows many similarities to the nematode research in both the approaches needed and the genes being used. The examples discussed here serve not only to illustrate some of the most promising genes under investigation but also to speculate about the potential of other, as yet unexploited, genes or gene classes.

2.1. ENZYME INHIBITORS

Plants have evolved a range of chemical defences to resist herbivorous pests. Most of the chemical defences involve toxic or limiting secondary metabolites, such as alkaloids or phenolics [73], but others are primary gene products (peptides or proteins) and are thus more amenable to gene transfer technologies. Many of the proteinaceous defence molecules are enzyme inhibitors and presumably function by inhibiting important enzymes of invading pests and pathogens.

2.1.1. Protease Inhibitors

Plants possess inhibitors of the four main mechanistic classes of proteolytic enzymes: serine, cystine, aspartic and metallo-proteases [58]. Protease inhibitors (PIs) are found in many plant tissues often in response to wounding or herbivory. They are particularly prevalent in seeds where their presence correlates with resistance to seed eating pests. Hilder *et al.* [31] isolated the gene for a Bowman-Birk type trypsin inhibitor (a serine proteinase inhibitor) from cowpea (*Vigna unguiculata*). When expressed in tobacco the cowpea trypsin inhibitor (CpTI) conferred resistance to browsing tobacco budworm (*Heliothis virescens*). Other PIs, such as soybean trypsin inhibitor, potato inhibitor-II, oryzacystatin and tomato inhibitor-I have also been shown to be useful as transgenes for conferring resistance to insects [reviewed in 58].

The value of PIs as anti-nematode genes is being investigated. This subject is reviewed in the chapter by McPherson *et al.* (this volume) and so will be discussed only briefly here. One of the first indications that PIs may be useful for controlling nematodes came with the observation that when CpTI is expressed in transgenic potatoes it reduces the fecundity of *Meloidogyne* and shifts the sex ratio of *G. pallida* to favour the production of males [1]. Alteration of the sex ratio by CpTI was also noted for *G. tabacum* on transgenic tobacco (P. R. Burrows unpublished data) and is presumably triggered by nutritional stress experienced by the juveniles soon after initiation of feeding. Recently, Urwin *et al.* [72] showed that an engineered cystine proteinase inhibitor derived from rice oryzacystatin-I affected markedly the development of *G. pallida* on hairy root cultures of tomato. Cystine PIs are likely to be a valuable class of anti-nematode genes, especially since cystine proteinases appear to be important in nematode digestion but not in mammalian guts [37, 58].

2.1.2. α-Amylase inhibitors

Inhibitors of α-amylases (α-AIs) are not as ubiquitous in the plant kingdom as protease inhibitors and most have been described from Leguminous and Graminaceous crops. The natural concentrations of α-AIs in seeds is related to resistance to seed pests such as bruchid beetles. α-amylases are the main carbohydrate digesting enzymes in insect guts and resistance is achieved presumably by interfering with carbohydrate assimilation from the diet. Schroeder *et al.* [60] transferred the α-amylase inhibitor gene from common bean (*Phaseolus vulgaris*) to pea (*Pisum sativum*) and demonstrated that the resultant transgenic peas were resistant to the pea weevil (*Bruchus pisorum*). Transgenic peas expressing common bean α-AI are also resistant to cowpea and azuki bean weevils [62]. Plant derived α-AIs tend to be strong inhibitors of mammalian amylases and as such are not suitable for expression in parts of crop plants that are eaten raw. However, normal cooking processes readily destroy the inhibitors and activity against animal or human enzymes should not be a great limitation to their use. Interestingly, an α-amylase inhibitor present in the seeds of *Amaranthus hypocondriacus*, a variety of Mexican crop plant, inhibits strongly α-amylase activity in insect larvae but apparently not the amylases of mammals [16]. Genes such as this are worthy of further investigation.

In a similar way to insect resistance the expression of α-amylase inhibitor genes in transgenic plants could confer resistance to plant nematodes. It is impossible to predict with confidence whether or not this is likely to be effective because little is known about the gut enzymes used by plant nematodes to digest carbohydrates. Indeed, the diets of sedentary nematodes may be rich in simple sugars and digestion of complex carbohydrates is not necessary. In the absence of artificial diets the most efficient way to test the efficacy of α-amylase inhibitors against nematodes is to screen existing α-AIs already expressed in transgenic plants.

2.2. ENZYMES

Plants are capable of expressing a range of enzymes that function by damaging or inhibiting invading pathogens. For this to be an effective means of defence, plants must exploit important structural/physiological differences between themselves and the invaders so as to damage or kill the pathogens without excessive damage to their own tissues. This general principle could be used to engineer nematode resistance.

2.2.1. *Chitinases*
Some PR-proteins exhibit chitinase activity and are thus thought to have an important role in defence against fungi by degrading chitin rich fungal cell walls [56]. Nematode eggshells are largely made of chitin and it is possible that the over expression of chitinase in plants is sufficient to damage the eggs of some species. The amount of damage need not be great. As long as it is sufficient to change indirectly the permeability of the egg membrane it could be enough to lead to the death of the embryo, inhibit hatch or even cause inappropriate hatch. If this was the case then it is unlikely that all species of plant parasitic nematodes would be effected equally. Those most at risk would be those that lay eggs within host tissues.

It has been demonstrated that chitin amendments to soil can suppress plant nematode populations and lead to increased crop yields [66, 67]. It was initially suspected that some of this suppressive effect was due directly to soil microorganisms releasing chitinase in order to take advantage of the chitin as a nitrogen source. However, Spiegel *et al.* [66] showed that this is probably incorrect and the real explanation is almost certainly biphasic. In the first phase the degradation of chitin releases nematicidal ammonia, while in the second phase there is a bloom of soil microorganisms with chitinase potential that are able to invade and parasitise nematode eggs. This does not necessarily exclude the possibility of using chitinases as anti-nematode genes in transgenic plants but the gene products would almost certainly have to be targeted for cellular export.

2.2.2. *Collagenases*
Collagens are major structural components of nematode cuticles and the expression of collagen degrading enzymes in plants may be an elegant way of engineering nematode resistance. The nematode cuticle is a multi-layered structure that completely encloses the organism and it also lines the anterior and posterior portions of the digestive tract. The cuticle is the interface with the environment and acts as a semipermeable barrier for the selective exchange of molecules. Clearly, the cuticle and associated collagens are extremely important for survival. Havstad *et al.* [29] reported the cloning of cDNAs for three human collagenase genes in transgenic tobacco in an attempt to confer resistance to

Meloidogyne and other nematodes. It is not clear whether these experiments were successful because the results have not been reported.

If collagenases are to be a valuable class of broad spectrum anti-nematode genes then two problems need to be addressed. First, there is some evidence that collagens of nematodes (sometimes referred to as pseudocollagen) are different in aspects of their structure from the collagens of vertebrates and are much less susceptible to digestion by collagenases from higher animals [19] has not been widely published. Careful selection of the most appropriate collagenase genes is required to ensure adequate activity against the target nematodes. In this respect, the most valuable source of suitable collagenases is likely to be soil and rhizosphere microorganisms, especially those that parasitise nematodes and are capable of penetrating the cuticle.

Second, to imagine the surface of plant nematodes as a barren sheet of collagen greatly underestimates its complexity. In most nematodes there are three basic layers to the cuticle that vary in composition and thickness between species and different life stages. The outermost layer is the epicuticle and its structure has been compared to a highly modified lipid based plasma membrane [51]. In many nematodes there is a fragile glycocalyx layer overlaying the lipid rich epicuticle. Little is known about the composition of plant nematode cuticles but based on detailed studies of other nematodes it seems that collagens are not part of the surface coat and are thus not exposed to the environment at the surface [50]. This has great relevance to using collagenases to combat plant nematodes because, even if appropriate collagenases are used, the enzymes will not have access to the underlying collagenous cuticular layers. Despite these reservations, collagenases should still be investigated for nematode resistance in transgenic plants. Although normally resistant to external collagenase activity nematodes may become susceptible during moulting. Alternatively, as we learn more about the structure/function of the surface coat it may become clear how we might expose the collagen containing layers, for example co-expression of a specific lipase with a collagenase, or even devise methods to deliver collagenases deeper into the cuticle.

2.2.3. *Ribosome Inactivating Proteins*
Ribosome-inactivating proteins (RIPs) are enzymes that interact with ribosomes in a highly specific way and thus interfer with protein synthesis [22, 28 for review]. The best characterised RIP is ricin, a highly toxic protein from the castor oil plant. Ricin, like other plant derived RIPs, is a RNA *N*-glycosidase that cuts out a single adenine residue from a conserved sequence in the large subunit ribosomal RNA of eukaryotic ribosomes. RIPs are generally classified as either type 1 or type 2. Type 1 RIPs are the most numerous and consist of a single enzymatically active polypeptide chain of about 30kDa. In contrast, type 2 RIPs are composed of an enzymatically active chain (like type 1) but linked to a lectin-like cell binding chain. Type 2 RIPs are by far the most toxic as they can gain entry into cells by binding to surface glycoproteins whereas type 1 RIPs are nearly always dependent on fluid endocytosis.

RIPs have been described from about 50 plant species covering 13 families. Their function within plants is still unclear but evidence is accumulating that they, at least in part, have a defence-related role especially against viruses and fungi [39, 40]. Some RIPs have also been shown to be effective *in vitro* and in artificial diets against insects such as

Carpophilus freemani beetles and *Spodoptera frugiperda* [11]. The insecticidal effects of two RIPs, Saporin (type 1) and ricin (type 2), were investigated on representative Lepidoptera and Coleoptera species in artificial diets [24]. Both RIPs were potent toxins towards the Coleoptera species but were largely ineffective against the two Lepidoptera species tested. This difference was attributed to the hydrolysis of the RIPs in the guts of Lepidopteran insects. The use of suitable nematicidal RIPs in transgenic plants has not yet been investigated thoroughly. However, to be useful, the general phytotoxicity of many RIPs must be avoided by appropriate sequestration or by using nematode responsive expression.

2.2.4. *Other Enzymes*

Two other types of enzymes have been investigated for insect control that might also be effective against nematodes. Patatin is a non-specific lipid acyl hydrolase from potato tubers. Strickland *et al.* [69] showed that patatin inhibits the growth of corn rootworm (*Diabrotica* spp.) in a dose-dependent manner. Tests on other herbivorous insects indicated that patatin is most effective in alkaline gut environments; typically those insects that use predominantly gut cystine proteinases (like some plant nematodes [37]) rather than serine proteinases. The mode of action of patatin may reside in its ability to promote autocatalysis of gut membranes and lipids. Similarly, cholesterol oxidase is another enzyme which leads to the disruption of gut membranes in some insects and could also be useful for nematode resistance [52].

2.3. LECTINS

Lectins are proteins other than enzymes and antibodies that bind carbohydrates (see [49] for review). They are widespread in nature and commonly found in many plant species. Lectins bind reversibly to carbohydrates in a specific manner with individual lectins showing a marked bias towards particular mono- or oligosaccharides. Several hundred lectins have been isolated and characterised from plants and they are loosely grouped depending on their sugar binding specificity. The physiological role of most plant lectins is unknown but there is growing evidence that they are principally involved in defence against pests and pathogens [49].

Some lectins have been demonstrated to be toxic to insects in artificial diets. For example, phytohaemagglutinin (PHA) from *P. vulgaris* and wheat germ agglutinin (WGA) are toxic to the larvae of the bruchid beetle *Callosobruchus maculatus* [33, 45]. Similarly, the larvae of the southern corn rootworm, *D. unidecimpunctata*, are killed by WGA and lectins from *Bauhinia purpurea* and *Phytolacca americana* (pokeweed). Mannose binding lectins from garlic (*Allium sativum*) and snowdrop (*Galanthus nivalis*) are active against browsing insects such as *C. maculatus* and the tobacco hornworm (*Spodoptera litoralis*). Importantly, the snowdrop lectin (GNA) is also effective against the brown planthopper (*Nilaparvata lugens*) which was the first record of lectins being effective against a sap-sucking insect (cited in [32]). The mechanism(s) by which lectins show toxicity to insects is not well understood but, in most cases, it is likely to be associated with the gut surface where there would be plenty of opportunity for lectins to interact with membrane glycoproteins, glycolipids or polysaccharides. The toxicity of many lectins towards mammals and birds is well documented [53, 54] but they are usually destroyed by cooking. Furthermore, a few lectins, such as P-LEC from pea (*P. sativum*) and GNA from snowdrops, are insecticidal to susceptible insects but have little effect on mammalian systems [53]. Lectins like these may be particularly valuable for engineering resistance in food plants. The toxic and inhibitory properties of plant lectins towards insects have also been demonstrated in transgenic plants that express foreign lectin genes. Edwards (cited in [25]) expressed P-LEC at high levels in transgenic tobacco and demonstrated that the resultant plants had enhanced resistance to the tobacco budworm *H. virescens*. In these experiments larval biomass and leaf damage were significantly reduced. Expression of the snowdrop lectin in tobacco plants also resulted in added protection against insects [32]. Additionally, expression of GNA directed toward phloem tissue by the rice sucrose synthase-1 promoter confers resistance to brown planthopper [63] and aphids [32].

Based on the qualified success of lectins as anti-insect genes their potential as anti-nematode genes is also being investigated. The anti-nematode properties of lectins are difficult to demonstrate *in vitro* due to the lack of artificial diets for phytophagous nematodes. The ways in which lectins interact with nematodes *in vivo* are, in the main, likely to be similar to those for insects e.g. gut associated activity. However, with microscopic root endoparasitic nematodes lectins would also have opportunity to bind to the surface/cuticle and to the primary chemosensory structures, the amphids. Lectin binding studies have revealed the presence of various carbohydrate residues associated with the amphidial secretions and on the surface of plant parasitic nematodes [23, 43, 68]. It has been suggested that the binding of lectins to amphids or amphidial secretions can interfere with nematode sensory perception and therefore the ability to locate food or a mate [74]. Furthermore, Marban-Mendoza *et al.* [42] found that concanavalin A applied as a soil drench reduced the galling caused by *Meloidogyne incognita* in tomato roots. If lectins can interfere significantly with the perception of external signals and stimuli by plant nematodes then the expression of suitable genes in plant roots, perhaps with the gene products directed for cellular/root export, may confer some resistance.

The potential of GNA for nematode resistance has been tested in a series of trials, mostly glasshouse based, using transgenic oilseed rape (OSR) and potatoes [13, 14]. In the first test, four lines of OSR were selected for the high levels at which they expressed GNA. Two of these expressed GNA alone and two expressed GNA together with CpTI. All four lines were challenged with *Heterodera schachtii* and the migratory nematode *Pratylenchus neglectus*. The results from this test were not clear-cut but they were, in part, encouraging (Fig. 1). The best resistance to sedentary and migratory nematodes was found in line 1 that expressed GNA and CpTI. This line reduced the number of *H. schachtii* females and *P. neglectus* (all stages) by approximately 25% and 75% respectively compared to control plants.

A second line (line 4, Fig. 1), which expressed GNA alone, conferred moderate levels of resistance to *P. neglectus* with 55% fewer nematodes being recovered relative to the controls. However, with this same line significantly ($p<0.05$) more *H. schachtii* females were recovered from the GNA plants than the controls. This same effect was also observed for *P. neglectus* in lines 2 and 3. The ability of GNA to confer resistance but also, in some cases, apparently increase nematode numbers is unlikely to be simply an

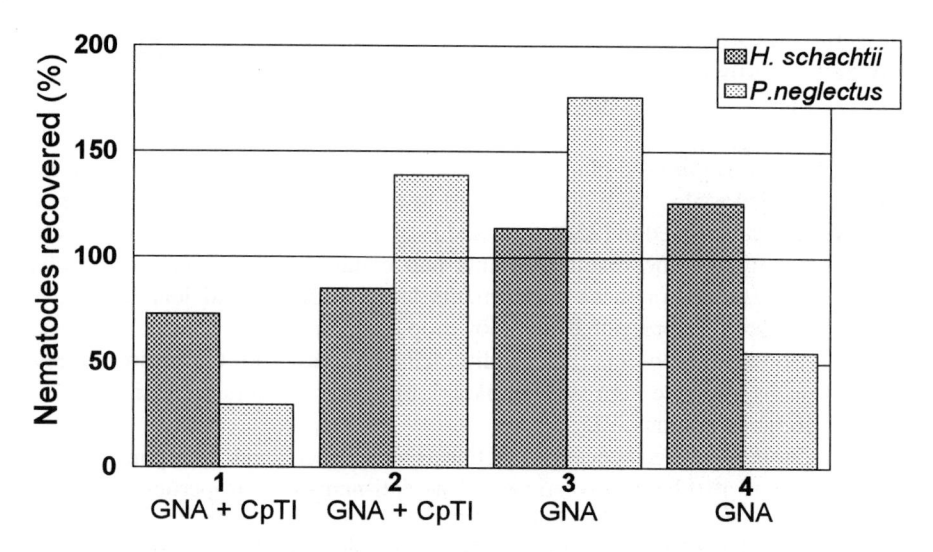

Figure 1. Reproduction of *Heterodera schachtii* and *Pratylenchus neglectus* on transgenic oilseed rape expressing either snowdrop lectin (GNA) alone or GNA in combination with cowpea trypsin inhibitor (CpTI). The nematodes recovered after the experiment are expressed as a percentage of those from control plant..

experimental artifact because it occurred in two independent tests with two very different nematodes. It does, however, illustrate the pitfalls that await when attempting to interfere with host parasite interactions that are poorly understood.

The observation that GNA could, in some instances, confer enhanced nematode resistance was investigated further using transgenic potatoes [13]. Four lines of transgenic potatoes, expressing GNA at different levels ranging from 0.1-0.5% total root protein, were challenged with *G. pallida*. Four criteria were considered to assess the effect of GNA on *G. pallida*: the numbers of females produced, the relative size of cysts, their egg content and the time taken to develop. Of these parameters, the number of females produced on the GNA potatoes was the most indicative of detrimental effect towards the nematodes. Figure 2 illustrates the numbers of females that developed on the GNA potatoes compared to control plants.

Three of the four lines showed little or no effect of GNA but line 21, which expressed GNA at approximately 0.25% of total root protein, reduced mean female numbers by 50%. Interestingly, the most resistant potato line was not the highest expresser of GNA.

Based on the above two experiments, it is possible to speculate that GNA is able to confer useful levels of nematode resistance in transgenic plants but the resistance achieved depends very much on an as yet unknown critical level of GNA. The GNA concentration that is most effective may vary for different nematodes due to differential susceptibility or exposure. Too much or too little GNA has a negligible effect or, in some cases, may even lead to increased nematode numbers. Much of this remains to be tested.

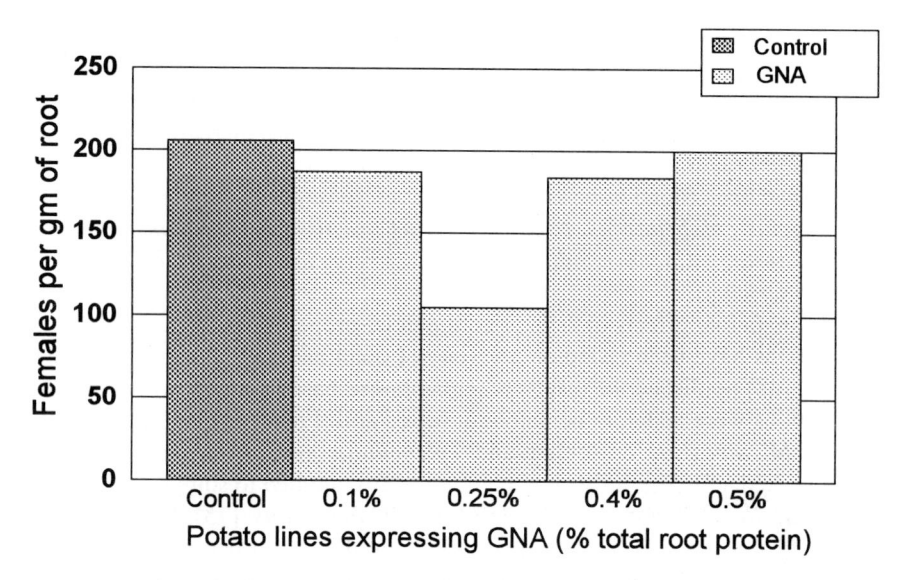

Figure 2. Reproduction of *Globodera pallida* on transgenic potatoes expressing snowdrop lectin (GNA) at 0.1-0.5% total root protein.

2.4. TOXINS FROM *BACILLUS THURINGIENSIS*

Bacillus thuringiensis (B.t.) has been used in various forms for over 20 years as a bio-insecticide [20]. B.t. is a gram-positive bacterium found commonly in soil and insect rich environments. The bacterium probably produces several toxic metabolites including both exotoxins and endotoxins, but it is the endotoxins that have attracted most attention as insecticidal or nematicidal genes. Under adverse conditions B.t sporulates and produces internal proteinaceous crystalline structures containing one or more pro-endotoxins. When B.t. spores are eaten by insects the pro-toxins are hydrolysed in the gut to produce active toxins. These endotoxins are potent insecticides that act by disrupting cell membranes in the midgut leading to ion imbalance, gut paralysis and eventually death. The majority of B.t. strains and their associated endotoxins are selective, each being active against one or a few insect groups.

Genes encoding B.t. endotoxins have been introduced into crops including tomato, cotton, tobacco, maize, rice and potatoes. In most cases resistance against a particular target group of insects has been achieved. Initial problems with low levels of expression of B.t. toxins in transgenic plants have been overcome by expressing truncated versions of the genes that encode only the active part of the protein and by altering the codon usage towards the G+C rich bias favoured by plants. The most successful genes therefore are highly modified DNAs that have low homology to the original sequences.

The anti-nematode properties of B.t. toxins have been investigated mainly with animal parasitic nematodes but some plant nematodes have also been included. B.t. exotoxins are nonspecific toxins released from the vegetative cells. The beta-exotoxin is a thermostable general cytotoxin with little or no specificity and is active against invertebrates and vertebrates [61]. Not surprisingly it is also effective against a range of animal and plant parasitic nematodes. The value of such generally cytotoxic peptides is likely to be small. Exposure of nematodes to natural B.t. strains, which after sporulation consist of mixtures of vegetative cells, spores and crystals, has increased the mortality of animal parasitic, plant parasitic and free living nematodes. To date over 20 species of nematodes have been found to be killed at one or more stages in the life-cycle, including *Trichostrongylus colubriformis, Nippostrongylus braziliensis, Haemonchus contortus, Ostertagia ostertagi, Caenorhabditis. elegans, C. briggsae, Panagrellus redivivus* and *Pratylenchus scribneri.*

The nematicidal effects of several commercial preparations of B.t. that are mostly delta-endotoxins have been tested on a variety of plant and animal parasitic nematodes. Osman *et al.* [48] published the results of a glasshouse-based assay in which commercial B.t. preparations were added to nematode-infected soil. Final population levels and egg viability of *M. incognita* on tomato and *Tylenchulus semipenetrans* on citrus were reduced in treated soil. Similarly, a preparation of B.t. applied to soil or seeds controlled *M. incognita, Rotylenchulus reniformis* and *P. penetrans* [75]. Treatment of *M. incognita* infected soil with B.t. resulted in increased yield of pepper and tomato. In a previous study (D. de Waele *et al.* unpublished data) transgenic tomatoes expressing the B.t. endotoxin CryIab were challenged with *M. incognita and M. javanica.* For both species the number of egg masses per gram of root was reduced by approximately 50% on the transgenic plants compared to the controls. Although interesting, it is emphasised that this work was only a preliminary study and would need to be repeated to gain confidence in the result.

2.4.1. *Species Specificity of Nematicidal Activity*
An important characteristic of the insecticidal delta-endotoxins is their specificity for particular species or groups of insects. The literature concerning species specificity of nematicidal action is often contradictory with some studies finding high degrees of specificity and others not. Ciordia and Bizzell [17] reported that one strain of B.t killed three different animal-parasitic nematodes. Likewise, a single isolate of *B.t israelensis* was lethal to eggs of six animal-parasitic nematodes and the free-living nematode *C. briggsae.* A strain used by Narva *et al.* [46] killed juveniles and adults of the free living *P. redivivus* and the plant parasitic *P. scribneri.* Conversely, Borgonie *et al.* [8] found marked specificity of their B.t. strains; out of 15 species of free living nematodes, belonging to three families (Rhabditidae, Panagrolaimidae and Cephalobidae), only two related species, *C. elegans* and *C. briggsae,* were sensitive. This high species specificity is comparable with that shown by the delta-endotoxins towards insects. The precise source of the nematicidal or nematostatic activity of commercial B.t. preparations is difficult to determine. Although commercial preparations are rich in crystal proteins, and therefore delta-endotoxins, they also contain vegetative cells which may be capable of producing non-specific exotoxins. In considering the use of B.t. endotoxin genes for nematode control the source of the nematicidal/nematostatic activity is extremely important.

2.4.2. *Mode of Nematicidal Activity of B.t Toxins.*
Compared to the mode of action of insecticidal delta-endotoxins very little is known about the mode of action of nematicidal B.t isolates. The ovicidal activity of some B.t. preparations and toxins appears to be due to some alteration to the permeability of the eggshell allowing the toxin to enter and interact with the membrane of the embryo. A detailed study of the mode of action of one B.t. isolate against the hatched juveniles and adults of *C. elegans* was made by Borgonie *et al.* [6, 7]. This study showed that the toxin acted directly against the nematode intestine and that the toxic process proceeded in two distinct phases. In the first 12 hours spores and crystals accumulated in the intestinal lumen and the four anterior-most intestinal cells were gradually destroyed. Beyond 12 hours the remainder of the intestine was destroyed and accumulated spores germinated resulting in complete colonisation of the nematode body [8]. These studies indicate that B.t. toxins may have different modes of action on nematodes resulting from ectopic contact or ingestion. In all cases the toxic effect seems to be related to membrane disruption/dysfunction of the egg, cuticle or intestine.

In view of the above work and the many reported examples of the nematicidal activity of some B.t. toxins or preparations, it is surprising that not more has been published regarding nematode resistance of transgenic plants expressing B.t. crystal proteins. Such plants are certainly not scarce in the biotechnology community so it is not lack of plants to test that has prevented progress. One possible explanation is that, despite the wealth of *in vitro* work, the great majority of the B.t plants available are constructed for insect resistance and the characteristic specificity of individual endotoxins probably precludes in most cases cross-protection against nematodes.

2.5. ANTIBODIES

The expression of functional mammalian antibodies or antibody fragments in transgenic plants has been achieved [3, 21, 30]. Antibodies produced in plants (the so-called `plantibodies') are being investigated as a means of conferring pest and disease resistance [59]. Initial success has largely been limited to a few reports of increased levels of viral resistance in plants engineered to express anti-viral coat protein antibodies [71]. Work is also under way to express antibody fragments that recognise oesophageal gland products of some sedentary plant parasitic nematodes such as root-knot and cyst nematodes ([57, 59] and Stiekema *et al.*, this volume). Oesophageal gland products from these worms are believed to play a central role in the induction and/or maintenance of the complex hypertrophic feeding sites induced in host roots. Expression of plantibodies to these antigens could disrupt feeding site development.

Over and above feeding site perturbation, plantibodies could have a more direct role as anti-nematode genes, for example by disrupting chemosensory functions or as enzyme inhibitors. Antibodies can be potent and specific inhibitors of a variety of enzymes including asparagine synthase, human pancreatic elastase 2, RNA polymerase and beta-lactamase (cited in [59]). Antibody fragments that inhibit important digestive enzymes in ways analogous to the plant-derived inhibitors (discussed above) are also likely to be nematotoxic. Similarly, if the binding of lectins to amphidial regions of nematodes can interfere with host location then so too could the binding of antibodies. Evidence is accumulating that plant nematodes such as *Meloidogyne* spp. use secreted enzymes such as pectinases to assist in invasion and migration through host plants (see chapter by von Mende, this volume). Plantibodies that selectively neutralise such enzymes could disrupt this aspect of the host parasite relationship. As we learn more about the various interactions between nematodes and their hosts then more targets for antagonistic plantibodies will be revealed.

2.6. OTHER SOURCES OF POTENTIAL ANTI-NEMATODE GENES

Any current discussion of potential anti-nematode genes necessarily involves an element of speculation. This section illustrates some sources or classes of genes that, with more work, may form the future basis of nematode resistance.

In addition to the work on B.t toxins some attention has focused on other microorganisms as possible sources of nematode toxins. Nematodes are an abundant source of nutrients for any soil microorganisms that possess the means to exploit them. A range of fungi and bacteria that colonize and kill plant parasitic nematodes has already been considered as biological control agents [18] and it is reasonable to suppose that many more are awaiting discovery. Investigation of these and similar microorganisms may yield nematode toxins or antagonists. One such nematode toxin was reported by Barron and Thorne [2] when they observed the destruction of nematodes by a species of the fungus *Pleurotus*. On water agar *Pleurotus ostreatus* produced tiny droplets of a toxin from spathulate secretory cells. Nematodes that touch these droplets show a sudden response in which the head region shrinks and the oesophagus is displaced. Within a few minutes the nematodes are immobilized but not killed and, stimulated by leakage of products from the nematodes, hyphae converge on the body orifices and invade the paralysed worms. No evidence was presented that the toxin responsible was a peptide, and thus amenable to gene transfer technologies, but observations like this are important to highlight potentially useful anti-nematode genes.

Various cytolytic peptides have been investigated as pest and disease control agents. Enterolobin is a 55 kDa haemolytic protein from the seeds of a Brazilian leguminous tree (*Enterolobium contortisiliquum*). In artificial diets it was toxic to the larvae of *C. maculatus* leading to 100% mortality at a concentration of 0.025% [65]. However enterolobin was innocuous to the Lepidopteran *S. litoralis*. This differential effect was probably due to the proteolytic activity in the gut of *S. litoralis* breaking down the protein before it had time to act. Another cytolytic peptide, Cecropin B, isolated from the giant silk moth *Hyalophora cecropia* has been expressed in transgenic tobacco and confers enhanced resistance to bacterial wilt caused by *Pseudomonas solanacearum* [35]. Conceivably some cytolytic peptides will be deleterious to nematodes when expressed in transgenic plants.

3. Practical Considerations of Using Anti-Nematode Genes

3.1. ANTI-NEMATODE GENES IN THE LABORATORY

3.1.1. *Identification of Leads*

Much of the preceding discussion concerning types of anti-nematode genes is based on proteins previously found to be toxic or inhibitory to insects. Leads for insecticidal proteins usually come either from simple literature searches or by investigating world collections of plants and wild germ plasm for insect resistant biotypes [9]. Often the resistance `factor' is a secondary metabolite but in cases where resistance can be correlated with a peptide or protein, then the gene may be isolated. In addition to screening plants, microbial sources of nematicidal genes are also likely to be productive. Selection of good candidate nematicidal genes for investigation is severely hampered by the absence of artificial diets or other feeding assays for plant parasitic nematodes in which possibly harmful peptides could be tested. Urwin *et al.* [72] attempted to circumvent the inherent screening problem by testing the inhibitory activity of an engineered oryzacystatin against *C. elegans* and extrapolating this to plant nematodes. Considering the very different food and feeding methods of these nematodes it is perhaps surprising that there was, in this instance, a good relationship between the efficacy against *C. elegans* and that against *G. pallida*. It remains to be seen whether or not *C. elegans* will be widely useful for screening enzyme inhibitors (and other gene products) or whether it is only a fortuitous but useful link in just a few cases. Other *in vitro* assays, such as gut enzyme location and characterisation in thin sections using fluorogenic substrates or other chromogenic methods, can give useful information as to the classes of digestive enzymes present in target nematodes [37]. Similarly, assays for the binding of lectins to nematode target tissues, such as the gut wall or amphids, can at least show whether or not a particular lectin (or antibody fragment) has the potential to be nematicidal.

Great care should be taken when selecting the most appropriate bioassay as the results obtained *in vitro* can be misleading. For example, an α-amylase inhibitor from wheat was equally effective at inhibiting enzyme activity from a pest and non-pest insect species *in vitro* but in feeding trials (*in vivo*) only the non-pest was affected [26]. The only true test for a potential anti-nematode gene is to put the gene into a plant expression system and screen for resistance. In this respect, *Arabidopsis thaliana* is a good model plant to use because it is quick and easy to transform and is an amenable host to a range of nematodes [64]. Wherever possible, nematode resistance conferred by expression of anti-nematode genes should be tested in soil-based trials as opposed to agar grown plants.

3.1.2. *Expression of Anti-Nematode Genes in Transgenic Plants*

Transgenic plants can, in general, tolerate high levels of constitutive expression of plant-derived enzyme inhibitors and lectins as insecticidal genes [9]. The expression of these transgenes in plants can be as high as 1-2% (or higher) of total soluble protein without significant effect on plant biomass or seed set. However, this clearly depends very much on the particular gene being used. For example, most RIPs are just as toxic to the plant cells that produce them as they are to other eukaryotic cells and, *in vivo*, need to be compartmentalised to prevent contact with ribosomes. The cellular fate of expressed nematicidal transgenes can be an important feature of their effective use. It seems reasonable to expect that the cellular export of chitinases to accumulate in the inter cellular spaces or to appear at the rhizosphere would give better access to nematode eggs. Likewise, collagenases, enzyme inhibitors, lectins etc could, if necessary, also be tagged for the most appropriate cellular or extracellular compartment.

The temporal and spatial expression patterns of nematicidal transgenes at the level of the whole plant is perhaps one of the most important considerations. Constitutive expression of a nematicidal gene product that only really needs to be present in the roots is inefficient. Furthermore, the cauliflower mosaic virus 35S gene promoter, that is used frequently in plant biotechnology to achieve constitutive expression of transgenes, is partly down regulated in the feeding sites of cyst and root-knot nematodes [27]. This makes delivery of anti-nematode gene products to sedentary nematodes via the feeding site considerably less than optimal. The use of nematode responsive promoters that drive gene expression inside the feeding sites of sedentary nematodes (see chapter by Fenoll *et al.* this volume) will overcome this problem. The advantage of using non-phytotoxic genes, like plant-derived enzyme inhibitors, is that expression in nematode feeding sites does not have to be very specific and expression elsewhere in the plant will be tolerated. This is not the case where cytotoxic genes, such as barnase, are used to destroy the feeding sites.

3.1.3. *Protein Engineering*
Site-directed mutagenesis and related technologies can lead to the introduction of specific changes in the amino acid sequences of a protein and thus alter its physical properties [4]. Protein engineering has been used to change the inhibitory spectrum and/or increase the power of plant derived enzyme inhibitors (for example [41]). Urwin *et al.* [72] engineered rice oryzacystatin to be more effective against *C. elegans* and *G. pallida* than the wild type inhibitor. In principle, protein engineering could be used to increase the nematicidal activity of any of the anti-nematode genes under consideration.

Another important role for protein engineering and manipulation may be in reducing the size of good nematicidal proteins whilst maintaining function. Microinjection of fluorescent dextrans into the feeding site of *H. schachtii* indicated that there is a molecular barrier between the plant cytosol and the nematodes digestive tract [5]. Dextrans of 10 kDa and 20 kDa were taken up by the nematodes but those of 40 kDa and 70 kDa were not. If a molecular weight cut-off of approximately 30 kDa exists with other sedentary nematodes then this will impose serious limitations on the size of any nematicidal gene products that could be ingested. Fortunately, most plant derived enzyme inhibitors are smaller than the predicted molecular exclusion and should be taken up freely. Nematotoxic antibodies (approximately 150 kDa) can be engineered and expressed as antibody fragments (Fv) that should retain the antigen binding characteristics of the original molecule. Expressed Fv molecules are generally in the region of 25-35 kDa. The suitability of other 'large' anti-nematode gene products for size reduction while still retaining their nematicidal activity is much less predictable and should be considered on a case by case basis. It is interesting to note that the snowdrop lectin (see 2.3.) is effective against two cyst nematodes in transgenic plants yet it is a molecule of approximately

50kDa. Either these nematodes are able to ingest molecules larger than the fluorescent dextran work indicates or the mode of action of GNA is not via ingestion. Clearly this needs to be investigated.

3.2. ANTI-NEMATODE GENES IN THE FIELD

There are three main considerations to address when using anti-nematode genes widely in transgenic crops in the field: 1. durability, 2. efficacy and 3. environmental impact.

3.2.1. *Durability and Efficacy*

It is extremely important that anti-nematode genes are managed effectively so as to reduce the selection of resistance breaking nematodes and thereby prolong field life. Resistance breaking insects capable of overcoming selected B.t endotoxin genes are already emerging [38, 44] and others are sure to follow. Expression of single transgenes, for example proteinase inhibitors, also has the disadvantage that the pests may adapt by inducing different classes of enzymes or isoenzymes that are insensitive to the inhibitory transgene [36]. Cultural practices such as non-transgenic refugia for non-resistant pest genotypes and seed mixtures are being evaluated as strategies to delay the emergence of resistance breaking populations of insects [70], but nematodes are far less mobile and it is not immediately obvious whether such practices would be of value. Another approach to increased durability of transgenes is gene pyramiding, in which two, three or more anti-nematode genes (preferably of different modes of action - mulitmechanistic) are expressed together in the same plant. Not only does this decrease the chance of selecting for resistance breaking nematodes but it could also increase efficacy and broaden the spectrum of nematode species controlled by any one transgenic line. Initial attempts at gene/protein combinations and pyramiding for pest and pathogen control have been reported [10, 34, 47].

3.2.2. *Environmental Impact*

Over and above the more general concerns expressed by some groups about the release of transgenic plants to the environment the expression and widespread use of anti-insect/nematode genes poses questions about impact on non-target or beneficial organisms. Aphids feeding on transgenic plants expressing the snowdrop lectin GNA produce honeydew containing GNA and are likely to hold levels of GNA in their gut and possibly other tissues. The effect of this GNA and the products of other insecticidal transgenes ingested indirectly by predatory insects, or even birds and animals further up the food chain, is unknown. Similarly, the effect, if any, on nectar/pollen feeding insects is also unclear. Many of these insects are vital for pollination. The consensus of scientific opinion is that any deleterious effects on non-target organisms are likely to be negligible (certainly much less than current pesticides), but experiments are under way to test this. However, where appropriate, constitutive expression of transgenes should be avoided and, in the case of nematodes, nematode responsive promoters can be used to direct gene expression more specifically to these pests.

4. Conclusions

Several classes of insecticidal genes, such as lectins, enzymes and enzyme inhibitors, are being evaluated as anti-nematode genes. Early indications are that these are likely to be effective for controlling nematodes when expressed in transgenic crops. More research is required to identify other potential nematicidal peptides and proteins and to test these in plants. As we learn more about the physiology, biochemistry and molecular biology of nematodes and their interactions with host plants, other potentially useful anti-nematode genes are sure to be highlighted.

Gene pyramiding is likely to be important to increase field durability and widen the spectrum of nematodes being controlled. Transgenic crops expressing anti-nematode genes will almost certainly be most effective and useful in integrated pest management schemes where they can be used to reduce greatly the input of chemical control agents and our over-defendence on current resistant cultivars.

ACKNOWLEDGEMENTS

The authors would like to acknowledge the considerable contribution made to the GNA studies by Drs W.D.O Hamilton and C. Newell from Pestax Ltd and Mr A. Barker IACR-Rothamsted. IACR-Rothamsted receives grant aided support from the Biotechnology and Biological Sciences Research Council of the United Kingdom.

References

1. Atkinson, H.J., Urwin, P.E., Hansen, E. and McPherson, M.J. (1995) Designs for engineered resistance to root parasitic nematodes, *Trends Biotechnol.* **13**, 369-374.
2. Barron, G.L. and Thorne, R.G. (1987) Destruction of nematodes by species of *Pleurotus, Can. J.Bot.* **65**, 774- 778.
3. Benvenuto, E., Ordas, R.J., Tavazza, R., Ancora, G., Biocca, S., Cattaneo, A. and Galeffi, P. (1991) Phytoantibodies: a general vector for the expression of immunoglobulin domains in transgenic plants, *Plant Mol. Biol.* **17**, 865-874.
4. Lundell, T.L. (1994) Problems and solutions in protein engineering - towards rational design, *Trends Biotechnol.* **12**, 145-148.
5. Böckenhoff, A. and Grundler, F.M.W. (1994) Studies on the nutrient uptake by the beet cyst nematode *Heterodera schachtii* by *in situ* microinjection of fluorescent probes into the feeding structures in *Arabidopsis thaliana, Parasitol.* **109**, 249-254.
6. Borgonie, G., Claeys, M., Leyns, F., Arnaut, G., de Waele, D. and Coomans, A. (1996) Effect of nematicidal *Bacillus thuringiensis* strains on free-living nematodes. 1. Light microscopic observations, speciesand biological stage specificity and identification of resistant mutants of *Caenorhabditis elegans, Fundam. Appl. Nematol.* **19**, 391-398.
7. Borgonie, G., Claeys, M., Leyns, F., Arnaut, G., De Waele, D. and Coomans A. (1997) Effect of nematicidal *Bacillus thuringiensis* strains on free-living nematodes. 2. ultrastructural analysis of the intoxication process in *Caenorhabditis elegans, Fundam. Appl. Nematol.* (in press).
8. Borgonie G., van Driessche R., Leyns F., Arnaut G., De Waele D. and Coomans A. (1995) Germination of *Bacillus thuringiensis* spores in bacteriophagous nematodes (Nematoda: Rhabditida), *J. Invertbr. Pathol.* **65**, 61-67.
9. Boulter, D. (1993) Insect pest control by copying nature using genetically modified crops, *Phytochemistry* **34**, 1453-1466.

10. Boulter, D., Edwards, G.A., Gatehouse, A.M.R., Gatehouse, J.A. and Hilder, V.A. (1990) Additive protective effects of different plant derived insect resistance genes in transgenic tobacco plants, *Crop Prot.* **9**, 351-354.
11. Brandhorst, T., Dowd, P.F. and Kenealy, W.R. (1996) The ribosome inactivating protein restrictocin deters insect feeding on *Aspergillus restrictus*, *Microbiol. UK.* **142**, 1551-1556.
12. Burrows, P.R. (1996) Modifying resistance to plant parasitic nematodes, in W.S Pierpoint and P.R. Shewry (eds), *Genetic Engineering of Crop Plants for Resistance to Pests and Diseases*, British Crop Protection Council, London, pp 38-65.
13. Burrows, P.R. (1997) Plant derived enzyme inhibitors and lectins for resistance to plant parasitic nematodes in transgenic crops, *Pestic. Sci.* (in press)
14. Burrows P.R., Barker, A., Merryweather, A., Newell, C.A. and Hamilton, W.D.O.: Engineering resistance against plant parasitic nematodes. Offered Papers in Nematology, Association of Applied Biologists (One day meeting) Linnean Society of London, 13th December 1995 (abs).
15. Burrows, P.R. and Jones, M.G.K. (1993) Cellular and molecular approaches to the control of plant parasitic nematodes, in K. Evans, D.L. Trudgill and J.M. Webster (eds), *Plant Parasitic Nematodes in Temperate Agriculture*, CAB International, Wallingford, Oxon, UK. University Press Cambridge, pp. 609-630.
16. Chagollalopez, A., Blancolabra, A., Patthy, A., Sanchez, R, and Pongor, S. (1994) A novel alpha-amylase inhibitor from Amaranth (*Amaranthus hypocondriacus*) seeds, *J. Biol. Chem.* **38**, 23675-23680.
17. Ciordia, H. and Bizzell, W.E. (1961) A preliminary report on the effects of *Bacillus thuringiensis* var *thuringiensis* Berliner on the development of the free-living stages of some cattle nematodes, *J. Parasitol.* **47**, 41 (abs).
18. Davies, K.G.. De Leij, F.A.A.M. and Kerry, B.R. (1991) Microbial agents for the biological control of plant parasitic nematodes in tropical agriculture, *Trop. Pest Manage.* **37**, 303-320.
19. Dawkins, H.J.S and Spencer, T.L. (1989) The isolation of nucleic acid from nematodes requires an understanding of the parasite and its cuticular structure, *Parasitol. Today* **5**, 73-76.
20. Dulmage, H.T. (1970) Insecticidal activity of HD-1, a new isolate of *Bacillus thuringiensis* var, *Alesti. J. Invert. Path.* **15**, 232-239.
21. During, K., Hippe, S., Kreuzaler, F. and Schell, J. (1990) Synthesis and self assembly of a functional monoclonal antibody in transgenic *Nicotiana tabacum, Plant. Mol. Biol.* **15**, 281-293.
22. Endo, Y., Mitsui, K., Motizuki, M. and Tsurugi, K. (1987) The mechanism of action of ricin and related toxic lectins on eukaryotic ribosomes - the site and the characteristics of the modification in 28S ribosomal RNA caused by the toxins, *J. Biol. Chem.* **262**, 5908-5912.
23. Forrest, J.M.S. and Robertson, W.M. (1986) Characterisation and localisation of saccharides on the head region of four populations of the potato cyst nematodes *Globodera rostochiensis* and *G. pallida, J. Nematol.* **18**, 23-26.
24. Gatehouse, A.M.R., Barbieri, L., Stirpe, F. and Croy, R.R.D. (1990) Effects of ribosome inactivating proteins on insect development differences between Lepidoptera and Coleoptera. *Entomol. Exp. Appl.* **54**, 43-51.
25. Gatehouse, A.M.R., Boulter, D. and Hilder, V.A. (1993) Potential of plant derived genes in the genetic manipulation of crops for insect resistance, in A.M.R Gatehouse, V.A Hilder and D. Boulter (eds), *Plant Genetic Manipulation for Crop Protection*, C.A.B International Wallingford, Oxon, UK. pp 155-181.
26. Gatehouse, A.M.R., Fenton, K.A., Jepson, I. and Pavey, D.J. (1986) The effects of α-amylase inhibitors on insect storage pests: inhibition of α-amylase *in vitro* and effects on development *in vivo, J. Sci. Food Agr.* **37**, 727-734.
27. Goddijn, O.J.M., Lindsey, K., van der Lee, F.M., Klap, J.C and Sijmons, P.C. (1993) Differential gene expression in nematode induced feeding structures of transgenic plants harbouring promoter *gus* A fusion constructs, *Plant J.* **4**, 863-873.
28. Hartley, M.R., Chaddock, J.J. and Bonness, M.S. (1996) The structure and function of ribosome inactivating proteins, *Trends Plant Sci.* **1**, 254-260.
29. Havstad, P., Sutton, D., Thomas, S., Sengupta-Gupalan, C. and Kemp, L.: Collagenase expression in transgenic plants: an alternative to nematicides. *Third International Congress of Plant Molecular Biology*, 6-11 October 1991. Tucson, Arizona, USA 345 (abs).

30. Hiatt, A., Cafferkey, R., Bowdish, K. (1989) Production of antibodies in transgenic plants, *Nature* **342**, 76-78.
31. Hilder, V.A., Gatehouse, A.M.R., Sheerman, S.E., Barker, R.F. and Boulter, D. (1987) A novel method for insect resistance engineered into tobacco, *Nature* **330**, 160-163.
32. Hilder, V.A., Powell, K.S., Gatehouse, A.M.R., Gatehouse, J.A., Gatehouse, J.N., Shi, Y., Hamilton, W.D.O., Merryweather, A., Newell, C.A., Timans, J.C., Peumans, W.J., van Damme, E. and Boulter, D. (1995) Expression of snowdrop lectin in transgenic tobacco plants results in added protection against aphids, *Transgenic Res.* **4**, 18-25.
33. Huesing, J.E., Murdock, L.L and Shade, R.E. (1991) Rice and stinging nettle lectins: insecticidal activity similar to wheat germ agglutinin, *Phytochem.* **30**, 3565-3568.
34. Jach, G., Gornhardt, B., Mundy, J., Logemann, J., Pinsdorf, P., Leah, R., Schell, J. and Maas, C. (1995) Enhanced quantitative resistance against fungal disease by combinatorial expression of different barley antifungal proteins in transgenic tobacco, *Plant J.* **8**, 97-109.
35. Jeynes, J.M., Nagpala, P., Destefano-Beltran, L., Hong Huang, J., Kim, J.H., Denny, T. and Cetiner, S. (1993) Expression of Cecropin B lytic peptide analog in transgenic tobacco confers enhanced to resistance to bacterial wilt caused by *Pseudomonas solanacearum, Plant Sci.* **89**, 43-53.
36. Jongsma, M.A., Bakker, P.L., Peters, J., Bosch, D. and Stiekema, W.J. (1995) Adaptation of *Spodoptera exigua* larvae to plant proteinase inhibitors by induction of gut proteinase activity insensitive to inhibition, *Proc. Nat. Acad. Sci.U.S.A.* **92**, 8041-8045.
37. Lilley, C.J., Urwin, P.E., McPherson, M.J. and Atkinson, H.J. (1996) Characterisation of intestinally active proteinases of cyst nematodes, *Parasitology* **113**, 415-424.
38. Liu, Y.B., Tabashnik, B.E. and Pusztaicarey, M. (1996) Field evolved resistance to *Bacillus thuringiensis* toxin CRYIC in diamondback moth (Lepidoptera, Plutellidae), *J. Econ. Ent.* **89**, 798-804.
39. Lodge, J.K., Kaniewski, W.K. and Tumer, N.E. (1993) Broad spectrum virus resistance in transgenic plants expressing pokeweed anti viral protein, *Proc. Nat Acad. Sci. U.S.A.* **90**, 7089-7093.
40. Logemann, J., Jach, G., Tommerup, H., Mundy, J. and Schell, J. (1992) Expression of a barley ribosome inactivating protein leads to increased fungal protection in transgenic tobacco plants. *Biotechnol.* **10**, 305-308.
41. Longstaff, C., Campbell, A.F. and Fersht,A.R. (1990) Recombinant chymotrypsin inhibitor-II - expression, kinetic analysis of inhibition with α-chymotrypsin and wild-type and mutant subtilisin BPN' and protein engineering to investigate inhibitory specificity and mechanism, *Biochemi.* **29**, 7339-7347.
42. Marban-Mendoza, N., Jeyaprakash, N., Jansson, H.B Jr and Zuckerman, B.M. (1987) Control of root-knot nematodes an tomato by lectins, *J. Nematol.* **19**, 331-335.
43. McClure, M.A and Stynes, B.A. (1988) Lectin binding sites on the amphidial exudates of *Meloidogyne, J. Nematol.* **20**, 321-326.
44. McGaughey, W.H. (1994) Implications of cross resistance among *Bacillus thuringiensis* toxins in resistance management, *Biocon. Sci. Tech.* **4**, 427-435.
45. Murdock, L.L., Huesing, J.E., Nielsen, S.S., Pratt, R.C. and Shade, R.E. (1990) Biological effects of plant lectins on the cowpea weevil, *Phytochem.* **29**, 85-89.
46. Narva, K.E., Payne, J., Schwab, G.E., Hickle, L.A., Galasan, T. and Sick, A.J. (1991) Novel *Bacillus thuringiensis* microbes active against nematodes and genes encoding novel nematode active toxins cloned from *Bacillus thuringiensis* isolates, *European Patent Application 91305047.2.*
47. Oppert, B., Morgan, T.D., Culbertson, C. and Kramer, K.J. (1993) Dietary mixtures of cystine and serine proteinase inhibitors exhibit synergistic toxicity towards the red flour beetle *Tribolium castaneum., Toxicol. Endocr.* **105**, 379-385.
48. Osman, G.Y., Salem, F.M. and Ghattas, A. (1988) Bio-efficacy of two bacterial insecticide strains of *Bacillus thuringiensis* as a biological control agent in comparison with a nematicide, Nemacur, on certain parasitic Nematoda, *Anz. Schadlinsk. Pflschutz. Umwschutz.* **61**, 35-37.
49. Peumans, W.J. and van Damme, E.J.M. (1995) Lectins as plant defence proteins, *Plant Physiol.* **109**, 347-352.
50. Politz, S.M and Philipp, M. (1992) *Caenorhabditis elegans* as a model for parasitic nematodes - a focus on the cuticle, *Parasitol. Today* **8**, 6-12.
51. Proudfoot, L., Kusel, J.R., Smith, H.V. and Kennedy, M.W. (1991).: Biophysical properties of the nematode surface, in M.W Kennedy (ed), *Parasitic Nematodes - Antigens, Membranes and Genes*, Taylor and Francis, London, New york, Philadelphia, pp 1-26.

52. Purcell, J.P., Greenplate, J.T., Jennings, M.G., Ryerse, J.S., Pershing, J.C., Sims, S.R., Prinsen, M.J., Corbin, D.R., Tran, M., Douglas Sammons, R. and Stonard, R.J. (1993) Cholesterol oxidase: a potent insecticidal protein active against boll weevil larvae, *Biochem. Biophys. Res. Com.* **3**, 1406-1413.
53. Pusztai, A., Ewen, S.W.B., Grant, G., Peumans, W.J., van Damme, E.J.M., Rubio, L. and Bardocz, S. (1990) The relationship between survival and binding of plant lectins during small intestine passage and their effectiveness as growth factors, *Digestion* **46**, 310-316.
54. Pusztai, A., Ewen, S.W.B., Grant, G., Brown, D.S., Stewart, J.C., Peumans, W.J., van Damme, E.J.M and Bardocz, S. (1993) Antinutritive effects of wheat germ agglutinin and other N-acetylglucosamine specific lectins, *Br. J. Nutr.* **70**, 313-321.
55. Roberts, P.A. (1992) Current status of the availability, development and use of host plant resistance to nematodes, *J. Nematol.* **24**, 213-227.
56. Roby, D. and Esquerre-Tugaye, M.T. (1987) Induction of chitinases and of translatable mRNA for these enzymes in melon plant infected with *Colletotrichum lagenarium, Plant Sci.* **52**, 175-185.
57. Rosso, M.N., Schouten, A., Roosien, J., Borstvrenssen, T., Hussey, R.S., Gommers, F.J., Bakker, J., Schots, A. and Abad, P. (1996) Expression and functional characterisation of a single chain FV antibody directed against secretions involved in plant nematode infection process, *Biochem. Bioph. Res. Co.* **220**, 255-263.
58. Ryan, C.A. (1990) Protease inhibitors in plants: genes for improving defence against insects and pathogens, *Annu. Rev. Phytopathol.* **28**, 425-449.
59. Schots, A., De Boer, J., Schouten, A., Roosien, J., Zilverentant, J.F., Pomp, H., Bouwman-Smits, L., Overmars, H., Gommers, F.J., Visser, B., Stiekema, W.J. and Bakker, J. (1992) Plantibodies: a flexible approach to design resistance against pathogens, *Neth. J. Pl. Path.* **98**, 183-191.
60. Schroeder, H.E., Gollasch, S., Moore, A,. Tabe, L.M., Craig, S., Hardie, D.C., Chrispeels, M.J., Spencer, D. and Higgins, T.J.V. (1995) Bean alpha-amylase inhibitor confers resistance to the pea weevil (*Bruchus pisorum*) in transgenic peas (*Pisum sativum* L), *Plant Physiol.* **109**, 1129.
61. Sebesta, K., Farkas, J., Horska, K. and Vankova, J.: Thuringensin, the beta exotoxin of *Bacillus thuringiensis*, in H.D Burges (ed) *Microbial Control of Pests and Plant Diseases*. New York, Academic Press. pp 271-281.
62. Shade, R.E., Schroeder, H.E., Pueyo, J.J.,Tabe, L.M., Murdock, L.L and Higgins, T.J.V. (1994) Transgenic pea seeds expressing the alpha-amylase inhibitor of the common bean are resistant to Bruchid beetles, *Biotechnol.* **8**, 793-796.
63. Shi, Y., Wang, M.B., Powell, K.S., van Damme, E., Hilder, V.A., Gatehouse, A.M.R., Boulter, D. and Gatehouse, J.A. (1994) Use of the rice sucrose synthase-1 promoter to direct phloem specific expression of beta glucuronidase and snowdrop lectin genes in transgenic tobacco plants, *J. Exp. Bot.* **45**, 623-631.
64. Sijmons, P.C., Grundler, F.M.W., von Mende, N., Burrows, P.R. and Wyss, U. (1991) *Arabidopsis thaliana* as a new model host for plant parasitic nematodes, *Plant J.* **1**, 245-254.
65. Sousa, M.V., Morhy, L., Richardson, M., Hilder, V.A. and Gatehouse, A.M.R. (1993) Effects of the cytolytic seed protein Enterolobin on Coleopteran and Lepidopteran insect larvae. *Entomol. Exp. Appl.* **69**, 231-238.
66. Spiegel Y., Chet I. and Cohn E.: Use of chitin for controlling plant parasitic nematodes. II. Mode of action. *Plant Soil* **98**, (1987), 337-345.
67. Spiegel, Y., Cohn, E. and Chet, I. (1986) Use of chitin for controlling plant parasitic nematodes. I. Direct effect on nematode reproduction and plant performance, *Plant Soil* **95**, 87-95.
68. Spiegel, Y., Inbar, J., Kahane, I. and Sharon, E. (1995) Carbohydrate recognition domains on the surface of phytophagous nematodes, *Exp. Parasitol.* **80**, 220-227.
69. Strickland, J.A., Orr, G.L. and Walsh, T.A. (1995) Inhibition of Diabrotica larval growth by patatin, the lipid acyl hydrolase from potato tubers, *Plant Physiol.* **109**, 667-674.
70. Tabashnik, B.E. (1994) Delaying insect adaption to transgenic plants - seed mixtures and refugia reconsidered, *P. Roy. Soc. Lond. B Bio.* **225**, 7-12.
71. Tavladoraki, P., Benvenuto, E., Trinca, S., Demartinis, D., Cattaneo, A. and Galeffi, P. (1993) Transgenic plants expressing a functional single chain FV-antibody are specifically protected from virus attack, *Nature* **366**, 469-472.
72. Urwin, P.E., Atkinson, H.J., Waller, D.A. and McPherson, M.J. (1995), Engineered oryzacystatin-I expressed in transgenic hairy roots confers resistance to *Globodera pallida, Plant J.* **8** 121-131.

73. Wink, M. (1988) Plant breeding: importance of plant secondary metabolites for protection against pathogens and herbivores, *Theor. Appl. Genet.* **75**, 225-233.
74. Zuckerman, B.M. (1983) Hypotheses and possibilities of intervention in nematode chemo- responses, *J. Nematol.* **15**, 173-182.
75. Zuckerman, B.M., Dicklow, M.B and Acosta, N. (1993) A strain of *Bacillus thuringiensis* for the control of plant parasitic nematodes, *Biocon. Sci. Technol.* **3**, 41-46.

ENGINEERING PLANT NEMATODE RESISTANCE BY ANTI-FEEDANTS

Proteinase inhibitors and nematode proteinases

Michael J. McPHERSON, Peter E. URWIN, Catherine J. LILLEY & Howard J. ATKINSON
Centre for Plant Biochemistry and Biotechnology, University of Leeds, Leeds, LS2 9JT, UK.

Abstract

Control of parasitic nematodes by transgenic plants will offer substantial benefits. Chemical nematicides and fumigants are highly toxic and environmentally harmful. Many of these compounds are being restricted by international agreement. Alternative mechanisms for control of nematodes must be implemented with some urgency. We briefly consider alternative approaches and describe in more detail one approach for achieving transgenic control against a range of nematode species. Our prototype resistance is based on the use of a plant proteinase inhibitor that has been improved by protein engineering. We also discuss studies of nematode proteinases that provide a foundation for understanding the resistance mechanism in molecular detail and will assist development of durable field resistance.

1. Introduction

Nematodes cause global crop losses valued at over $100 billion per year [46] and their control is associated with widespread use of highly toxic nematicides with consequent environmental and health risks. In northern Europe, potato cyst-nematode (PCN; *Globodera* spp.) and sugarbeet cyst-nematode (BCN; *Heterodera schachtii*) represent major problems. In Southern Europe, and in all subtropical and tropical regions of the world, the most significant problems arise due to root-knot nematodes (*Meloidogyne* spp). Unlike the cyst-nematodes, which have very narrow host ranges, some root-knot nematodes are capable of infecting and reproducing on a wide range of plant species.

1.1. CURRENT CONTROL MEASURES

Control of nematodes currently relies on three principal approaches, chemicals, cultural practices and resistant varieties, often used in an integrated manner [22]. Chemical control is not only costly and therefore largely inaccessible to most farmers in the developing world but it involves application of environmentally unacceptable compounds. Examples include, the oximecarbamate, Aldicarb, one of the most toxic and environmentally hazardous pesticides in widespread use [21], and nematicidal fumigants such as methyl bromide and 1,3-dichloropropene (Telone II). Toxicological problems and environmental damage caused by nematicides has resulted in their withdrawal or in severe restrictions on their use. Measures for planned reduction or withdrawal of many of these chemicals exist in countries such as The Netherlands and USA. The prospect of severe crop losses within major sectors of the agricultural industry may, however, limit such withdrawal plans until replacement crop protection strategies become available.

Cultural measures such as crop rotation, although widely practised, rarely provide an adequate solution alone and are not practical for growers that are specialists either from choice or necessity.

Resistant cultivars have been a commercial success for a limited range of crops but the approach is unable to control many nematode problems for various reasons [14, 44]. To give one example, the natural resistance gene H1 present in potato cultivars such as Maris Piper has proved extremely successful in controlling *G. rostochiensis* populations in the United Kingdom. Unfortunately, these cultivars do not control *G. pallida* which has now become the prevalent potato cyst-nematode in Britain as a direct consequence of selection through widespread growing of resistant H1-cultivars [49]).

The limitations of conventional control approaches provide an important opportunity for plant biotechnology to produce effective and durable forms of nematode control. Designs for such novel plant defences should aim to minimise environmental, producer and consumer risk while providing significant economic and social benefits for both the developed and developing world [4].

2. Developing a Strategy for Engineered Resistance

It is important to identify a clear strategy for developing engineered resistance and this will be governed by economic need. If control of a specific nematode species (*e.g.* potato cyst-nematode) or group (e.g. root-knot nematode), is sought then a strategy designed to disrupt a characteristic feature of the target plant/nematode interaction may be appropriate. For example, control of potato cyst-nematode could be achieved by mimicking the natural hypersensitive response leading to syncytium collapse [42]. However, the very specificity of this approach presents a potential problem of selection for non-prevalent nematode species. Changes in pest status have occurred between *Globodera* spp. following use of H1-based resistant potato cultivars in the UK [49] and in the relative importance of nematodes on pineapple in Hawaii as a consequence of changes in agricultural practice [12]). An ability to counter such changes in pest species prevalence is an important consideration for those designing engineered plant resistance.

If control of several nematodes that exhibit different invasion and feeding habits is required a second more generic anti-nematode defence is preferable. Such an approach should seek to target a common feature required for the growth and development of all nematode species. The effector molecule(s) should ideally be nematode specific, have no adverse effect on the plant and be present in all, principally root, cells from which a nematode may feed. For several years the group at Leeds has been developing both nematode-specific and generic anti-nematode strategies as both approaches offer distinct and important merits.

2.1. PLANT PARASITISM AND TARGETS FOR ENGINEERED RESISTANCE

There are several stages in the host parasite interaction at which an engineered plant resistance response could be mounted against a nematode.

2.1.1. *Invasion and migration*
The invasion and root migration by the nematode before selection of a feeding site provides an early stage for disrupting the potential parasitism [5]. Initial interaction with the plant relies upon chemotactic sensory perception by the nematode [56]. Exogenous application of lectins can disrupt the normal invasion processes [37] so expression of certain lectins as transgenes may also disrupt the invasion process (see chapter by Burrows and De Waele, this volume).

Differences in the nature of the migratory phase of nematodes as they move through the plant root prior to feeding site induction [see 47] may make it difficult to devise a universal anti-migratory strategy. Root-knot nematodes migrate between cells without damaging them and therefore do not elicit significant wound responses. By contrast, cyst-nematodes move through plant cells and elicit a pronounced wound response during the migration phase [23]. Therefore expression of a defence during this phase of the parasitism must be tailored to the target nematode. Wound-inducible promoters such as *wun-1* [36] could prove useful for cyst-nematode and even migratory endo- and ecto-parasite control when the period of interaction exceeds the promoter response time. It would not be suitable for root-knot species and others that fail to induce wound responses or have only transient interactions with the plant at one locale [23].

Effectors expressed from appropriate promoters could be directed at either the feeding site or the parasite. A wound-specific promoter and an appropriate effector could induce localised plant cell death so mimicking natural hypersensitive reactions. Anti-nematode defences could include lectins or antibodies which might interfere with sensory perception. Enzymes, such as collagenases that degrade structural proteins, could be considered (see chapter by Burrows and De Waele, this volume). The key to an anti-migratory approach is the rapidity of the plant response to ensure the nematode induces and then encounters the effector. Ideally a promoter should provide a root "systemic" response to ensure effector expression in cells before they are subject to nematode attack. Such expression characteristics are seen with the *wun-1* gene [23].

2.1.2. *Nematode Feeding*
The most appropriate target for nematode control is feeding. Several economically important nematodes including *Meloidogyne, Globodera, Heterodera, Rotylenchulus, Nacobbus* and *Xiphinema*, form feeding cells within their plant hosts [47]. While the process of formation and the physiological and morphological characteristics of these feeding cells differ their ultimate role is to provide the nutritional requirements of the

240

growing female. By contrast, damaging migratory ecto- and endoparasitic nematodes such as *Radopholus*, *Pratylenchus*, *Hirschmanniella* and *Trichodorus* feed intermittently from unmodified plant cells [47].

The importance of the feeding cell to cyst- and root-knot nematodes has provided a focus for significant research activity aimed at providing engineered resistance. One approach is to elicit feeding cell death or its metabolic attenuation. A second approach is to use the feeding cell to deliver anti-nematode effectors which are ingested by and directly interfere with the development of the feeding nematode.

3. Disruption of the Feeding Cell

3.1. PROMOTERS

Targeting the feeding cell for either destruction or metabolic attenuation demands the availability of promoters for genes that are specifically induced in the feeding cells. Plant genes that respond to nematodes have been isolated by various cDNA cloning [19, 20, 33, 7, 54] and gene tagging procedures [18, 35]. However, it seems unlikely that a plant will harbour a gene whose only function is to allow nematode parasitism. Given the high metabolic activity of the feeding cells it is not surprising that many of the genes up-regulated in feeding cells are expressed in other actively dividing and expanding cells [7].

An alternative approach is to select nematode-responsive promoters from previously characterised genes. For example *Tob*RB7, a tobacco root-specific putative water channel-encoding gene shows expression at root meristems in healthy plants [13, 55] but is also active in the giant cells induced by *Meloidogyne* [40].

It seems likely that in some cases the expression characteristics of candidate promoters may be amenable to modification to provide greater feeding cell specificity. In the case of *Tob*RB7 deletion studies have identified a 300 bp promoter fragment that directs expression within the nematode feeding cells but not in root meristems [40].

There remains a need to select a subset of genes specifically induced in feeding cells in response to nematodes and to provide detailed developmental maps of their expression patterns. This will provide a resource from which promoter selections can be made to suit the planned defence strategy.

3.2. EFFECTORS

An approach that has been adopted to demonstrate the concept of feeding cell disruption is based on the use of barnase, a bacterial RNase, and its inhibitor protein, barstar [24]. Barnase was successfully used in a plant cell suicide system for engineered emasculation in maize [38] and barstar restoration of fertility has also been achieved [39]. However, a significant difference between this system and a root feeding cell system, is that male-specific genes can be expected to have promoters with little "leaky" expression in other cells, unlike the situation for nematode responsive promoters discussed above. This limitation must be overcome to prevent a toxic effector, such as barnase, destroying non-target cells within the plant. Ohl and colleagues describe elsewhere in this volume the use of constitutive expression of the barstar protein to protect normal plant cells from leaky expression of the potent barnase effector, itself expressed from a predominantly feeding cell specific promoter.

Development of alternative feeding cell attenuation mechanisms, based on the co-ordinate expression of inactive components which interact to yield an active effector is currently underway at Leeds. Selection of promoters giving overlapping patterns of expression only within the feeding cells allows delivery of active effector molecule only to these cells. Protection of the remainder of the plant does not rely upon the constitutive expression of another protein.

4. Anti-Nematode Strategies

Our principal motivation for developing an anti-nematode approach is the need to control nematodes with differing feeding habits. Many crops are targeted by more than one nematode species with dissimilar feeding processes. Even when a crops major nematode pest induces a specialised feeding site, its specific control may reduce competition thereby raising another species with a distinct feeding habit to major pest status. An added advantage of an anti-nematode approach is that the grower does not need to be aware of the specific nematode pests, a situation that has particular relevance to subsistence farmers in the developing world [5].

4.1. PROMOTERS

Restricting effector expression to the feeding cell is not necessary for certain defence strategies, such as delivery of an anti-feedant. In fact more wide ranging expression can be desirable to provide a standing generic defence capable of controlling a range of nematodes. Such an approach is suitable where the effector is not hazardous to the plant, the consumer or other non-pest organisms. An example is when it is based on a naturally occurring plant protein that is consumed in traditional foods. For the prototype defence that we describe below, we have used the *CaMV 35S* promoter derived from the pBI-vector series. A range of promoters we have developed with more root specific expression are currently being evaluated and compared. It is important to understand the promoter expression dynamics both in feeding regions and elsewhere within the transgenic plant. As a first step in this direction we have used the modified Green Fluorescent Protein (mGFP) reporter [25] to monitor levels of CaMV 35S expression in nematode feeding sites. This promoter is down-regulated by *H. schachtii* in *Arabidopsis* but remains sufficiently active to deliver effector proteins [52]. The GFP system may prove useful for mapping expression patterns throughout transgenic lines at all stages during plant development.

4.2. EFFECTORS

4.2.1. *Nematode proteinases and proteinase inhibitor effects*

Our objective is to identify targets in the feeding process that are likely to be common to all plant parasitic nematodes. We have chosen to focus on proteinases in the first instance. Proteinases are involved in a broad range of processes in eukaryotic cells such as intra- and extra-cellular protein metabolism and digestion of dietary protein. Cysteine proteinases have been reported from animal parasitic nematodes [15] and *in situ* hybridisation has demonstrated transcripts of such genes are abundant in the intestine of the microbivorous nematode *Caenorhabditis elegans* [41]. By contrast, little was known about the proteinases of plant parasitic nematodes and their roles in the host-parasite interaction before we initiated our current research programme.

Proteinases are known to be inhibited by proteinase inhibitors (PIs) many of which are natural, defence-related, proteins induced by wounding and herbivory in aerial and certain other parts of plants [45]. Although nematode parasitism of roots leads to systemic induction of PIs in the aerial tissues [10] it is surprising that they are not induced in roots.

Cowpea trypsin inhibitor (CpTI), a serine proteinase inhibitor, has been shown to provide some control against insects when expressed as a transgene [27]. However, some insects can to some extent circumvent natural plant PI-mediated defence mechanisms by adjusting the relative expression levels of mechanistically distinct gut proteinases [29]. A similar protective feedback operates in mammals [28]. For nematodes the situation may be somewhat different. Lack of exposure to PIs in roots means nematodes have not been subjected to evolutionary selective pressure to overcome PI-mediated plant defences. This increases the potential of PIs for nematode control.

We have shown that CpTI expressed in transgenic plants suppresses early growth and development of *G. pallida* [26, 3]. Subsequently we showed that cyst-nematodes possess a major cysteine proteinase [32, 34]. PIs are control agents of high potential for several reasons; they are naturally occurring plant proteins, they are consumed in significant quantities by humans in many plant foods and mammals lack cysteine proteinases in their intestine further enhancing the safety aspects for this group of PIs. Therefore our recent research has centred on developing effective nematode resistance using cysteine proteinase inhibitors of plant origin.

4.2.2. *Cystatins*

Cystatins are small protein inhibitors of cysteine proteinases. They have been identified in a range of plants [45, 43] and several genes have been sequenced including those from rice (*oc-I* and *oc-II*; [2, 31]), cowpea (*cpci*; [16]) and maize (*zmci*; [1], Urwin *et al.*, in preparation). The phytocystatins are a distinct class from type I and II cystatins representing a hybrid between these classes [30]. We have worked principally with oryzacystatin I (OC-I) as this is present in the seed of rice, a very widely consumed food. Plant cystatins are less efficient inhibitors than certain animal cystatins but we made the decision to focus upon plant proteins for transgenic control. The key question is whether improvements can be made to increase the efficacy of a plant cystatin.

OC-I has been expressed in *E. coli* using the QIAexpress system (QIAGEN) to provide large quantities of inhibitor protein for bioassays, biochemical assays and toxicological testing. We have prepared antibodies against recombinant OC-I for ELISA and western blotting screening and for quantifying expression levels in transgenic plants [50].

Molecular modelling of a cysteine proteinase/OC-I interaction. An understanding of the molecular interactions between two molecules such as a proteinase inhibitor and its target proteinase can provide useful information to guide modifications to improve the efficacy of the inhibitor. Structures are not yet available for plant cystatins or nematode cysteine proteinases. We have therefore used X-ray crystallographic structures for chicken egg white cystatin (CEWC; [9]) and the complex of human stefin B with papain [48] to build a three dimensional model of OC-I. The modelled structure was subjected to energy minimisation using the program Xplor [11] to ensure reasonable stereochemistry. The cystatin model was docked into the active site of papain to provide an unrefined model for the inhibitor/proteinase complex [50]. In addition, alignment of 28 cysteine proteinase inhibitor sequences of both the cystatin and stefin classes allowed comparison of sequence features with the reported K_i values for inhibitors for the design of mutagenesis experiments [50].

Mutagenesis of OC-I. Two proteinases, papain and GCP-1, were used for *in vitro* assays to measure efficacy of inhibitor variants. GCP-1 is the digestive cysteine proteinase from the nematode *C. elegans*. The *gcp-1* gene was PCR amplified, cloned into pQE, expressed in *E. coli* and GCP-1 protein purified by nickel-chelate affinity chromatography.

Based on our model and sequence alignments, we initially focused on amino acid insertions and deletions. Several mutations did not affect K_i while others led to an increase in K_i suggesting decreased inhibitory capacity. However, deletion of an aspartic acid at position 86 (OC-IΔD86) resulted in a K_i some 13-fold lower than for OC-I against both papain and GCP-1. This suggested OC-IΔD86 had enhanced inhibitory activity against both proteinases compared with the parent OC-I. Asp86 was substituted by 12 other amino acids to test if it is the absence or chemical nature of the side chain that is important in improved proteinase binding. All substitutions increased K_i indicating it was the absence of the residue that was critical for increased efficacy. This study demonstrated that protein engineering can enhance the efficacy of a plant cystatin [50]. Further progressive improvements in OC-I efficacy have been made using mutagenesis approaches (Urwin *et al.*, in preparation). The mutagenesis studies have also clarified the nature of involvement of the N-terminal region of OC-I in cysteine proteinase inhibition [51].

4.2.3. *Proteinase activity of plant nematodes*

A standard, histochemical assay with a specific peptide substrate of cysteine proteinases (Z-Ala-Arg-Arg-4MNA) was used with cyst-nematode cryosections to localise a cysteine proteinase to the animals intestine. Yellow fluorescent crystals form when proteolytically cleaved 4-MNA reacts with added 4-hydroxy-5-nitrobenzaldehyde. The use of a range of specific synthetic substrates led to classification of the cysteine proteinase activity as cathepsin L type [34]. This activity was completely inhibited by pretreating the sections with OC-IΔD86, but was not affected by pretreatment with either BSA or an inactive OC-I variant. Similarly, by using different substrates the presence of a serine proteinase activity was demonstrated, and this could be inhibited by pre-treatment with the serine proteinase inhibitor CpTI [34].

4.2.4. *Cloning proteinase genes*
Characterisation of target proteinases and their expression for use in biochemical assays and structural studies underpins our understanding of inhibitor/proteinase interaction and protein engineering to optimise PI redesign.

Degenerate PCR primers designed from conserved regions of proteinases were used to amplify short fragments of cysteine proteinase genes from *H. glycines, G. pallida, M. javanica* and *M. incognita*. The *Globodera* and *Meloidogyne* products show most homology to cathepsin B-like sequences while a product from *H. glycines* shows homology to a cathepsin L-like sequence [34]. This *H. glycines* product was used to screen a cDNA library prepared from feeding females and full-length clones have been isolated. A similar approach was used to isolate full-length cDNA clones for serine proteinases. The finding that serine as well as cysteine proteinases are active in the intestine of cyst-nematodes suggests that a combination of inhibitors will provide a more durable defence system in plants than use of a single PI as a transgene.

4.3. TESTING ENGINEERED CYSTATINS

Having established that our engineered inhibitor OC-IΔD86 was able to a) inhibit cysteine proteinase activity *in vitro* and b) inhibit proteinase activity *in situ* in nematode cryosections we then tested its effects on nematode growth and development. We were able to show a lethal effect on *C. elegans* feeding on cystatin-containing media and this effect was more pronounced with OC-IΔD86 than with the unmodified OC-I [50].

4.3.1. *Growth of Globodera pallida on transgenic hairy roots*
For an initial test of the effect on plant parasitic nematodes we used a transgenic hairy root assay. Expression of cystatins was directed by the *CaMV 35S* promoter derived from pBI121. Transformed tomato hairy root lines expressing either OC-I or OC-IΔD86 at approximately 0.5% total soluble protein were selected for comparative studies. Nematode growth on OC-I and OC-IΔD86 expressing roots was significantly lower than on control roots. This effect was most pronounced on OC-IΔD86 roots where there was no significant increase in nematode size between weeks 4 and 6 in contrast to the OC-I and control lines.

4.3.2. *Parameters for rapid analysis of nematode size and fecundity*
We developed an image analysis approach to allow rapid classification of nematode developmental state and fecundity in whole plant assays. The shape parameters have been established for *G. pallida* and *M. incognita* on tomato hairy roots and for *H. schachtii* on cabbage roots. Cross-sectional area, length and roundness measurements provided values for distinguishing three body shapes; fusiform (J2, J3 of both sexes plus J4 and adult

male), saccate (J4 and adult female prior to egg formation) and enlarged saccate (gravid females) [6, 53]. A range of cysts of various sizes were examined for egg content to define the relationship between cross-sectional area and egg content for *G. pallida* and *H. schachtii*. The smallest *H. schachtii* cyst recovered had a cross-sectional area of 0.075 mm^2 and an egg content of 32 eggs. This may approximate to the lower size limit for new *H. schachtii* cysts with viable eggs [53].

4.3.3. Resistance assays on OC-IΔD86-transgenic Arabidopsis

We have tested the effects on nematode growth and development of OC-IΔ86 expressed as a transgene in *Arabidopsis* plants. A homozygous line carrying a single copy of the *oc-IΔD86* gene was selected and shown by quantitative western blot analysis to be expressing OC-IΔD86 as 0.4 % total soluble protein. This line was used in resistance assays with both *H. schachtii* and *M. incognita* [53].

On control *Arabidopsis* plants most *H. schachtii* females became enlarged saccate in shape and were of comparable size to those recovered from cabbage plants, indicating normal growth and development. By comparison the nematodes on OC-IΔD86 expressing *Arabidopsis* were significantly smaller. Most nematodes were fusiform and the few saccate females were all below the size of the smallest *H. schachtii* cyst recovered from cabbage plants. Expression of the cystatin prevented egg-laying individuals from developing, growth was arrested and some animals showed abnormal development. There was no evidence that expression of the cystatin in *Arabidopsis* prevented development of males of normal size [53].

For *M. incognita* the effects were similar to those observed for *H. schachtii*. This suggests that growth of females of both species is similarly inhibited by expression of cystatin [53]. Three effects occur; few females reach an egg-laying size, those that do have a greatly diminished size and fecundity, and there is evidence that some saccate *H. schachtii* are developmentally compromised.

4.3.4. Cystatin ingestion by feeding nematodes

Using the previously described fluorogenic assay in nematode cryosections (section *5.2.3*), proteinase activity was detected in saccate females recovered from control *A. thaliana* C24 but only exceptionally in those from plants expressing OC-IΔD86. Further evidence of ingestion is the detection by western blot analysis of OC-IΔD86 in extracts of nematodes collected from the cystatin expressing *Arabidopsis*. Using a similar approach we have demonstrated uptake of mGFP by *M. incognita* but not by *H. schachtii* from transgenic *Arabidopsis* [52]. This may reflect a difference in the architecture of the nematode feeding tubes between root-knot and cyst-nematodes. Previous studies involving microinjection of fluorescent dextran probes into the syncytia of *H. schachtii*, feeding on *Arabidopsis*, have indicated a size exclusion limit of between 20 and 40 kDa [8]. Protein uptake from transgenic plants by *H. schachtii* is in agreement with these values with limits of 11 kDA (OC-IΔD86) and 28 kDa (mGFP). By contrast the *M. incognita* feeding tube must have an upper size exclusion limit of at least 28 kDa [52]. The upper exclusion limit of the feeding tube is an important parameter that will govern the size of effector molecules that may be delivered as part of an anti-feedant defence strategy.

5. Conclusions

We have provided the first demonstration of transgenic control of two major groups of agronomically important nematode parasites using a single effector transgene. The resistance is based upon delivery of a proteinase inhibitor *via* the plant cells from which the nematode feeds. Transgenic resistance against plant nematodes will be of most value when it is effective against several species with differing feeding habits. In this respect the size of a generic effector is dictated by the maximum size molecule that can be ingested by any of the target nematodes to be controlled. Our prototype defence uses the *CaMV 35S* promoter to provide constitutive expression of the OC-IΔD86 effector. Development of a generic resistance relies upon selection of the correct promoter and effector(s) combination and we are now selecting a root-specific promoter for delivery of an effective defence against all root-feeding nematodes. Our assays are being extended to test several groups of nematodes, including *Pratylenchus*, *Hirschmanniella* and *Rotylenchulus*.

The defence is now being introduced for testing in crop plants, including rice (Paul Christou, John Innes Centre), sugarbeet (Hilleshög NK) and potato, as the next stage towards field resistance. In this context it is important to ensure that a transgene does not adversely affect other agronomic characteristics such as vigour and yield. We are examining toxicological consequences of expression of a cystatin in transgenic plants and are conducting environmental impact studies on non-target invertebrates.

Durability of transgenic defence remains a key concern and we are developing a variety of approaches designed to provide a pyramided defence. Although plant nematodes probably have had no evolutionary exposure to cysteine proteinase inhibitors, it is important to anticipate any future breakdown in resistance and to devise strategies for countering the consequences. Our rational approach is to develop protein engineered PIs targeted at characterised proteinases. This assures a molecular understanding of the transgenic resistance. This strategy will allow definitive analysis of any breakdown in resistance and provide a basis for further rational design. However, a key to durable plant defence is likely to involve gene pyramiding [17].

The potential of our approach for the developing world is also being examined for rice and a range of other crops with support from ODA Plant Sciences Programme. A basis has been developed to assure free distribution of the technology to subsistence crops in the developing world [4].

ACKNOWLEDGEMENTS

We are grateful for financial support from the Biotechnology and Biological Sciences Research Council, Advanced Technologies (Cambridge), Hilleshög NK, Nickerson BIOCEM, The Overseas Development Administration and The Scottish Office Agriculture, Environment and Fisheries Department. We thank J. Goodall for excellent technical support and Simon Møller, Gaelle Richard, Paul Jones and Mark Sutton for helpful discussions.

References

1. Abe, M., Abe, K., Kuroda, M., and Arai, S. (1992) Corn kernel cysteine proteinase inhibitor as a novel cystatin superfamily member of plant origin. Molecular cloning and expression studies, *Eur J Biochem* **209**, 933-937.
2. Abe, K., Emori, Y., Kondo, H., K. Suzuki, and Arai, S. (1987) Molecular cloning of a cysteine proteinase inhibitor of rice (oryzacystatin). Homology with animal cystatins and transient expression in the ripening process of rice seeds, *J Biol Chem* **262**, 16793-16797.
3. Atkinson, H.J. (1993) Opportunities for improved control of plant parasitic nematodes via plant biotechnology, in Beadle, D.J., et al. (eds.) *Opportunities for Molecular Biology in Crop Production*, British Crop Protection Council, Farnham, pp. 257-266.
4. Atkinson, H.J. (1995) Can genetic engineering help reduce eelworm problems on rice? *Landmark* **11**, 10.
5. Atkinson, H.J., Urwin, P.E., Hansen, E. and McPherson, M.J. (1995) Designs for engineered resistance to root-parasitic nematodes, *Trends in Biotechnology* **13**, 369-374.
6. Atkinson, H.J., Urwin, P.E., Clarke, M. and McPherson, M.J. (1996) Image analysis of the growth of *Globodera pallida* and *Meloidogyne incognita* on transgenic tomato roots expressing cystatins, *J. Nematology* **28**, 209-215.
7. Bird, D.M. and Wilson, M.A. (1994) DNA sequence and expression analysis of root-knot nematode-elicited giant cell transcripts, *Molecular Plant-Microbe Interactions* **7**, 419-424.
8. Bockenhoff, A. and Grundler, F.M.W. (1994) Studies on the nutrient-uptake by the beet cyst-nematode *Heterodera schachtii* by *in situ* microinjection of fluorescent probes into the feeding structures in *Arabidopsis thaliana*, *Parasitology*, **109**, 249-254.
9. Bode, W., Engh, R., Musil, D., Thiele, U., Huber, R., Karshikov, A., Brzin, J., Kos, J. and Turk, V. (1988) The 2.0 Å X-ray crystal structure of chicken egg white cystatin and its possible mode of interaction with cysteine proteases, *EMBO J.* **7**, 2593-2599.
10. Bowles, D.J., Gurr, S-J., Scollan, C., Atkinson, H.J. and Hammond-Kosack, K. (1991) Local and systemic changes in plant gene expression, in Smith, C.J. (ed.) *Biochemistry and Molecular Biology of Plant-Pathogen Interaction*, Oxford University Press, pp. 225-236.
11. Brunger, A.T., Kuriyan, J. and Karplus, M. (1987) Crystallographic R-factor refinement by molecular dynamics, *Science*, 235, 458-460.
12. Caswell, E.P., Sarah, J-L. and Apt, W.J. (1990) Nematode parasites of pineapple, in M. Luc, R.A. Sikora and J. Bridge (eds.), *Plant Parasitic Nematodes in Subtropical and tropical Agriculture* CAB International, Oxon, pp. 519-537.
13. Conkling, M.A., Cheung, C.-L., Yamamoto, Y.T. and Goodman, H.M. (1990) Isolation of transcriptionally regulated root-specific genes from tobacco, *Plant Physiology* **93**, 1203-1211.
14. Cook, R., Evans, K. (1987) Resistance and tolerance, in R.H. Brown and B.R. Kerry (eds.), *Principles and Practice of Nematode Control in Crops*, Sydney Academic Press, pp. 179-231.
15. Cox, G.N., Pratt, D., Hageman, R. and Boisvenue, R.J. (1990) Molecular cloning and primary sequence of a cysteine protease expressed by *Haemonchus contortus* adult worms, *Molecular and Biochemical Parasitology* **1**, 25-34.
16. Fernandes, K.V., Sabelli, P.A., Barratt, D.H., Richardson, M., Xavier-Filho, J., and Shewry, P.R. (1993) The resistance of cowpea seeds to bruchid beetles is not related to levels of cysteine proteinase inhibitors, *Plant Mol Biol* **23**, 215-219.
17. Gatehouse, A.M.R., Shi, Y., Powell, K.S., Brough, C., Hilder, V.A., Hamilton, W.D.O., Newell, C.A., Merryweather, A., Boulter, D. and Gatehouse, J.A. (1993) Approaches to insect resistance using transgenic plants, *Philosophical Transactions of the Royal Society of London B*, **342**, 279-286.
18. Goddijn, O.J.M., Lindsey, K., van der Lee, F.M., Klap, J.C. and Sijmons, P.C. (1993) Differential gene expression in nematode-induced feeding structures of transgenic plants harbouring promoter-*gusA* fusion constructs, *Plant J.* **4**, 863-873.
19. Gurr, S.J. McPherson, M.J. Atkinson, H.J. and Bowles, D.J. (1992) Plant Parasitic Nematode Control, International Patent Application Number, PCT/GB91/01540; International Publication Number, WO 92/04453.
20. Gurr, S.J., McPherson, M.J., Scollan, C., Atkinson, H.J. and Bowles, D.J. (1991) Gene expression in nematode-infected plant roots, *Molecular and General Genetics*, **226**, 361-366.
21. Gustafson, D.I. (1993) *Pesticides in drinking water*, Chappell Hill, N. Carolina, USA.

22. Hague, N.G.H. and Gowen, S.R. (1987) Chemical control of nematodes, in R.H. Brown and B.R. Kerry (eds.), *Principles and Practice of Nematode Control in Crops*, Sydney Academic Press, pp. 131-178.
23. Hansen, E., Harper, G., McPherson, M.J. and Atkinson, H.J. (1996) Differential expression patterns of the wound-inducible transgene *wun-1-uidA* in potato roots following infection with either cyst or root-knot nematodes, *Physiological and Molecular Plant Pathology* **48**, 161-170.
24. Hartley, R.W. (1988) Barnase and barstar. Expression of its cloned inhibitor permits expression of a cloned ribonuclease, *J. Molecular Biology* **202**, 913-915.
25. Hasseloff, J. and Amos, B. (1995) GFP in plants, *Trends in Genetics* **11**, 328-329
26. Hepher, A. and Atkinson, H.J. (1992) Nematode Control with Proteinases Inhibitors, European Patent Application Number, 92301890.7, Publication Number 0 492 730 A1.
27. Hilder, V.A., Gatehouse, A.M.R., Sheerman, S.E., Barker, R.F. and Boulter, D. (1987) A novel mechanism of insect resistance engineered into tobacco, *Nature* **330**, 160-163.
28. Holm, H., Jorgensen, A. and Hanssen, L.E. (1991) Raw soy and purified proteinase inhibitors induce the appearance of inhibitor-resistant trypsin and chymotrypsin activities in Wistar rat duodenal juice, *Journal of Nutrition* **121**, 532-538.
29. Jongsma, M.A., Bakker, P.L., Peters, J., Bosch, D. & Stiekema, W.J. (1995) Adaptation of *Spodoptera exigua* larvae to plant proteinase inhibitors by induction of proteinase activity insensitive of inhibition, *Proceedings of the National Academy of Sciences, USA* **92**, 8041-8045.
30. Kondo, H., Abe, K., Emori, Y. and Arai, S. (1991) Gene organisation of oryzacystatin-II, a new cystatin superfamily member of plant origin, is closely related to that of oryzacystatin-I but different from those of animal cystatins, *FEBS Letters* **278**, 87-90.
31. Kondo, H., Abe, K., Emori, Y., and Arai, S. (1991) Gene organization of oryzacystatin-II, a new cystatin superfamily member of plant origin, is closely related to that of oryzacystatin-I but different from those of animal cystatins, *FEBS Lett* **278**, 87-90.
32. Koritsas, V.M. and Atkinson, H.J. (1994) Proteinases of females of the phytoparasite *Globodera pallida*, *Parasitology* **109**, 1-9.
33. Lambert, K.N. and Williamson, V.M. (1993) cDNA library construction from small amounts of RNA using paramagnetic beads and PCR, *Nucleic Acids Research* **3**, 775-776.
34. Lilley, C.J., Urwin, P.E., McPherson, M.J. and Atkinson, H.J. (1996) Characterisation of intestinally active proteinases of cyst-nematodes, *Parasitology* (in press).
35. Lindsey, K., Wei, W., Clarke, M.C., McArdle, H.F., Rooke, L.M. and Topping, J.F. (1993) Tagging genomic sequences that direct gene expression by activation of a promoter trap in plants, *Transgenic Research* **2**, 33-47.
36. Logemann, J. and Schell, J. (1989) Nucleotide sequence and regulated expression of a wound-inducible potato gene, *Molecular and General Genetics* **219**, 81-88.
37. Marban-Mendoza, N., Jeyaprakash, A., Jansson, H.-B., Damon, Jr., R.A. and Zuckerman, B.M. (1987) Control of root-knot nematodes on tomato by lectins, *J. Nematol.* **19**, 331-335.
38. Mariani, C., Beuckeleer, M., Truettner, J., Leemans, J. and Goldberg, R.B. (1990) Induction of male sterility in plants by a chimaeric ribonuclease gene, *Nature* **347**, 737-741.
39. Mariani, C., Gossele, V., Beuckeleer, M., De Block, M., Goldberg, R.B., De Greef, W. and Leemans, J. (1992) A chimaeric ribonuclease-inhibitor gene restores fertility to male sterile plants, *Nature* **357**, 384-387.
40. Opperman, C.H., Taylor, C.G. and Conkling, M.A. (1994) Root-knot nematode-directed expression of a plant root-specific gene, *Science* **263**, 221-223.
41. Ray, C. and McKerrow, J.H. (1992) Gut-specific and developmental expression of a *Caenorhabditis elegans* cysteine protease gene, *Molecular and Biochemical Parasitology* **51**, 239-49.
42. Rice, S.L., Leadbeater, B.S.C and Stone, A.R. (1985) Changes in cell structure in roots of resistant potatoes parasitised by potato cyst-nematodes I. Potatoes with resistance gene H1 derived from *Solanum tuberosum* ssp. *andigena*, *Physiol. Plant Pathol.* **27**, 219-234.
43. Richardson, M. (1991) Seed storage proteins: the enzyme inhibitors, *Methods in Plant Biochemistry* **5**, 259-305.
44. Roberts, P.A. (1992). Current Status of the Availability, Development and Use of Host Plant Resistance to Nematodes, *J. Nematology* **24**, 213-227.
45. Ryan, C.A. (1990) Protease inhibitors in plants: Genes for improving defenses against insects and pathogens, *Annual Review of Phytopathology* **28**, 425-49.
46. Sasser, J.N. and Freckman, D.W. (1987) A world perspective on nematology: the role of the society, in

J.A.Veech and D.W.Dickerson (eds.), *Vistas on Nematology*, Society of Nematologists, pp.7-14.
47. Sijmons, P. C., Atkinson, H. J. and Wyss, U. (1994) Parasitic strategies of root nematodes and associated host cell responses. *Annual Review of Phytopathology* **32**, 235-259.
48. Stubbs, M.T., Laber, B., Bode, W., Huber, R., Jerala, R., Lenarcic, B. and Turk, V. (1990) The refined 2.4 Å X-ray crystal structure of recombinant human stefin B in complex with the cysteine protease papain; a novel type of protease inhibitor interaction, *EMBO J.* **9,** 1939-1947.
49. Trudgill, D.L. (1991) Resistance to and tolerance of plant parasitic nematodes in plants, *Annual Review of Phytopathology* **29**, 167-192.
50. Urwin, P.E., Atkinson, H.J., Waller, D.A. and McPherson, M.J. (1995a) Engineered oryzacystatin-I expressed in transgenic hairy roots confers resistance to *Globodera pallida*, *Plant J.* **8**, 121-131.
51. Urwin, P.E., Atkinson, H.J., and McPherson, M.J. (1995b) Involvement of the N-terminal region of oryzacystatin-I in cysteine proteinase inhibition *Protein Engineering* **8**, 1303-1307.
52. Urwin, P.E., Moller, S.G., Lilley, C.J., Atkinson, H.J. and McPherson, M.J. (1996a) Continual GFP monitoring of CaMV35S promoter activity in nematode-induced feeding cells in *Arabidopsis thaliana* (*submitted*).
53. Urwin, P.E., Lilley, C.J., McPherson, M.J. and Atkinson, H.J. (1996b) Resistance to both cyst- and root-knot nematodes conferred by transgenic *Arabidopsis* expressing a modified plant cystatin (*submitted*).
54. Van der Eycken, W., Engler, J.D., Inze, D., Van Montagu, M. and Gheysen, G. (1996) Molecular study of root-knot nematode-induced feeding sites, *Plant J.* **9**, 45-54.
55. Yamamoto, Y.T., Taylor, C.G., Acedo, G.N., Cheng, C-L. and Conkling, M.A. (1991) Characterisation of *cis*-acting sequences regulating root-specific gene expression in tobacco, *Plant Cell* **3**, 371-382.
56. Zuckerman, B.M. and Jansson, H.-B. (1984) Nematode chemotaxis and possible mechanisms of host/prey interactions, *Ann. Rev. Phytopathol.* **22**, 95-113.

ANTI-FEEDING STRUCTURE APPROACHES TO NEMATODE RESISTANCE

Stephan A. OHL[1], Frédérique M. van der LEE[1], Peter C. SIJMONS[2]
[1]*MOGEN International nv Einsteinweg 97, 2333 CB Leiden, The Netherlands*
[2]*Agrotechnological Research Institute (ATO-DLO), P.O. Box 17, 6700 Wageningen, The Netherlands*

Abstract

Sedentary nematodes are a very successful and economically damaging group of plant parasites found throughout the world in modern agriculture. Since agrochemical nematode control measures are viewed as increasingly problematic due to their high toxicity for environment and farmers and since other conventional methods such as crop rotation and breeding for natural resistance are not considered effective long term solutions there is a growing demand for novel molecular approaches to engineering a more durable and broad range resistance to nematodes. This chapter discusses resistance strategies directed against nematode feeding structures. One of the most advanced approaches uses a two component system consisting of the cytotoxic RNase barnase and its specific inhibitor barstar whereby barnase expression is directed towards the feeding structures and barstar protects the plant from aberrant barnase activity in uninfected tissues.

1. Introduction

Plant-parasitic nematodes are an important factor causing damage to modern agriculture throughout the world. Their impact on yield losses has been estimated to over US$ 75 billion annually [39]. Current nematode control measurements include application of nematicides and soil fumigants, crop rotation schemes and the introduction into crops of natural resistance traits by conventional breeding. However, almost all nematicidal agrochemicals are very toxic to the environment, crop rotation restricts farmers in often difficult agronomical situations and offers very few choices against some broad host range nematodes. Natural resistances tend to break down due to their high pathogen-specificity which provides no protection against new virulent nematode populations. These limitations create a strong demand for the novel molecular resistance strategies described in this book.

Most plant-parasitic nematodes feed on root tissue and damage their host mainly by inhibiting root growth resulting in reduced water uptake, and by promoting microbial infections through wound sites or by serving as vectors for pathogenic viruses. Depending on their parasitic adaptations they are classified in sedentary or migratory endo- or ectoparasites. The most advanced and successful group is formed by the sedentary endoparasitic nematodes, particularly root-knot and cyst nematodes.

Sedentary nematodes have developed a very complex mode of parasitism which requires a prolonged interaction with their host plants. The two most characteristic features that evolved as a result of this interaction are functionally related. After a short migratory phase in the plant root the infectious juveniles select a feeding site and manipulate via an unknown inducer a few cells at this site to redifferentiate into a specialized sink organ (syncytium, giant cells) which is able to meet the nematode's nutrient requirements until reproduction is accomplished. This adaptation allows the sedentary nematodes to degenerate their locomotive apparatus and become immobile during feeding. However, sedentary endoparasites may be particularly exposed to the plant's defense system and a crucial aspect of their strategy must be to either avoid plant recognition or to somehow counteract the defense response. In developing strategies for the genetic engineering of nematode resistance one might learn from the natural resistance responses.

2. Natural Resistance

Genetic resistances have been identified in many plant species [35]. Some of these resistance traits are conferred by major dominant genes. Well known examples include the Mi gene from tomato conferring resistance to *M. incognita* (see Liharska and Williamson, this volume) and the H1 gene from potato conferring resistance to some *G. rostochiensis* pathotypes. The fact that a single plant gene confers resistance to a small range of nematode pathotypes suggests that resistance (R) genes are involved in a specific recognition process which triggers a local defense response. Resistance reactions are frequently accompanied by the hypersensitive death of cells in the immediate vicinity of the nematode during migration or after establishment of a feeding site [9, 13, 34, 36]. These hypersensitive reactions (HR) are a special manifestation of apoptosis, a regulated suicide program on the cellular level which is also found during normal development of most organisms [14, 18]. Due to the difficulties in performing genetic studies on plant-parasitic nematodes which relates to their obligate parasitism, and the parthenogenetic propagation of some *Meloidogyne* species, nematode (a)virulence factor(s) have not been identified yet and a true gene-for-gene relationship has only been established for the incompatible interaction between *Globodera rostochiensis* and potato containing the H1 resistance gene [17].

Despite the frequent occurrence of resistance traits it is evident that particularly root-knot nematodes are extremely successful in infecting an enormous number of plant species. Videomicroscopic studies of the infection process in living root tissue show that they are very careful to minimize damage during their complex intercellular migration. This behaviour certainly helps the nematodes to evade plant recognition. In addition, sedentary nematodes may have developed mechanisms to actively suppress the plant defense response. Some experimental evidence suggests that this is indeed the case. The expression of several pathogenesis-related (PR) proteins was found to be induced in leaves of potato plants infected with *Globodera* but not in the roots of these plants [33]. The expression of phenylalanine ammonia lyase (PAL), a key enzyme of phenylpropanoid metabolism leading to the biosynthesis of defense-related phenolics in many plants, was found to be down-regulated at the infection site in compatible interactions [12] but upregulated in incompatible interactions [8]. Further evidence that nematodes are able to modify plant defense responses comes from split root experiments with successive inoculations of virulent and avirulent nematodes [30]. Inoculation of one half of the root with a virulent population resulted in decreased resistance to a normally avirulent population which is able to successfully infect the second half of the root a few days later. An avirulent population, on the other hand, can increase the plant's resistance to a subsequent inoculation with a virulent population resulting in significantly reduced infection levels. The latter phenomenon is comparable to the 'systemic acquired resistance' (SAR) described for other plant-pathogen interactions [37]. These findings suggest that strategies which attempt to trigger or enhance the plant's endogenous defense system and which have been proposed for resistance against fungi [7, 22] may not work against nematodes.

The use of resistance traits in plant breeding has shown that the agricultural benefits are very often limited in time as new virulent pathotype populations emerge. The goal of the novel approaches described below is to engineer a resistance system which is independent of the plants endogenous defense response. Obviously the feeding structure is a suitable target to interfere with nematode development as exemplified by natural resistance responses. Finding the right trigger to induce the system by a broad range of nematodes will be crucial for success.

3. Feeding Structures

Nematode-induced feeding structures are complex specialized organs the development and structure of which is described in more detail in the chapters written by Bleve-Zacheo and Golinowski. Cyst nematodes start feeding on a single parenchyma cell which soon begins to fuse with surrounding cells as a result of partial cell wall dissolution. In this way a multinucleate unicellular structure, a syncytium, is formed. Root-knot nematodes in contrast feed alternating on several parenchyma cells in the developing vascular cylinder. These cells become enlarged and multinucleate through mitoses in the absence of cytokinesis, and are called giant cells [19]. Mitotic divisions of cortical cells give the feeding structures of root-knot nematodes their characteristic gall shape. Despite their contrasting ontogeny syncytia and giant cells share a number of morphological characteristics which reflect functional similarities and classify them as transfer cells [19]. An electron-dense cytoplasm containing many organelles and enlarged nuclei indicate high metabolic activity. Cell wall invaginations increase the cell surface thus facilitating the exchange of nutrients. The nematodes withdraw large amounts of water and solutes from the cytoplasm during their periodic feedings which the feeding cells replenish from the vascular system.

It is quite obvious that the transformation of root cells into feeding structures and the new function of providing large amounts of nutrients requires changes in the expression of a large number of genes. To identify and characterize these genes is crucial for our understanding of feeding structure development and function on a molecular level. However, the very small size of feeding structures relative to the rest of the root tissue has long hampered the identification of genes involved in this process. Recent technical advances allow to make PCR-based cDNA libraries for differential screening [2, 15, 48] or to detect differences in gene expression by differential display.

In this way a number of genes have been reported to be upregulated in or close to feeding structures. Gurr and coworkers [15] reported the identification by differential cDNA screening of a potato gene which is specifically upregulated in syncytia after PCN infection. The same approach led to the identification of *Lemmi9*, a tomato gene which is expressed in most healthy plant tissues and up-regulated in giant cells after *Meloidogyne incognita* infection. This gene is homologous to the cotton late embryogenesis abundant gene *lea14-A* and could have a function in osmoprotection of giant cells [44].

Another class of genes which was reported to be upregulated in galls of root-knot nematodes encodes extensins [29, 44]. In situ hybridization in tomato galls showed expression mainly in the gall tissue early in infection and predominantly in giant cells 4 weeks after inoculation . Extensins play a role in growing tissues to increase cell wall rigidity and may help the gall cells to counteract the high turgor pressure inside the giant cells.

Another interesting gene which appears to be specifically up-regulated in giant cells is *tobRB7*. This root-specific gene from tobacco encodes a water channel protein. In healthy root tissue the protein accumulates in the meristem and in the plasma membrane of young endodermis [31, 49]. Since the casparian strip seals the extracellular space between the root cortex and the vascular cylinder radial water transport in the root has to occur via the cytoplasm of the endodermis and presumably these water channel proteins facilitate water flow across the plasma membrane of both endodermis and giant cells. Surprisingly the *tobRB7* gene is not induced in syncytia.

The expression of known genes can be analyzed by fusing their promoter to a reporter gene such as *gus* (encoding ß-glucuronidase). In this way *hmgr*, encoding a key enzyme of isoprenoid metabolism, was found to be up-regulated in tobacco giant cells [46]. The function of this increased gene activity is not very clear but a possible explanation could be nematodes requirement for sterols which they cannot produce themselves [3].

Using promoter-*uid*A fusions (*uid*A is the gene encoding ß-glucuronidase - GUS) also the cell cycle genes *cdc2a* and *cyc1At* from Arabidopsis were found to be strongly induced in the early stages of syncytium and giant cell development [28].

Not all of the differentially expressed genes are up-regulated in feeding structures. Differential screening of cDNA libraries [44] and analyzing a variety of plant, bacterial and viral promoters using *uid*A fusion constructs in cyst and root-knot nematode feeding structures revealed that several genes including some which encode enzymes of the primary (malate synthase, plasma membrane H+ ATPase) and secondary plant metabolism (PAL), but also *Agrobacterium* genes and the cauliflower mosaic virus 35S gene, are downregulated in syncytia and giant cells [12].

Screening of mutants for aberrant phenotypes followed by a molecular characterization of the mutant locus is another approach to identify novel genes, which has proven extremely successful in developmental biology [20] and plant pathogenesis [5]. However, screening of Arabidopsis-mutagenized seeds has not resulted in the identification of mutants with altered susceptibility to *H. schachtii* infection (Grundler, personal communication).

4. Anti-Feeding Structure Approaches

Molecular resistance strategies against plant parasitic nematodes can be divided into two classes depending on their immediate target. Anti-nematode strategies aim to directly interfere with the pathogen by expressing a nematicidal gene product and are discussed in the chapters by P. Burrows and M.J. McPherson. Anti-feeding structure strategies are specific for sedentary nematodes and aim to destroy the nematode's nutrient supply. The main difficulty of this type of strategy is its requirement to interact with a modified plant tissue in a highly specific manner, its main advantage may lie in a high durability since nematodes cannot influence the resistance-conferring process directly.

For a genetic engineer there are several principal ways to selectively destroy a specific plant tissue or cell type using a transgenic approach. Probably the most straight forward one is the tissue-specific expression of a toxic protein. Another approach is to down-regulate a component necessary for the development or maintenance of this tissue by anti-sense inhibition or cosuppression. Still another option would be to activate the plant's internal capacity for 'programmed cell death' or apoptosis, which occurs during normal development and associated with environmental stress [38] and pathogen attack [18, 26, 45].

4.1. THE APOPTOSIS APPROACH

Recently apoptosis has gained considerable attention in plant physiology. Apopotosis leads to a regulated ordered dying of cells with nuclear DNA fragmentation by a Ca^{2+} activated endonuclease and the formation of apopoptotic bodies as characteristic hallmarks of this process. Indicators of this process in animals are the accumulation of 180bp DNA multimers which appear as a ladder after gel electrophoresis. The 3' hydroxyls of these oligonucleosomal fragments can be detected by in situ labelling with UTP conjugated to a radioactive or fluorescent marker.

The characteristic DNA fragments and the formation of apoptotic bodies, which are membrane-bound vesicles containing these fragments, were found during embryogenesis, xylem differentiation and the development of unisexual flowers [18], as well as after treatment of plants with KCN [38] and associated with plant defense against fungal infection, where it is called HR [18, 38, 45]. In tomato this pathway can be activated by the host-selective AAL toxins secreted by *Alternaria alternata* [48] and the transgenic expression of a bacterial proton pump [26].

To date a transgenic approach to nematode resistance by triggering apoptosis selectively in nematode-induced feeding structures has not been reported yet but as our understanding of the signals and components in this process increases such an approach may become feasible in the near future. One way to induce an HR is via expression of a natural resistance gene in combination with an avirulence factor. Several plant resistance genes against microbial attack and some of their corresponding avirulence genes have been cloned [4]. Combined expression in a tomato cell of the resistance gene *cf9* and the corresponding avirulence gene *avr9* from the fungus *Cladosporium fulvum* induce a hypersensitive response. De Wit suggested that the combined expression of these two factors can be used in a resistance strategy against fungi [6]. The specificity of such an engineered resistance system thus resides entirely in the promoters used to drive expression of the two components. At least one of these promoters should be specific for the target tissue. Of course the system will require the presence in the target crop of a signal transduction chain responding to the inducers and leading to apoptosis. With the cloning of new resistance genes and their transfer to more distantly related plant species we will soon know more about how conserved the signalling pathways leading to an HR are and consequently how broadly applicable such an approach would be. As one of the first examples for the functioning of resistance genes in heterologous systems the tobacco N gene conferring resistance to tobacco mosaic virus (TMV) infection has been transferred to tomato and found to be functional [47].

4.2. THE GENE SILENCING APPROACH

Gene silencing, often referred to as 'antisense inhibition' or 'cosuppression', is a technology which utilizes the fact that transformation of a plant cell with several copies of a transgene which is homologous to an endogenous gene can cause a dramatic reduction in the transcript accumulation of both the foreign and the endogenous gene [25]. Although this process is not yet fully understood it appears to involve the direct or indirect interaction of at least two highly homologous non-allelic DNA sequences. Crossing experiments showed that silent loci usually suppress non-silent homologous loci, indicating that silencing is a dominant property. Transgene-mediated silencing is a stochastic process the frequency of which appears to depend on the locus and the copy number of transgenes present in the genome. Transcription of the homologous repeats appears to be required for silencing but not translation, since it is not influenced by stop codons or translation inhibitors that prevent protein synthesis. As some silencing events manifest themselves at the transcriptional and others at the post-transcriptional level there may be more than one mechanism involved. In many cases local changes in DNA methylation were found associated with silencing. An attractive hypothesis suggests a defense mechanism against viruses as the biological role for this process [10, 27].

Although the molecular mechanism of gene silencing has not been elucidated yet the phenomenon has been applied frequently in genetic engineering. In fact among the very first transgenic plant products reaching the market are tomatoes engineered for delayed fruit ripening via suppression of polygalacturonase [42].

Down-regulating the expression of a gene with an important function in feeding structure development or maintenance may prevent the nematodes from reaching maturity, strongly reduce fertility or cause a significant shift in the sex ratio of some promoter fragment. Surprisingly, both promoters did not confer detectable GUS expression in all syncytia of a root system. The percentage of GUS positive syncytia appeared to be somewhat dependent on the infection stage with the highest percentage (70%) around 7-10 days after inoculation with J2 larvae of *H. schachtii*. We do not know yet if this variability in the GUS staining is due to quantitative differences in GUS activation in different syncytia or results from transient promoter activation at variable times after inoculation in an unsynchronised infection system.

None of the two characterized promoters nor sequences flanking the other side of the T-DNA insertions showed homology to entries in the EMBL/GenBank database. We also could not find larger open reading frames close to the T-DNA insertions. Therefore it is still unknown if functional genes have been tagged in the two lines characterized in more detail. The fact that both promoters are induced in feeding structures of both cyst and root-knot nematodes suggests that a number of genes may be required for the development or maintenance of both of these morphologically distinct but functionally related cell types. For engineering purposes these promoters offer the potential to develop a system which confers a very broad type of resistance to both groups of sedentary nematodes. Moreover, since genes which are controlled by promoters such as #1164 or #25.1 are likely to play a basic role in feeding structure development or maintenance, this type of resistance should be very durable.

4.3.4. *Increased resistance to H. schachtii in Arabidopsis transformed with barnase/barstar constructs*

Promoter #1164 was cloned in front of barnase in a binary construct containing also a barstar gene linked to the rolD promoter. This construct was transformed to *Arabidopsis*. The resulting 106 lines were analyzed in the next generation in three separate experiments. Four weeks old plants (15 per line) grown on soil in translucent plastic tubes were inoculated with J2 larvae of *H. schachtii* and the number of female nematodes visible on the surface of the root systems was determined 4 weeks after inoculation. For the untransformed controls and the majority of the transgenic lines an average of 50-60 females per root were counted (with considerable variation between lines and even controls). However, in several lines infection levels were reduced by up to 70% compared to the wt controls.

A difficulty in this system is that due to the high toxicity of free barnase transgene expression levels cannot be measured directly and compared to the resistance data. However, an observation which may relate to the nematode-induced activation of barnase in these plants was a slight reduction (10-20%) in root development in some lines which appeared to be correlated with reduced infection levels but which was too small to account for the reduced number of cysts measured in these lines. This root inhibiting effect may be caused by nematode-induced barnase activation at infection sites but more experiments will be required to confirm this hypothesis.

5. Conclusions

Increasing environmental concerns and regulatory restrictions cause a demand for novel strategies to nematode resistance. Natural defense responses demonstrate that nematode feeding structures are an effective target for engineering resistance. The most straight forward approach aims to express a toxic gene product in the target tissue. A potential draw back of resistance strategies directed against feeding structures is their requirement for a highly specific promoter. One way to circumvent the problem of negative side effects caused by leaky expression of the toxin in healthy tissue is to include an inhibitor in the system which neutralizes low levels of toxin up to a threshold. The barnase/barstar system, which offers these features, is currently developed for the engineering of nematode-resistant plants. First results indicate that the system is able to confer a measurable reduction in infection levels in phenotypically normal plants but further improvements will be necessary to obtain levels of resistance which allow an effective nematode control.

ACKNOWLEDGEMENT

We would like to thank J. Klap for nematode resistance assays and O. Goddijn for critically reading the manuscript. Some of the work described was funded in part by the dutch company AVEBE.

References

1. Atkinson, H.J., Urwin, P.E., Hansen, E. and McPherson, M.J. (1995) Designs for engineered resistance to root-parasitic nematodes, *Trends in Biotechnology* **13**, 369-374.
2. Bird, D.M. and Wilson, M.A. (1994) DNA sequence and expression analysis of root-knot nematode-elicited giant cell transcripts, *Molecular Plant-Microbe Interactions* **7**, 419-424.
3. Chitwood, D.J. and Lusby, W.R. (1991) Metabolism of plant sterols by nematodes, *Lipids* **26**, 619-627.
4. Dangl, J.L. (1995) Piece de resistance: Novel classes of plant disease resistance genes. *Cell* **80**, 363-366.
5. Delaney, T.P., Friedrich, L., Ryals, J.A. (1995) Arabidopsis signal transduction mutant defective in chemically and biologically induced disease resistance. *Proc. Natl. Acad. Sci. USA* **92**, 6602-6606.
6. De Wit, P. J. G.M. (1995) Fungal avirulence genes and plant resistance genes: Unraveling the molecular basis of gene-for-gene interactions. in: Andrews, J. H., Tommerup, I. C. Advances in Botanical Research, Vol 21. : 147-185. Academic Press, London
7. Dong, X.N., Mindrinos, M., Davis, K.R., Ausubel, F.M. (1991) Induction of *Arabidopsis* defense genes by virulent and avirulent *Pseudomonas syringae* strains and by a cloned avirulence gene, *Plant Cell.* **3**, 61-72.
8. Edens, R. M., Anand, S. C., Bolla, R. I. (1995) Enzymes of the phenylpropanoid pathway in soybean infected with *Meloidogyne incognita* or *Heterodera glycines*. *J. Nematology* **27**, 292-303.
9. Endo, B.Y. (1991) Ultrastructure of initial responses of susceptible and resistant soybean roots to infection by *Heterodera glycines*, *Rev. Nématol.* **14**, 73-94.
10. English, J.J., Mueller, E. and Baulcombe, D.C. (1996) Suppression of virus accumulation in transgenic plants exhibiting silencing of nuclear genes, *Plant Cell* **8**, 179-188.
11. Gheysen, G. and van Montagu, M. (1995) Invited paper: Plant/nematode interactions, a molecular biologist's approach, *Nematologica* **41**, 366-384.
12. Goddijn, O.J.M., Lindsey, K., van der Lee, F.M., Klap, J.C. and Sijmons, P.C. (1993) Differential gene expression in nematode-induced feeding structures of transgenic plants harbouring promoter-*gusA*

fusion constructs, *Plant J.* **4**, 863-873.
13. Golinowski, W. and Magnusson, C. (1991) Tissue response induced by *Heterodera schachtii* (Nematoda) in susceptible and resistant white mustard cultivars, *Can. J. Bot.* **69**, 53-62.
14. Greenberg, J. T., Guo, A. L., Klessig, D. F., Ausubel, F. M. (1994) Programmed cell death in plants: a pathogen-triggered response activated coordinately with multiple defense functions, *Cell* **77**, 551-563.
15. Gurr, S.J., McPherson, M.J., Scollan, C., Atkinson, H.J. and Bowles, D.J. (1991) Gene expression in nematode-infected plant roots, *Molecular and General Genetics* **226**, 361-366.
16. Hartley, R.W. (1988) Barnase and barstar. Expression of its cloned inhibitor permits expression of a cloned ribonuclease, *J. Molecular Biology* **202**, 913-915.
17. Janssen, R., Bakker, J., and Gommers, F. (1991) Mendelian proof for a gene-for-gene relationship between virulence of *Globodera rostochiensis* and the H1 resistance gene in *Solanum tuberosum* spp. andigena CPC 1673, *Revue Nematol.* **14**, 213-219.
18. Jones, A.M. and Dangl, J.L. (1996) Longjam at the Styx: programmed cell death in plants, *Trends Plant Sci.* **1**, 114-119.
19. Jones, M.G.K. and Northcote, D.H. (1972) Nematode-induced syncytium —a multinucleate transfer cell. *J. Cell Sci.* **10**, 789-809.
20. Jurgens, G. (1995) Axis formation in plant embryogenesis: Cues and clues, *Cell* **81**, 467-470.
21. Koltunow, A.M., Truettner, J., Cox, K.H., Wallroth, M. and Goldberg, R.B. (1990) Different temporal and spatial gene expression patterns occur during anther development, *Plant Cell* **2**, 1201-1224.
22. Lawton, K.A., Friedrich, L., Hunt, M., Weymann, K., Delaney, T., Kessmann, H., Staub, T. and Ryals, J. (1996) Benzothiadiazole induces disease resistance in Arabidopsis by activation of the systemic acquired resistance signal transduction pathway, *Plant J.* **10**, 71-82.
23. Mariani, C., Beuckeleer, M., Truettner, J., Leemans, J. and Goldberg, R.B. (1990) Induction of male sterility in plants by a chimaeric ribonuclease gene, *Nature* **347**, 737-741.
24. Mariani, C., Gossele, V., Beuckeleer, M., De Block, M., Goldberg, R.B., De Greef, W. and Leemans, J. (1992) A chimaeric ribonuclease-inhibitor gene restores fertility to male sterile plants, *Nature* **357**, 384-387.
25. Meyer, P. and Saedler, H. (1996) Homology-dependent gene silencing in plants. *Ann. Rev. Plant Physiol. Plant Mol. Biol.* **47**, 23-48.
26. Mittler, R., Shulaev, V. and Lam, E. (1995) Coordinated activation of programmed cell death and defense mechanisms in transgenic tobacco plants expressing a bacterial proton pump, *Plant Cell* **7**, 29-42.
27. Mueller, E., Gilbert, J., Davenport, G., Brigneti, G. and Baulcombe, D.C. (1995) Homology-dependent resistance: Transgenic virus resistance in plants related to homology-dependent gene silencing, *Plant J.* **7**, 1001-1013.
28. Niebel, A. (1994) Molecular and genetic approaches to plant-nematode interactions, *Ph.D. thesis, Gent, Universiteit Gent.*
29. Niebel, A., de Almeida Engler, J., Tiré, C., Engler, G., Van Montagu, M. and Gheysen, G. (1993) Induction pattern of an extensin gene in tobacco upon nematode infection, *Plant Cell* **5**, 1697-1710.
30. Ogallo, J. L. and McClure, M. A. (1996) Systemic acquired resistance and susceptibility to root-knot nematodes in tomato, *Phytopathology* **86**, 498-501.
31. Opperman, C.H., Taylor, C.G. and Conkling, M.A. (1994) Root-knot nematode-directed expression of a plant root-specific gene, *Science* **263**, 221-223.
32. Opperman, C.H. and Conkling, M.A. (1996) Analysis of transgenic root-knot nematode resistant tobacco, *Abstract, Third International Nematology Congress, Guadeloupe.*
33. Rahimi, S., Perry, R.N. and Wright, D.J. (1996) Identification of pathogenesis-related proteins induced in leaves of potato plants infected with potato cyst nematodes, *Globodera* species, *Physiol. Mol. Plant Pathol.* **49**, 49-59.
34. Rice, S.L., Leadbeater, B.S.C and Stone, A.R. (1985) Changes in cell structure in roots of resistant potatoes parasitised by potato cyst-nematodes I. Potatoes with resistance gene H1 derived from *Solanum tuberosum* ssp. *andigena, Physiol. Plant Pathol.* **27**, 219-234.
35. Roberts, P.A. (1992). Current Status of the Availability, Development and Use of Host Plant Resistance to Nematodes, *J. Nematology* **24**, 213-227.
36. Robinson, M. P., Atkinson, H. J., and Perry, R. N. (1988) The association and partial characterisation of a fluorescent hypersensitive response of potato roots to the potato cyst nematodes *Globodera rostochiensis* and *G. pallida., Révue Nematol.* **11**, 99-107.

37. Ryals, J., Uknes, S. and Ward, E. (1994) Systemic Acquired Resistance, *Plant Physiology* **104**, 1109-1112.
38. Ryerson, D.E. and Heath, M.C. (1996) Cleavage of nuclear DNA into oligonucleosomal fragments during cell death induced by fungal infection or by abiotic treatments, *Plant Cell* **8**, 393-402.
39. Sasser, J.N. and Freckman, D.W. (1987) A world perspective on nematology: the role of the society, in J.A.Veech and D.W.Dickerson (eds.), *Vistas on Nematology*, Society of Nematologists, pp.7-14.
40. Sijmons, P. C., Atkinson, H. J. and Wyss, U. (1994) Parasitic strategies of root nematodes and associated host cell responses. *Annual Review of Phytopathology* **32**, 235-259.
41. Sijmons, P.C., Goddijn, O.J.M., van den Elzen, P.J.M. (1994) Plants with reduced susceptibility to plant-parasitic nematodes. Patent WO 94/10320.
42. Smith, C.J.S., Watson, C.F., Ray, J., Bird, C.R., Morris, P.C. et al. (1988) Antisense RNA inhibition of polygalacturonase gene expression in transgenic tomatoes, *Nature* **334**, 724-726.
43. Sobczak, M. (1996) Investigations on the structure of syncytia in roots of *Arabidopsis thaliana* induced by the beet cyst nematode *Heterodera schachtii* and its relevance to the sex of the nematode. PhD Thesis, University of Kiel.
44. Van der Eycken, W., Engler, J.D., Inze, D., Van Montagu, M. and Gheysen, G. (1996) Molecular study of root-knot nematode-induced feeding sites, *Plant J.* **9**, 45-54.
45. Wang, H., Li, J., Bostock, R.M. and Gilchrist, D.G. (1996) Apoptosis: a functional paradigm for programmed plant cell death induced by a host-selective phytotoxin and invoked during development, *Plant Cell* **8**, 375-391.
46. Weissenborn, D.L., Zhang, X., Eisenback, J.D., Radin, D.N. and Cramer, C.L. (1994) Induction of the tomato *hmg2* gene in response to endoparasitic nematodes. Abstract, *Proc. 4th Intl. Congress Plant Mol. Biol, Amsterdam.*
47. Whitham, S., Mccormick, S., Baker, B. (1996) The N gene of tobacco confers resistance to tobacco mosaic virus in transgenic tomato. *Proc. Natl. Acad. Sci. USA* **93**, 8776-8781.
48. Wilson, M.A., Bird, D.McK. and van der Knaap, E. (1994) A comprehensive substractive cDNA cloning approach to identify nematode-induced transcripts in tomato, *Phytopathology* **84**, 299-303.
49. Yamamoto, Y.T., Taylor, C.G., Acedo, G.N., Cheng, C-L. and Conkling, M.A. (1991) Characterisation of *cis*-acting sequences regulating root-specific gene expression in tobacco, *Plant Cell* **3**, 371-382.

TOWARDS PLANTIBODY-MEDIATED RESISTANCE AGAINST NEMATODES

Willem J. STIEKEMA[1], Dirk BOSCH[1], Annemiek WILMINK[1], Jan M. DE BOER[2], Alexander SCHOUTEN[2], Jan ROOSIEN[3], Aska GOVERSE[2], Gert SMANT[2], Jack STOKKERMANS[3], Fred J. GOMMERS[2], Arjen SCHOTS[3], Jaap BAKKER[2]
[1]Department of Molecular Biology, DLO-Centre for Plant Breeding and Reproduction Research (CPRO-DLO), P.O. Box 16, 6700 AA Wageningen, The Netherlands.
[2]Department of Nematology, Agricultural University, P.O. Box 8123, 6700 ES Wageningen, The Netherlands.
[3]Laboratory for Monoclonal Antibodies, P.O. Box 9060, 6700 GW Wageningen, The Netherlands.

Abstract

The lack of available resistance genes severely hampers the introduction of durable disease resistance in plants. Therefore we explored the feasibility to obtain resistance by the expression in plants of monoclonal antibodies. The rational behind this idea is that by binding to its antigen a monoclonal antibody is capable to inactivate the biological activity of that antigen. As a model system we have chosen the interaction between potato and potato cyst nematodes while saliva proteins of nematodes served as antigen. These proteins are thought to play an important role in this interaction. Inhibition of the biological activity of these proteins might interupt the interaction which results in resistance.

The isolation of genes coding for antibodies is facilitated by the application of the polymerase chain reaction (PCR). A small set of primers was designed which allowed the amplification of such genes. Subsequently genes which coded for full-length heavy and light chains of an antibody against nematodal saliva proteins were isolated and transferred to potato. Also genes encoding single-chain antibodies against the same antigen were constructed and transferred to potato. Analysis of transgenic potato plants showed that high level expression of such antibodies in roots is conceivable. These data will be discussed in the light of the introduction of nematode resistance in potato.

1. Introduction

The expression of functional antibodies in transgenic plants has been considered highly promising for plant disease control and manipulation of metabolic pathways [26]. This promise is fulfilled with reports on plant protection against virus infection [28, 29], aberrant phytochrome-dependent germination of plants [22] and a wilty phenotype [1] obtained through expression of antibody derivatives against viral coat protein, phytochrome and abscisic acid, respectively.

Our research focuses on the introduction of resistance against potato cyst nematodes in potato. These sedentary plant parasites feed from their hosts by exploiting a syncytium of metabolically active root cells [17]. This causes severe damage to potato roots that make potato cyst nematodes one of the most prominent pathogens of potato. Accordingly farmers use large amounts of nematicides which is a severe threat to the environment. The presence of resistance against cyst nematodes in potato should avoid the use of these detrimental agrochemicals. Most commercial potato varieties however lack resistance. Therefore we pursue the introduction of such a resistance by the expression in potato of antibodies against proteins of nematodes that are essential in the interaction between this pathogen and its host.

Antibodies are heteromultimeric proteins composed of two heavy and two light chains connected by disulfide bridges. Consequently, the production of functional antibodies in plants requires the proper oligomerization of antibody subunits. In addition proper expression of the genes encoding both antibody chains is a requisite. To promote antibody assembly and reduce the size of the molecule, antibodies are engineered to form a single-chain, in which the variable, antigen-binding domain of the light and the heavy chain are connected by a linker polypeptide [4]. However, the affinity of such a single-chain antibody (scFv) for the antigen is often lower than the parent antibody, which may be a serious drawback in applications requiring a high affinity.

To obtain transgenic plants producing antibodies, [14] initially expressed light and heavy chains in separate plants, which were subsequently crossed. However, this procedure is time-consuming. Co-transformation of the genes encoding both antibody chains is faster but suffers from a large proportion of low antibody expressors [8]. In addition, in both these approaches segregation will occur in the progeny if the two T-DNA's are not genetically linked. Introduction of heavy and light chain genes on a single T-DNA circumvents the aforementioned disadvantages, but puts additional demands on the promoter and terminator sequences employed. These sequences should preferably be different to avoid a potential lack of coordination of expression of the two transgenes [9] or silencing of transgene expression in the progenies [2].

Because we aim for resistance against root pathogens, antibodies have to be expressed in the roots of plants. On the basis of the above mentioned considerations, we decided to transfer whole antibody genes on a single T-DNA controlled by two different promoters that both have high activity in roots. To anticipate to possible problems inherent to the post translational modification of complete antibodies such as processing of the four subunits, intermolecular disulfide formation and glycosylation, we have also studied the secretion of scFv. In addition we studied the possibilities to expres scFv in the cytosol. As a model we first employed the monoclonal antibody 21C5 that was raised against a cutinase produced by the fungus *Botrytis cinerea* [25]. The expression of full-size antibodies and scFv was studied in tobacco. We also identified nematode proteins that most likely play a key role in the interaction between potato and potato cyst nematodes. Monoclonal antibodies were raised against these proteins, and the expression of one of these was studied in potato.

Our data showed that high level expression of both these antibodies is conceivable in roots of tobacco as well as potato. Both antibodies could be targeted to the apoplast and were shown to bind to their antigen. The expression of scFv against cutinase appeared to be dependent on the targeting signals present in the gene construct. Without any targeting signal, expression of scFv protein could not be detected *in planta*. However if provided with a signal sequence for translocation into the secretory pathway and an endoplasmic reticulum (ER) retention signal (KDEL), also scFv proteins are expressed to a high level in tobacco leaves. Surprisingly, the KDEL sequence had also a positive effect on a construct lacking a secretory signal sequence. These results will be discussed in relation to the introduction of nematode resistance in potato.

2. Results

2.1. EXPRESSION OF A MODEL ANTIBODY IN TOBACCO

2.1.1. *Cloning of full-size and single chain 21C5 antibody cDNAs*

The 21C5 monoclonal antibody employed in our research binds to a cutinase produced by *Botrytis cinerea* and is used as a model to study the feasibility of antibody expression in plants. The cDNA sequences encoding the full-size heavy and light chain of 21C5 were amplified by PCR using cDNA prepared from total 21C5 hybridoma mRNA as depicted in Figure 1. Specific 3'-primers were designed for the constant domains of light and heavy chains, C_L respectively C_{H3}. To increase the fidelity of 5'-end amplification by PCR eight different primers based on the 5'-end of variable domain sequences of both heavy and light chains as found in the Kabat database [18] were designed. The amplified 21C5 cDNAs were fused to a fragment encoding a slightly modified antibody signal sequence [11] and cloned into a binary plant expression vector. In the resulting construct a *35S* promoter with duplicated enhancer [19] directed the expression of the heavy chain, while the light chain expression is controlled by the *TR2'* promoter. This construct was used for *Agrobacterium tumefaciens*-mediated transformation of tobacco [11].

A scFv gene was assembled containing the variable domains of the 21C5 antibody heavy and light chain genes in the 5'-V_L-linker-V_H-3' orientation [27]. The end of the V_H region was fused to the c-myc tag coding sequence [20] for detection and purification purposes as depicted in Figure 1. To enable translocation of the 21C5 scFv to the lumen of the ER it was preceded by the same signal peptide as used for the full-size antibody gene constructs [11] while for retaining the scFv antibody in the ER a C-terminal retention signal KDEL [23] was added. Also two cytosolic versions of the scFv antibody with and without the KDEL extension were constructed which both lacked the ER translocation signal sequence. These constructs were cloned behind the doubled *CaMV*

35S promoter in a binary *Agrobacterium tumefaciens* vector for transformation of tobacco [11].

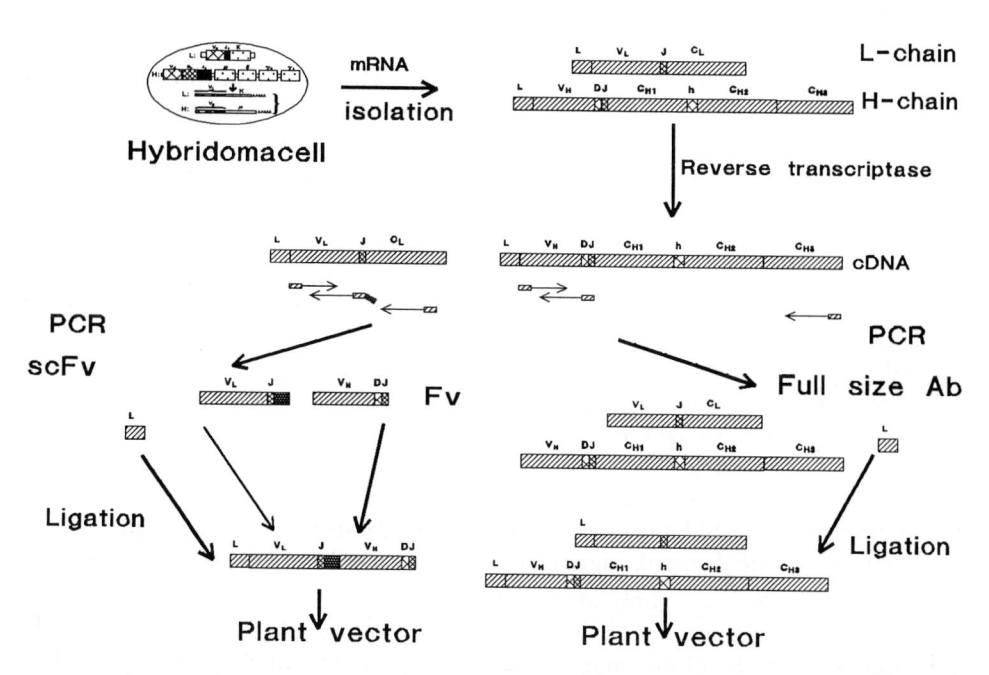

Figure 1. Schematic drawing of the cloning of full-length or single chain antibody genes from hybridoma cells. cDNA is synthesized on total hybridoma cell mRNA by reverse transcriptase using specific primers. For the construction of single chain antibody genes the variable domains of the light and heavy chains were amplified by the polymerase chain reaction. After ligation and addition of a translocation signal peptide encoding DNA fragment, the chimeric genes are cloned in a binary plant transformation vector and transferred to plants by *Agrobacterium tumefaciens*.

2.1.2. *Expression and functionality of 21C5 full-size and scFv antibody cDNAs in tobacco*

Immunoblotting of root extracts of a randomly selected set of axenically grown transgenic tobacco plants accommodating the full-size antibody gene construct revealed a protein of the size of a fully assembled antibody [11]. Expression levels of over 1% of total leaf protein were detected in independent transgenic plants. In addition to full-size antibodies another prominent protein band with an estimated size of 110 kDa was detected in the plants that was shown to be most likely an F(ab')$_2$ fragment, a dimer of two Fab fragments held together by disulfide bridges. Both these forms of the plant-formed antibodies appeared to be functional as binding of the antibodies to immobilized cutinase revealed that both specifically bound to the cutinase antigen [11].

Immunoblotting of total protein extracts of leaves of transgenic tobacco plants showed that the scFv construct accommodating the ER translocation signal sequence was only poorly expressed to level of 0.01% of total leaf protein [27]. Also in plants containing the scFv construct lacking the signal sequence, scFv protein could not be detected. In contrast 90% of the transgenic tobacco plants accommodating the scFv construct containing in addition to the ER translocation signal sequence, the C-terminal ER retention signal KDEL, were expressing the scFv protein: the highest expression level was 1% of total leaf protein. The scFv construct lacking the ER translocation signal sequence but containing the C-terminal KDEL sequence also showed a high expression level of up to 0.2% of total leaf protein. Analysis by an ELISA assay with immobilized cutinase showed that the scFv protein contained antigen binding capabilities. Immunoblotting experiments further showed that both the scFv construct with signal sequence and C-terminal KDEL sequence as well as the scFv construct only containing the C-terminal KDEL sequence resulted in the synthesis of scFv protein that was capable to specifically bind to the cutinase antigen.

2.1.3. *Targeting of complete and scFv 21C5 antibodies to different cell compartments in tobacco*

To study the secretion of full-size antibodies to the apoplast, protoplasts from leaves of a transgenic tobacco plant with a high antibody expression level were put to use [11]. Culturing of protoplasts and analysis of the culture medium as well as the cells by immunoblotting revealed that full-size antibodies accumulated in the culture medium with time, while they were not detectable inside the protoplasts. Thus, full-size antibodies were secreted by the tobacco protoplasts. Furthermore, mainly full-size (150 kDa) antibodies were detected, suggesting that the formation of the 110 kDa $F(ab')_2$-like fragment *in planta* occurs only after secretion and is most likely due to proteolytic processing in the apoplast. The fate of the antibodies after secretion was investigated in a cell suspension culture of the same tobacco plant. Both the full-size antibodies and the 110 kDa $F(ab')_2$-like fragments were observed in the culture medium whereas only full-size antibodies were in addition also associated with the cells in this culture. This indicates that full-size antibodies are partially retained by the cell walls while the $F(ab')_2$-like fragments can freely diffuse through the cell walls of the suspension cells.

Transient expression assays showed that the translocation signal sequence and the KDEL retention signal had the predicted effects on scFv protein translocation [27]. The scFv containing the ER translocation signal was secreted into the incubation medium suggesting that the signal peptide was indeed functional. The C-terminal KDEL extension directed scFv predominantly inside the protoplasts. This appeared also to be true *in planta* as shown in protoplasts prepared from transgenic plants expressing the scFv construct containing the C-terminal KDEL extension. Such plants only showed intracellular accumulation of the scFv protein. Thus the KDEL ER retention signal appears to be functional too.

The results obtained with constructs made for cytosolic expression are puzzling and need further investigation [27]. The KDEL ER retention signal had a positive effect on the expression levels of scFv constructs without a secretory signal sequence. It is well known, that a KDEL sequence increases the expression of constructs having a secretory sequence, but this has never been reported for cytosolic proteins.

2.2. EXPRESSION OF AN ANTIBODY RECOGNIZING A NEMATODE PROTEIN IN POTATO

2.2.1. *Selection of targets in nematodes to raise antibodies against*

Secretory products from the esophageal glands of sedentary plant parasitic nematodes are essential for the development. These secretions are thought to play a key role in the penetration and colonization of the roots [17, 15]. The production of these secretions takes place in two subventral gland cells and one dorsal gland cell. There are indications that some secretions from the dorsal gland are responsible for the formation of feeding tubes in the cytoplasm of the feeding cells [24, 30;16, 31]. The function of the subventral glands, however, is still unclear. They may be involved in the release of cell wall degrading enzymes during root invasion [32], in feeding site induction [3] , in mobilization of lipid reserves [31, 32] or in internal food digestion [6].

If subventral gland secretions are indeed involved in one of these processes they are of extreme importance for a successful completion of the lifecycle of sedentary plant parasitic nematodes. The inhibition of the function of these secretions either following their release in the apoplast or in the cytoplasm of the host cells, may arrest the growth and development of the juvenile nematodes.

To determine the role of subventral gland secretions we decided to raise monoclonal antibodies against subventral gland proteins of J_2 of the potato cyst nematode *Globodera rostochiensis* and express such monoclonal antibodies in potato roots. A classical way to study the function of a gene or protein, is to block its function. Binding of an antibody to its antigen often results in the inhibition of the function of the antigen [26], and the expression of these antibodies in potato might inhibit the penetration, growth or development of cyst nematodes. These experiment may disclose the essential role of subventral gland proteins, and in addition may reveal if antibodies raised against these proteins can be used to confer resistance. Because not all antibodies will inhibit the function of an antigen, these experiments have to be carried out with a panel of monoclonal antibodies. It is noted, that when raising monoclonal antibodies against a protein antigen, a relatively high proportion of the antibodies interfere with the function of an antigen. This proportion may range from 25 to 75%. A possible explanation is that the antigenicity of protein regions is determined by their accessibility, flexibility or both. Since the functional regions of a protein are among the most accessible and flexible parts of proteins, this might explain the high incidence of inhibiting monoclonal antibodies.

In addition we follow also an alternative approach to elucidate the function of the subventral glands. Antibodies against the subventral glands are used to purify the antigen by immunoaffinity chromatography and to determine the amino acid sequence of the NH_2-termini. These sequences are used to design primers and to clone the gene by RT-PCR. Comparison of the cloned genes with known genes may provide a clue for the function of the subventral glands and may lead to enzyme assays or other in vitro assays to select inhibiting monoclonal antibodies.

2.2.2. *Expression of full-size MGR 48 cDNAs in potato and the functionality of plant-encoded MGR 48 antibodies.*

To identify secretory proteins from the subventral gland of *Globodera rostochiensis* SDS-extracted proteins from J_2 were fractionated by preparative continuous flow electrophoresis as described by de Boer *et al.* [5]. Monoclonal antibodies were raised against the 38 to 40.5 kDa protein fraction and 12 monoclonal antibodies that bound specifically to the subventral gland were identified. One of these monoclonal antibodies, MGR 48, identified four major protein bands in the electrophoresis pattern of J_2 of *Globodera rostochiensis*. Most likely these proteins are structurally related.

Immunoelectron microscopy with MGR 48 showed intense labeling of secretory granules in the subventral gland cells of J_2 showing that one or more of the proteins are located within these cells [5].

The cDNA sequences encoding the full-size heavy and light chain of MGR 48 were amplified by PCR and fused to a fragment encoding a slightly modified antibody signal sequence as described before for the heavy and light chains of monoclonal antibody 21C5 [11]. Subsequently the construct was cloned into a binary plant expression vector in such a way that a 35S CaMV promoter with duplicated enhancer directed the expression of the heavy chain, while the light chain expression is controlled by the TR2' promoter. This construct was used for *Agrobacterium tumefaciens* transformation of potato [21].

Immunoblotting of root extracts of a randomly selected set of axenically grown potato plants accommodating the full-size antibody gene constructs revealed a protein with the size of a fully assembled antibody (data not shown) as found earlier for the 21C5 antibody construct [11]. Again expression levels of over 1% of total root protein were detected in independent transgenic potato plants (see Figure 2). In addition the antibodies formed in potato roots appeared to be functional because an ELISA assay showed that they bound to an immobilized nematode homogenate (data not shown).

3. Discussion

Antibody-mediated inactivation of antigens provides new possibilities for resistance breeding in plants. We have demonstrated high level expression of functional antibodies against two different antigens in roots of tobacco and potato plants. Controlled by the *CaMV 35S* and the *TR2'* promoters on a single T-DNA, expression of both light and heavy chain genes is coordinated and balanced giving rise to expression levels of around 1% of total soluble root protein in more than half of the transgenic plants obtained.

The proper targeting of antibodies will be crucial for their effectiveness against pathogens. Therefore we studied the secretion of antibodies in transgenic plants. Murine secretory signal sequences appear well suited to direct antibodies to the apoplast of plant cells. This was previously shown [13] and confirmed by our data that show the accumulation of fully assembled immunoglobulins in protoplast culture medium. Once in the apoplast, full-size antibodies appear to be subject to proteolytic processing as indicated by the occurrence of a 110 kDa F(ab')$_2$-like antibody fragment. Similar results have been obtained by de Neve *et al.* [8] and Hiatt *et al.* [14]. Although antibodies are

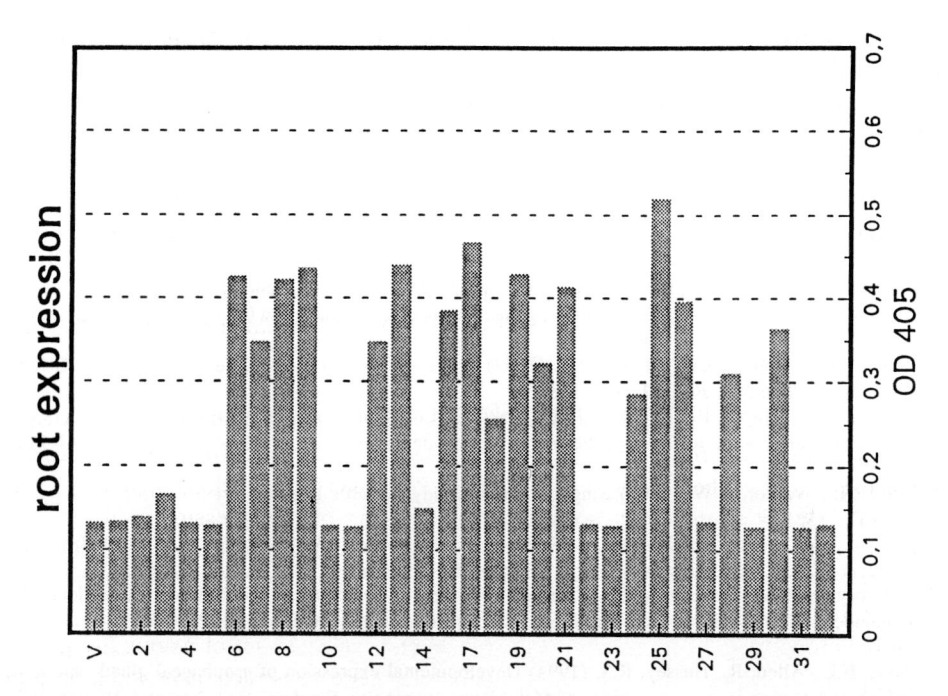

Figure 2. Expression of full-length monoclonal antibody MGR 48 in roots of a number of independent transgenic potato plants. The level of expression is depicted in OD 405 units as determined in root extracts by ELISA using alkaline phosphatase-conjugated sheep-anti-mouse IgG antibodies against F(ab')2 fragments. c: control roots; 1-32: independent transgenic potato plants.

susceptible to proteolytic processing, the products are functional and stably maintained in plant roots.

Also scFv antibodies can be targeted to particular subcellular compartments. Tavladoraki *et al.* [28] described the successful expression of an scFv directed to the cytosol while Firek *et al.* [12] showed a significant increase in the expression level of a scFv against phytochrome when secreted instead of expressed in the cytosol. In contrast we were not able to express the scFv construct to high levels in plants if it accommodated only the ER translocation signal [27]. This might reflect a difference in stability between the two scFv proteins applied, which may depend on the amino acid constitution. Our data also show that scFv can be expressed to a high level inside the ER by adding next to a translocation signal an ER retention signal. Enhanced expression of scFv antibodies in the cytosol seems possible by adding only the ER KDEL retention signal.

4. Conclusions

The possibility to express antibodies in potato roots may prove to be useful for the control of cyst nematodes. Assays are under way to test this hypothesis. We anticipate that resistance obtained in this way will be durable in the field because the reproduction capacity of potato cyst nematodes is too low to break resistance by mutation. This has also been shown in practice in Great Britain for the monogenic resistance locus H_1 present in *Solanum tuberosum* spp. *andigena*. This resistance against *G. rostochiensis* has been proven to be very effective and durable over the years.

References

1. Artsaenko, O., Peisker, M., zur Nieden, U., Fiedler U., Weiler, U. W., Muntz, K. and Conrad U. (1995) Expression of a single-chain Fv antibody against abscisic acid creates a wilty phenotype in transgenic tobacco. Plant Journal 8 745-750.
2. Assaad, F.F., Tucker, K.L., Signer, E.R. (1993) Epigenetic repeat-induced gene silencing (RIGS) in *Arabidopsis*, *Plant Mol. Biol.* **22**, 1067-1085.
3. Atkinson, H.J., Harris, P.D., Halk, E.J., Novitski, C., Leighton-Sands, J., Nolan, P., Fox, P.C. (1988) Monoclonal antibodies to the soya bean cyst nematode *Heterodera glycines*, *Ann. Appl. Biol.* **112**, 459-469.
4. Bird, R.E., Walker, B.W. (1991) Single chain antibody variable regions, *Trends Biotechnology* **9**, 132-137.
5. Boer de, J.M., Smant, G., Goverse, A., Davis, E.L., Overmars, H.A., Pomp, H., van Gent-Pelzer, M., Zilverentant, J.F., Stokkermans, J.P.W.G., Hussey, R.S., Gommers, F.J., Bakker, J., Schots, A. (1996) Secretory granule proteins from the subventral esophageal gland of the potato cyst nematode identified by monoclonal antibodies to a protein fraction from second-stage juveniles, *Mol. Plant Microbe Inter.* **9**, 39-46.
6. Davis, E.L., Allen, R., Hussey, R.S. (1994) Developmental expression of esophageal gland antigens and their detection in stylet secretions of *Meloidogyne incognita*, *Fundam. Appl. Nematol.* **17**, 255-262.
7. Boer de, J.M. (1996) Towards identification of oesophageal gland prteins in Globodera rostochiensis. Thesis Wageningen Agricultural University, 144 pp.
8. De Neve, M., Deloose, M., Jacobs, A., Vanhoudt, H., Kaluza, B., Weidle, U., van Montagu, M. Depicker, A. (1993) Assembly of an antibody and its derived antibody fragment in Nicotiana and Arabidopsis, *Transgen. Res.* **2**, 227-237.
9. Dean, C., Favreau, M., Tamaki, S., Bond-Nutter, D., Dunsmuir, P., Bedbrook, J. (1988) Expression of tandem gene fusions in transgenic tobacco plants, *Nucl. Acids Res.* **16**, 7601-7616.
10. During, K., Hippe, S., Kreuzaler, F., Schell, J. (1990) Synthesis and self-assembly of a functional monoclonal antibody in transgenic *Nicotiana tabacum*, *Plant Mol. Biol.* **15**, 281-293.
11. Engelen van, F.A., Schouten, A., Molthoff, J., Roosien, J., Salinas, J., Dirkse, W.G., Bakker, J., Gommers, F.J., Jongsma, M.A., Bosch, D., Stiekema, W.J. (1994) Coordinate expression of antibody subunit genes yields high levels of functional antibodies in roots of transgenic tobacco, *Plant Mol. Biol.* **26**, 1701-1710.
12. Firek, S., Draper, J., Owen, M.R.L., Gandecha, A., Cockburn, B., Whitelam, G.C. (1993) Secretion of a functional single-chain Fv protein in transgenic tobacco plants and cell suspension cultures, *Plant Mol. Biol.* **23**, 861-870.
13. Hein, M.B., Tang, Y., McLeod, D.A., Janda, K.D., Hiatt, A. (1991) Evaluation of immunoglobulins from plant cells, *Biotechnol. Prog.* **7**, 455-461.
14. Hiatt, A., Cafferkey, R., Bowdish, K. (1989) Production of antibodies in transgenic plants, *Nature* **342**, 76-78
15. Hussey, R.S. (1989) Disease-inducing secretions of plant-parasitic nematodes, *Ann. Rev. Phytopathol.* **27**, 123-141.
16. Hussey, R.S. and Mims, C.W. (1991) Ultrastructure of feeding tubes formed in giant-cells induced in plants by the root-knot nematode *Meloidogyne incognita*, *Protoplasma* **162**, 99-107.
17. Jones, M.G.K. (1981) Host cell response to endoparasitic nematode attack: structure and function of giant cells and syncytia, *Ann. Appl. Biol.* **97**, 353-372.
18. Kabat, E.A., Wu, T.T., Perry, H.M., Gottesman, K.S., Foeller, C. (1991) Sequences of proteins of immunological interest, 5th ed. US Dept. of Health and Human Services, Washington DC., USA.
19. Kay, R., Chan, A., Daly, M., McPherson, J. (1987) Duplication of CaMV 35S promoter sequences

creates a strong enhancer for plant genes, *Science* **236**, 1299-1302.
20. Munro, S. and Pelham, H.R.B. (1986) An Hsp-70-like protein in the ER: identity with the 78 kD glucose-regulated protein and immunoglobulin heavy chain binding protein, *Cell* **46**, 291-300.
21. Nap, J.P., van Spanje, M., Dirkse, W.G., Baarda, G., Mlynarova, L., Loonen, A., Grondhuis, P., Stiekema, W.J. (1993) Activity of the promotor of the potato *lhca3.St.1* gene, encoding the potato apoprotein 2 of the light-harvesting complex of Photosystem I, in transgenic potato and tobacco plants, *Plant Mol. Biol.* **23**, 605-612
22. Owen, M., Gandecha, A., Cockburn, B., Whitelam, G. (1992) Synthesis of a functional anti-phytochrome single-chain Fv protein in transgenic tobacco, *Biotechnology*, **10**, 790-794.
23. Pelham, H.R.B. (1989) Heat shock and the sorting of luminal ER proteins, *EMBO J.* **8**, 3171-3176.
24. Rumpenhorst, H.J. (1984) Intracellular feeding tubes associated with sedentary plant parasitic nematodes, *Nematologica*, **30**, 77-85.
25. Salinas, J. (1992) Function of cutinolytic enzymes in the infection of *Gerbera* flowers by *Botrytis cinerea*. Ph. D. thesis, University of Utrecht, Utrecht, The Netherlands.
26. Schots, A., de Boer, J., Schouten, A., Roosien, J., Zilverentant, J.F., Pomp, H., Bouwman-Smits, L., Overmars, H., Gommers, F.J., Visser, B., Stiekema, W.J., Bakker, J. (1992) Plantibodies: a flexible approach to design resistance against pathogens, *Neth. J. Pl. Path. suppl.* **2**, 183-191.
27. Schouten, A., Roosien, J., van Engelen, F.A., de Jong, G.A.M., Borst-Vrenssen, A.W.M., Zilverentant, J.F., Bosch, D., Stiekema, W.J., Gommers, F.J., Schots, A., Bakker, J. (1996) The C-terminal KDEL sequence increases the expression level of a single-chain antibody designed to be targeted to both the cytosol and the secretory pathway in transgenic tobacco, *Plant Mol. Biol.* **30**, 781-793.
28. Tavladoraki, P., Benvenuto, E., Trinca, S., Demartinis, D., Cattaneo, A., Galeffi, P. (1993) Transgenic plants expressing a functional single-chain Fv-antibody are specifically protected from virus attack, *Nature* **366**, 469-472.
29. Voss, A., Niersbach, M., Hain, R., Hirsch, H.J., Liao, Y.C., Kreuzaler, F., Fischer, R. (1995) Reduced virus infectivity in *N. tabacum* secreting TMV-specific full-size antibody. *Mol. Breeding* **1**: 39-50
30. Wyss, U. and Zunke, U. (1986) Observations on the behavior of second stage juveniles of *Heterodera schachtii* inside host roots. Rev. Nematol. **9**, 153-165.
31. Wyss, U. (1992) Observations on the feeding behavior of *Heterodera schachtii* throughout development, including events during molting, *Fundam. Appl. Nematol.* **15**, 75-89.
32. Wyss, U. and Grundler, F.M.W. (1992) Feeding behavior of sedentary plant parasitic nematodes, *Neth. J. Plant Pathol.* **98**, Suppl. 2, 165-173.
33. Wyss, U., Grundler, F.M.W. and Munch, A. (1992) The parasitic behavior of second-stage juveniles of *Meloidogyne incognita* in roots of *Arabidopsis thaliana*, *Nematologica* **38**, 98-111.

EPILOGUE

Paul R. BURROWS
IACR-Rothamsted, Harpenden, Herts, AL5 2JQ, UK.

Even a cursory glance at this book should convince the reader that sedentary plant parasitic nematodes are remarkable animals. Worldwide there are more than twenty genera of plant nematodes that contain species which at some point in their life cycle induce and maintain specialised feeding structures in host roots. Depending on the nematode species these feeding sites can range from a single enlarged cell to complex structures derived from several highly modified cells within the vascular cylinder.

Ten years ago plant nematologists knew little about the nature of these feedings sites apart from the structural details revealed by light and electron microscopy. Today, as this book testifies, this situation is changing rapidly and immunology and molecular biology are being used as powerful tools with which to dissect the interactions between sedentary nematodes and their hosts. Gene expression studies using promoter/*UidA* fusions, cDNA libraries and differential display of mRNA have shown us that feeding site formation is associated with gross local changes in host gene expression. Host genes are not only turned on (up-regulated) but also many are probably specifically switched off (down-regulated). Concomitant with this is the re-entry of developing feeding site cells into a modified form of the cell cycle. Furthermore, we should not forget that feeding site formation is only the culmination of a series of host interactions that lead from hatching and host location through to root penetration and endoparasitic migration. The research presented in the preceding chapters illustrates the progress being made and the value of multidisciplinary collaboration between laboratories.

So far we have only just started to scratch the surface in our exploration of nematode/host interactions and many interesting questions remain to be answered. One of the most intriguing questions concerns the molecular basis of feeding site induction. It has passed into nematology folklore that the cyst and root-knot nematodes use proteinaceous secretory components from oesophageal glands to initiate feeding sites, but there is not a shred of direct evidence in the literature to support this assumption. Indeed some researchers believe that the active effector molecules may not originate from the oesophageal glands or even, for that matter, be proteins. Similarly, we should also be asking at what level does the effector(s) interact with plant gene regulatory pathways? It is easy to imagine a membrane or cytoplasmic receptor that translates a nematode signal into changes in plant gene expression, most likely provoking a regulatory cascade. It is also possible that a nematode effector molecule acts directly as a plant transcription factor or inhibitor. Alternatively, it may be none of the above! Whatever the mechanism of feeding site induction, it will certainly have implications beyond far plant nematology and could tell us much about basic plant gene regulation.

On a purely practical note, some chapters in this book present the progress being made in the area of engineering resistance against plant parasitic nematodes in plants. The first transgenic plants with enhanced nematode resistance have already been made and tested. Although these plants contain only the early generations of constructs other improved versions are being developed. Suitable promoters will be key components of all nematode resistance approaches. If feeding site destruction is the ultimate goal then specific nematode responsive promoters are required to facilitate the expression of transgenes in the feeding sites with little expression elsewhere in the plant, e.g. the two component system (see chapter by Ohl this volume. More work is needed to identify additional promoters and to get them into prototype resistance constructs for testing. Resistance mechanisms that target the nematodes directly, such as those dependant on protease inhibitors or lectins, can be less strict about the promoters used. These could confer high levels of expression in roots as long as they were not down regulated (which many appear to be) in the feeding sites of sedentary nematodes. Even so, much development work still remains to be done especially in the characterisation of nematode responsive promoter subdomains and transcription factor binding sites. Looking further ahead, a greater knowledge at the molecular level of the fundamental interactions between nematodes and their host plants will highlight other, as yet, unexploited ways in which resistance could be engineered. This is an exciting and fast moving area of research that will make a significant contribution to reducing our dependence on toxic and expensive nematicides.

The collaboration generated through the EU Concerted Action Programme *'Resistance Mechanisms Against Plant Parasitic Nematodes'*, which ultimately resulted in this book, will continue. Twelve European laboratories are now linked through a new EU project *'Basis and Development of Molecular Approaches to Nematode Resistance'* (acronym ARENA) funded under Framework IV Biotechnology. This project aims to produce the technology necessary to engineer durable resistance to a broad spectrum of plant parasitic nematodes. In the process of this work we will also certainly gain a greater insight to the intriguing world of plant nematodes.

INDEX

2-D gel electrophoresis 135

A
AFLP 158-159, 161, 170, 181, 197
Agrobacterium tumefaciens 138, 264-265, 268
Agrochemical 250, 263
Allelochemical 38-39, 45-46
Amphids 38-40, 45, 54, 56-57, 61, 68, 98, 102, 223, 229
Amphimixis 170
Anti-sense inhibition 254
Apoptosis 251, 254-255
Artificial promoter 134
Avirulence gene 255

B
Barnase 230, 240, 250, 256, 258-259
Barstar 240, 250, 256, 258-259
Beet cyst nematode 176-178
Botrytis cinerea 264
Breeding 167, 176-180, 182-183, 192-194, 217-218, 250, 252, 268

C
Caenorhabditis elegans 99, 171, 242
CaMV 35S 29-30, 128, 241, 244, 246, 256-257, 264, 268
Cdc2aAt 73, 124, 126-127, 129, 136, 141, 143
Cell cycle gene 73, 120, 127, 130, 143, 253
Cell differentiation 25, 35, 67, 134, 141, 143, 145
Cell fate 27-28, 31, 35, 134
Chemoreceptor 45, 61
Chemitaxis 38, 68

Chromosome walking 168, 196
Cloning strategy 172, 176, 180
Collagen 58, 220-221
Collagenase 40, 220-221, 230, 239
Crop rotation 176-177, 238, 250
Cuticle 15, 54, 56, 58, 81, 98, 102-103, 220-221, 223
Cutinase 264-266
CyclAt 73, 124, 127, 253
Cyclin 36, 124, 126, 136
Cystatin 242-246
Cysteine proteinase 204, 242-244
Cytodifferentiation 93, 128
Cytoskeleton 75, 127

D
Developing world 238, 241, 246
Differential display 26, 133, 136, 161, 253, 273
Differential screening 135, 172, 253
Dormancy 39, 41, 43
Durability 194, 217, 232-233, 246, 254

E
Endoplasmic reticulum 15, 19, 59, 67, 70, 75, 80, 85, 91, 109, 128, 264
Engineered resistance 238-240, 255
Extensin 60, 73, 136, 141, 253
Exudate 68, 72, 102

F
Fecundity 204, 219, 244-245
Feeding tube 8, 11-12, 14-15, 17, 19, 69-70, 72, 83, 85, 98, 100, 109-110, 210, 245, 267
Fumigant 237-238, 250

G

Gall 8-9, 12-13, 53, 55, 60, 67-68, 108, 120, 124, 127-128, 140, 142, 206, 252-253
Genetic engineering 114, 116, 218, 251, 255
Glycoprotein 45, 56, 68, 101-103, 221-222
Green Fluorescent Protein 241
GUS 26, 31, 73-74, 124, 126-127, 136-137, 139, 144, 253, 257-258

H

Hairy root 155, 176, 185-187, 219, 244
Hatching 38-44, 46, 67, 103, 177, 273
HMG1 74, 136, 138, 142-145
Host location 38-39, 45, 227, 273
Hypersensitive response 80, 191, 193, 201, 205, 238, 255

I

Immunocytochemistry 65
Initial syncytial cell 14-15, 80-83, 108

K

Kinesin 72, 75, 141-142

L

Lectin 45, 68, 99, 103, 217, 221-225, 227, 229-230, 232, 239, 274
Lemmi9 135-136, 140-141, 143, 145, 253
Linkage Map Analysis 168

M

Marker assisted selection 180, 182, 187
Microinjection 15, 110-112, 114, 139, 173, 230, 245
Mitochondria 8, 10, 67, 85-89, 91-92, 128
Mitosis 18, 69, 121-122, 124-129, 141
Monoclonal antibody 68, 70, 72, 101

Moulting 17, 58, 68, 86, 103, 108, 221
Mutagenesis 139, 230, 243
MYB-like protein 136, 141

N

Necrosis 12, 60, 88, 90-91, 193-194, 201
Nematicide 1-2, 45, 202, 217, 237-238, 250, 263, 274
Nematode feeding site 60, 80-81, 86, 92, 99, 128, 130, 133-134, 207, 230, 241
Nematode responsive element 134, 140

O

Oesophageal gland 5, 9, 16, 39, 56-58, 70, 72, 75, 83, 98-101, 109, 120, 227, 273
Oryzacystatin 204, 219, 229-230, 242

P

Parthenogenesis 168, 171
PCR-based library 135
Pericycle cell 83, 116, 125
Phloem unloading 18, 112, 114
Plantibody 227
Polyploidy 124, 127-128
Positional cloning 168, 176, 180, 191, 197
Potato cyst nematode 39, 42, 46, 155, 169, 262-264, 267, 270
PR protein 204-205
Procambial cell 18, 83, 85, 87-88, 92-94, 116
Promoter element 133-134, 146
Promoter trap 25, 29, 35-36, 137-138
Protein engineering 230, 237, 243-244
Proteinase inhibitor 202, 204, 219, 231, 237, 242-244, 246
Protoplasts 9, 12, 14, 16, 18, 59, 81, 85, 87, 125-126, 266, 268

R
RAPD 158-160, 169, 171, 181, 197
Resistance gene 66, 154, 167-170, 173, 176-181, 183-187, 192, 198, 238, 251, 255, 262
Resistance response 65, 107, 136, 192, 201, 239, 251-252
Resistance strategy 250, 252-253, 259
Root diffusate 14, 39-43, 46, 103
Root morphogenesis 81, 93

S
SCAR 160
Secretory system 70, 98, 101
Sensory perception 38-39, 45, 222, 239
Serine proteinase 222, 244
Sex determination 107, 115-116
Signal sequence 264-266, 268
Signal transduction 69, 198, 210, 255
Single-chain antibody 262-263
Stylet secretions 70, 72, 100, 203
Systemic acquired resistance 252

T
TobRB7 73, 115, 138, 141, 145, 240, 253, 256
Tolerance 153-155, 177
Transcription factor 69-70, 134, 136, 139-141, 145-146, 256, 273-274
Tubulin 72, 75, 124, 136, 141-142

V
Vacuole 15, 17, 66, 80, 83, 85-89, 91-93, 121, 128, 181, 208-209
Vascular tissue 29-31, 73, 85-86, 107, 121, 131, 141, 257
Virulence 61, 153-162, 167-173, 251

W
Wun-1 239

X
Xylem 18, 29, 53-55, 66, 68-69, 83, 85, 89, 93, 111, 114, 205, 254

Y
Yeast artificial chromosome 176

Authors

ABAD, Pierre
INRA, Laboratoire de Biologie des Invertébrés
BP 2078
F-06606 Antibes CEDEX
FRANCE
Phone: +33 (4) 9367 8800
Facsimile: +33 (4) 9367 8955
E-mail: abad@antibes.inra.fr

ARISTIZABAL, Fabio A.
Laboratorio de Bioquimica y Biol. Molecular Vegetal
Centro Internacionalde Fisica (CIF)
Apartado Aereo 4948
Santafe de Bogota D.C.
COLOMBIA
Phone: +57 (1) 368 1517
Facsimile: +57 (1) 368 1335
E-mail: fisica@latino.net.co

ATKINSON, Howard J.
Centre for Plant Biochemistry and Biotechnology
University of Leeds
Leeds, LS2 9JT
UNITED KINGDOM
Phone. +44 113 233 2900
Facsimile. +44 113 233 3144
E-mail: h.j.atkinson@leeds.ac.uk

BAKKER, Jaap
Wageningen Agricultural University
Department of Nematology
P.O. Box 8123
NL-6700 ES Wageningen
THE NETHERLANDS
Phone: +31 (317) 482197

Facsimile: +31 (317) 484 254
E-mail: jaap.bakker@medew.nema.wau.nl

BLEVE-ZACHEO, Teresa
Istituto di Nematologia Agraria, CNR
Via Amendola 165/A
I-70126 Bari
ITALY
Phone: +39 (80) 5583 377
Facsimile: +39 (80) 5580 468
E-mail: Zacheo @ cnrarea. unile. it

BLOK, Vivian C.
Department of Nematology
Scottish Crop Research Institute
Invergowrie, Dundee
UNITED KINGDOM
Phone: +44 1382 562731
Facsimile +44 1382 562426
E-mail vblok@scri.sari.ac.uk

BÖCKENHOFF, Annette
Institut für Phytopathologie
Christian-Albrechts-Universität Kiel
D-24098 Kiel
GERMANY
Phone: +49 (431) 880 4669
Facsimile: +49 (431) 880 1583
E-mail: fgrundler@phytomed.uni-kiel.de

BOSCH, D.
Department of Molecular Biology
Centre for Plant Breeding and Reproduction Research (CPRO-DLO)
P.O. Box 16
NL- 6700 AA Wageningen
THE NETHERLANDS
Phone: +31 (317) 477001

Facsimile: +31 (317) 418094
E-mail: h.j.bosch@cpro.dlo.nl

BURROWS, Paul. R.
Entomology and Nematology Department
IACR-Rothamsted
Harpenden, Herts AL5 2JQ
UNITED KINGDOM
Phone: +44 (1582) 763133 ext 2280
Facsimile: + 44 (1582) 760981
E-mail: paul.burrows@bbsrc.ac.uk

CAI, Daguang
Institute of Crop Science and Plant Breeding
Christian-Albrechts-University of Kiel,
Olshausenstr. 40
D-24118 Kiel
GERMANY
Phone: +49 (431) 880 2580
Facsimile: +49 (431) 880 2566
E-mail: dcai@plantbreeding.uni-kiel.de

CASTAGNONE-SERENO, Philippe
INRA, Laboratoire de Biologie des
Invertébrés
BP 2078
F-06606 Antibes CEDEX
FRANCE
Phone: +33 (4) 9367 8800
Facsimile: +33 (4) 9367 8955
E-mail: pca@antibes.inra.fr

DALMASSO, Antoine
INRA, Laboratoire de Biologie des
Invertébrés
BP 2078
F-06606 Antibes CEDEX
FRANCE
Phone: +33 (4) 9367 8800
Facsimile: +33 (4) 9367 8955
E-mail: dalmasso@antibes.inra.fr

DE ALMEIDA ENGLER, JANICE
Laboratorium voor Genetica
Universiteit Gent
K.L. Ledeganckstraat 35
B-9000 Gent
BELGIUM
Phone. +32 (9) 264 5170
Facsimile: +32 (9) 264 5349
E-mail: mamon@gengenp.rug.ac.be

DE BOER, J.M.
Department of Nematology
Wageningen Agricultural University
P.O. Box 8123
NL-6700 ES Wageningen
THE NETHERLANDS
Phone: +31 (317) 482197
Facsimile: +31 (317) 484254
E-mail: j.m.deboer@medew.nem.wau.nl

DEL CAMPO, Francisca F.
Departamento de Biología
Universidad Autónoma de Madrid,
Cantoblanco
E-28049 Madrid
SPAIN
Phone: +34 (1) 397 8198
Facsimile: +34 (1) 397 8344
E-mail: francsica.fernandez@uam.es

DE VRIEZE, Geert
State University Utrecht
Dpt. of Molecular Cell Biology
Padualaan 8
NL-3584 CH Utrecht
THE NETHERLANDS
Phone: +31 (30) 253 3581
Facsimile: +33 (30) 251 3665

DE WAELE, Dirk
Laboratory of Tropical Crop Improvement
Katholieke Universiteit Leuven

Kardinaal Mercierlaan 92
B-3001 Heverlee
BELGIUM
Phone: +32 1632 1693
Facsimile: +32 1632 1993
E-mail: dirk.waele@agr.kuleuven.ac.be

FENOLL, Carmen
Departamento de Biología
Universidad Autónoma de Madrid,
Cantoblanco
E-28049 Madrid
SPAIN
Phone: +34 (1) 397 8198
Facsimile: +34 (1) 397 8344
E-mail: carmen.fenoll@uam.es

GHEYSEN, Godelieve
Laboratorium voor Genetica
Universiteit Gent
K.L. Ledeganckstraat 35
B-9000 Gent
BELGIUM
Phone: +32 (9) 264 5170
Facsimile: +32 (9) 264 5349
E-mail: mamon@gengenp.rug.ac.be

GOMMERS, Fred J.
Wageningen Agricultural University
Department of Nematology
PO Box 8123
NL-6700 ES Wageningen
THE NETHERLANDS
Phone: +31 (317) 482197
Facsimile: +31 (317) 484 254
E-mail: fred.gommers@medew.nema.wau.nl

GOLINOWSKI, Wladyslaw
Department of Botany
Warsaw Agricultural University
Rakowiecka 26/30
P-02528 Warszawa

POLAND
Phone: +48 (22) 492 251
Facsimile: +48 (22) 471 561
E-mail: golinowski@delta.sggw.waw.pl

GOVERSE, A.
Department of Nematology
Wageningen Agricultural University
P.O. Box 8123
NL-6700 ES Wageningen
THE NETHERLANDS
Phone: +31 (317) 482197
Facsimile: +31 (317) 484254
E-mail: a.goverse@medew.nem.wau.nl

GRUNDLER, Florian M.W
Institut für Phytopathologie
Christian-Albrechts-Universität Kiel
D-24098 Kiel
GERMANY
Phone: +49 (431) 880 4669
Facsimile: +49 (431) 880 1583
E-mail: fgrundler@phytomed.uni-kiel.de

GRYMASZEWSKA, Grazyna
Department of Botany
Warsaw Agricultural University
Rakowiecka 26/30
P-02528 Warszawa
POLAND
Phone: +48 (22) 492 251
Facsimile: +48 (22) 471 561
E-mail: golinowski@delta.sggw.waw.pl

HARLOFF, Hans
Institute of Crop Science and Plant Breeding
Christian-Albrechts-University of Kiel,
Olshausenstr. 40
D-24118 Kiel
GERMANY
Phone:+49 (431) 880 3212
Facsimile: +49 (431) 880 2566

E-mail: hharloff@plantbreeding.uni-kiel.de

JONES, John T.
Nematology Department
Scottish Crop Research Institute
Invergowrie, Dundee, DD2 5DA
UNITED KINGDOM
Phone: +44 (382) 562 731
Facsimile: +44 (382) 562 426
E-mail: jjones@scri.sari.ac.uk

JUNG, Christian
Institute of Crop Science and Plant Breeding
Christian-Albrechts-University of Kiel,
Olshausenstr. 40
D-24118 Kiel
GERMANY
Phone: +49 (431) 880 3210
Facsimile: +49 (431) 880 2566
E-mail: cjung@plantbreeding.uni-kiel.de

McKHANN, Heather
State University Utrecht
Dpt. of Molecular Cell Biology
Padualaan 8
NL-3584 CH Utrecht
THE NETHERLANDS
Phone: +31 (30) 253 3581
Facsimile: +33 (30) 251 3665

KIFLE, Sirak
Institute of Crop Science and Plant Breeding
Christian-Albrechts-University of Kiel,
Olshausenstr. 40
D-24118 Kiel
GERMANY
Phone: +49 (431) 880 2580
Facsimile: +49 (431) 880 2566
E-mail: skifle@plantbreeding.uni-kiel.de

KLEINE, Michael
Institute of Crop Science and Plant Breeding
Christian-Albrechts-University of Kiel,
Olshausenstr. 40
D-24118 Kiel
GERMANY
Phone: +49 (431) 880 3210
Facsimile: +49 (431) 880 2566
E-mail: mkleine@plantbreeding.uni-kiel.de

KLEIN-LANKHORST, Rene M.
DLO-Centre for Plantbreeding and
Reproduction Research (CPRO-DLO)
Droevendaalsesteeg 1
NL-6700 AA Wageningen
THE NETHERLANDS
Phone: +31 (317) 477001
Facsimile: +31 (317) 418094
E-mail: R.M.Kleinlankhorst@cpro.dlo.nl

KUREK, Wojciech
Department of Botany
Warsaw Agricultural University
Rakowiecka 26/30
P-02528 Warszawa
POLAND
Phone: +48 (22) 492 251
Facsimle: +48 (22) 471 561
E-mail: kurek@delta.sggw.waw.pl

LIHARSKA, Tsvetana B.
Free University of Amsterdam
Faculty of Biology
De Boelelaan 1087
1081 HV Amsterdam
THE NETHERLANDS
Phone: +31 (20) 444 7148
Facsimile: +31 (20) 444 7137
E-mail: tbl@bio.vu.nl

LILLEY, Catherine J.
Centre for Plant Biochemistry and
Biotechnology
University of Leeds

Leeds, LS2 9JT
UNITED KINGDOM
Phone. +44 113 233 3035
Facsimile. +44 113 233 3144
E-mail: c.j.lilley@leeds.ac.uk

MARCKER, Kjeld A.
Department of Molecular Biology
University of Aarhus
Gustav Wieds Vej 10
DK-8000 Aarhus
DENMARK
Phone: +45 (86) 202 000
Facsimile: +45 (86) 201 222
E-mail: marcker@biobase.dk

MELILLO, Maria Teresa
Istituto di Nematologia Agraria, CNR,
Via Amendola 165/A,
I-70126 Bari
ITALY
Phone: +39 (80) 5583 377
Facsimile: +39 (80) 5580 468
E-mail Zacheo @ cnrarea. unile. it

OHL, Stephan A.
MOGEN International nv
Einsteinweg 97
NL-2333 CB Leiden
THE NETHERLANDS
Phone: (31) 71 525 8282
Facsimile: (31) 71 522 1471
E-mail: sohl@mogen.nl

PERRY, Roland N.
Entomology and Nematology Department
IACR- Rothamsted
Harpenden, Hertfordshire, AL5 2JQ
UNITED KINGDOM
Phone: +44 (1582) 763 133
Facsimile: +44 (1582) 760 981
E-mail: Roland.Perry@bbscr.ac.uk

McPHERSON, Michael J.
Centre for Plant Biochemistry and
Biotechnology
University of Leeds,
Leeds, LS2 9JT,
UNITED KINGDOM
Phone. +44 113 233 2595
Facsimile. +44 113 233 3144
E-mail: m.j.mcpherson@leeds.ac.uk

PHILLIPS , Mark S.
Department of Nematology
Scottish Crop Research Institute,
Invergowrie, Dundee
UNITED KINGDOM
Phone: +44 1382 562731
Facsimile +44 1382 562426
E-mail: mphill@scri.sari.ac.uk

ROBERTSON, Walter M.
Nematology Department
Scottish Crop Research Institute
Invergowrie, Dundee, DD2 5DA
UNITED KINGDOM
Phone: +44 (382) 562 731
Facsimile: +44 (382) 562 426
E-mail: wrober@scri.sari.ac.uk

ROOSIEN, J.
Laboratory of Monoclonal Antibodies
P.O. Box 9060
NL-6700 GW Wageningen
THE NETHERLANDS
Phone: +31 (317) 482197
Facsimile: +31 (317) 484254
E-mail: j.roosien@medew.nem.wau

SALENTIJN, Elma M. J.
DLO-Centre for Plantbreeding and
Reproduction Research (CPRO-DLO)
Droevendaalsesteeg 1
NL-6700 AA Wageningen

THE NETHERLANDS
Phone: +31 (317) 477001
Facsimile: +31 (317) 418094
E-mail: e.m.j.salentijn@cpro.dlo.nl

SANDAL, Niels N.
Department of Molecular Biology
University of Aarhus
Gustav Wieds Vej 10
DK-8000 Aarhus
DENMARK
Phone: +45 (86) 202 000
Facsimile: +45 (86) 201 222
E-mail: sandal@biobase.dk

SANZ-ALFEREZ, Soledad
Departamento de Biología
Universidad Autónoma de Madrid,
Cantoblanco
E-28049 Madrid
SPAIN
Phone: +34 (1) 397 8198
Facsimile: +34 (1) 397 8344
E-mail: soledad.sanz@uam.se

SCHERES, Ben
State University Utrecht
Dpt. of Molecular Cell Biology
Padualaan 8
NL-3584 CH Utrecht
THE NETHERLANDS
Phone: +31 (30) 253 3133
Facsimile: +31 (30) 251 3665
E-mail: B.Scheres@cc.ruu.nl

SCHOTS, A.
Laboratory of Monoclonal Antibodies
P.O. Box 9060
NL-6700 GW Wageningen
THE NETHERLANDS
Phone: +31 (317) 482197
Facsimile: +31 (317) 484254
E-mail: a.schots@medew.nem.wau.nl

SCHOUTEN, A.
Department of Nematology
Wageningen Agricultural University
P.O. Box 8123
NL-6700 ES Wageningen
THE NETHERLANDS
Phone: +31 (317) 482197
Facsimile: +31 (317) 484254
E-mail: a.schouten@medew.nem.wau.nl

SMANT, G.
Department of Nematology
Wageningen Agricultural University
P.O. Box 8123
NL-6700 ES Wageningen
THE NETHERLANDS
Phone: +31 (317) 482197
Facsimile: +31 (317) 484254
E-mail: g.smant@medew.nem.wau.nl

SOBCZAK, Miroslaw
Department of Botany
Warsaw Agricultural University
Rakowiecka 26/30
P-02528 Warszawa
POLAND
Phone: +48 (22) 492 251
Facsimile: +48 (22) 471 561
E-mail: sobczak@delta.sggw.waw.pl

STIEKEMA, Willem J.
Department of Molecular Biology
Centre for Plant Breeding and Reproduction
Research (CPRO-DLO)
P.O. Box 16
NL- 6700 AA Wageningen
THE NETHERLANDS
Phone: +31 (317) 477 130
Facsimile: +31 (317) 418 094
E-mail: w.j.stiekema@cpro.dlo.nl

STOKKERMANS, J.
Department of Nematology
Wageningen Agricultural University
P.O. Box 8123
NL-6700 ES Wageningen
THE NETHERLANDS
Phone: +31 (317) 482197
Facsimile: +31 (317) 484254
E-mail: j.stokkermans@medew.nem.wau.nl

SIJMONS, Peter C.
ATO-DLO
P.O. Box 17
NL-6700 AA Wageningen
THE NETHERLANDS
Phone: (31) 317 475026
Facsimile: (31) 317 412260
E-mail: psijmons@euronet.nl

URWIN, Peter E.
Centre for Plant Biochemistry and Biotechnology
University of Leeds
Leeds, LS2 9JT
UNITED KINGDOM
Phone. +44 113 233 3035
Facsimile. +44 113 233 3144
E-mail: p.e.urwin@leeds.ac.uk

VAN DEN BERG, Claudia
State University Utrecht
Dpt. of Molecular Cell Biology
Padualaan 8
NL-3584 CH Utrecht
THE NETHERLANDS
Phone: +31 (30) 253 3581
Facsimile: +33 (30) 251 3665

VAN DER LEE, Frédérique M.
MOGEN International nv
Einsteinweg 97
NL-2333 CB Leiden
THE NETHERLANDS
Phone: (31) 71 525 8282
Facsimile: (31) 71 522 1471
E-mail: flee@mogen.nl

VAN MONTAGU, Marc
Laboratorium voor Genetica
Universiteit Gent
K.L. Ledeganckstraat 35
B-9000 Gent
BELGIUM
Phone: +32 (9) 264 5170
Facsimile: +32 (9) 264 5349
E-mail: mamon@gengenp.rug.ac.be

VON MENDE, Nicola
Entomology and Nematology Department
IACR Rothamsted
Harpenden, Herts AL5 2JQ
UNITED KINGDOM
Phone: +44 (1582) 763 133
Facsimile: +44 (1582) 760 981
E-mail: nicola.vonmende@bbsrc.ac.uk

WILLEMSEN, Viola
State University Utrecht
Dpt. of Molecular Cell Biology
Padualaan 8
NL-3584 CH Utrecht
THE NETHERLANDS
Phone: +31 (30) 253 3581
Facsimile: +33 (30) 251 3665

WILLIAMSON, Valerie M.
University of California
Department of Nematology
Davis, CA 95616-8668
U.S.A.
Phone: +1 (916) 752 3502
Facsimile: +1 (916) 752 5809
E-mail: vmwilliamson@ucdavis.edu

WILMINK, J.M.J.
Department of Molecular Biology
Centre for Plant Breeding and Reproduction
Research (CPRO-DLO)
P.O. Box 16
NL- 6700 AA Wageningen
THE NETHERLANDS
Phone: +31 (317) 477001
Facsimile: +31 (317) 418094
E-mail: j.m.j.wilmink@cpro.dlo.nl

WOLKENFELT, Harald
State University Utrecht
Dpt. of Molecular Cell Biology
Padualaan 8
NL-3584 CH Utrecht
THE NETHERLANDS
Phone: +31 (30) 253 3581
Facsimile: +33 (30) 251 3665

WYSS, Urs
Institut für Phytopathologie
Universität Kiel
Hermann-Rodewald-Str. 9
D-24118 Kiel
GERMANY
Phone: +49 (431) 880 4558
Facsimile: +49 (431) 880 1583
E-mail: uwyss@phytomed.uni-kiel.de

ZACHEO, Giuseppe
Istituto di Ricerca sulle Biotecnologie
Agroalimentari, CNR,
Prov.le Lecce-Monteroni
I-73100 Lecce
ITALY
Phone: +39 (80) 5583 377
Facsimile: +39 (80) 5580 468
E-mail Zacheo @ cnrarea. unile. it

ZIJLSTRA, Carolien
Research Institute for Plant Protection
(IPO-DLO),
P.O. Box 9060,
NL-6700 GW Wageningen
THE NETHERLANDS
Phone: +31 (317) 476155
Facsimile: +31 (317) 410112
E-mail C.Zijlstra@IPO.DLO.NL

Developments in Plant Pathology

1. R. Johnson and G.J. Jellis (eds.): *Breeding for Disease Resistance.* 1993
   ISBN 0-7923-1607-X
2. B. Fritig and M. Legrand (eds.): *Mechanisms of Plant Defense Responses.* 1993
   ISBN 0-7923-2154-5
3. C.I. Kado and J.H. Crosa (eds.): *Molecular Mechanisms of Bacterial Virulence.* 1994
   ISBN 0-7923-1901-X
4. R. Hammerschmidt and J. Kuć (eds.), *Induced Resistance to Disease in Plants.* 1995
   ISBN 0-7923-3215-6
5. C. Oropeza, F.W. Howard, G. R. Ashburner (eds.): *Lethal Yellowing: Research and Practical Aspects.* 1995
   ISBN 0-7923-3723-9
6. W. Decraemer: *The Family Trichodoridae: Stubby Root and Virus Vector Nematodes.* 1995
   ISBN 0-7923-3773-5
7. M. Nicole and V. Gianinazzi-Pearson (eds.): *Histology, Ultrastructure and Molecular Cytology of Plant-Microorganism Interaction.* 1996
   ISBN 0-7923-3886-3
8. D.F. Jensen, H.-B. Jansson and A. Tronsmo (eds.): *Monitoring Antagonistic Fungi Deliberately Released into the Environment.* 1996
   ISBN 0-7923-4077-9
9. K. Rudolph, T.J. Burr, J.W. Mansfield, D. Stead, A. Vivian and J. von Kietzell (eds.): *Pseudomonas Syringae Pathovars and Related Pathogens.* 1997
   ISBN 0-7923-4601-7
10. C. Fenoll, F.M.W. Grundler and S.A. Ohl (eds.): *Cellular and Molecular Aspects of Plant-Nematode Interactions.* 1997
    ISBN 0-7923-4637-8

KLUWER ACADEMIC PUBLISHERS – DORDRECHT / BOSTON / LONDON